D0098037

Technician's Guide to HVAC Systems

Gary K. Skimin, P.E.

McGraw-Hill

New York San Francisco Washington, D.C. Auckland Bogotá
Caracas Lisbon London Madrid Mexico City Milan
Montreal New Delhi San Juan Singapore
Sydney Tokyo Toronto

Library of Congress Cataloging-in-Publication Data

Skimin, Gary K.
Technician's guide to HVAC systems / by Gary K. Skimin.
p. cm.
Includes index.
ISBN 0-07-057914-8 (H)
1. Heating—Equipment and supplies. 2. Ventilation—Equipment and supplies. 3. Air conditioning—Equipment and supplies. I. Title.
TH7345.S55 1995
697—dc20 95-17261
CIP

hc 2 3 4 5 6 7 8 9 DOC/DOC 9 9 8 7 6 5

ISBN 0-07-057914-8 (H)

The sponsoring editor for this book was April D. Nolan. The executive editor was Joanne Slike. The managing editor was Lori Flaherty. Laura Bader was the manuscript editor. The director of production was Katherine G. Brown. This book was set in ITC Century Light. It was composed in Blue Ridge Summit, Pa.

Printed and bound by Donnelley of Crawfordsville, Indiana.

MH95
0579148

To my family. Everyone, especially my wife, Miriam, stood with me throughout this project with support and enthusiasm. It was truly appreciated.

Contents

Acknowledgment

Reference to the System Syzer Calculator courtesy of ITT Fluid Technology Corporation.

Introduction

This book is for the practicing HVAC technician and those learning the trade. Covering residential and light commercial systems, the focus is on those types of systems technicians work with every day. All too often technical books on this subject are heavy on math and engineering details. These leave the technician looking for a practical, easy-to-apply source of information he or she can use. This book provides that information. It explains the design, construction, and operation of HVAC systems enabling the technician to provide more efficient and effective service.

The intent of this book is to provide the technician with a sound understanding of HVAC system operating principles and their components. This book provides information that makes it easier to service and select systems that meet the customer's needs.

Knowledge of the operating details of HVAC equipment allows the technician to make better decisions when making repairs. It is easier to diagnose a problem if you have a good understanding of how the system works. The fundamentals are not hard to learn, but the technician must have the proper information. Wrong decisions can be made when servicing, sizing, or designing new systems. It is unfortunately true that there is a body of misinformation regarding heating and cooling system operation. This book gives the real facts.

A note about math: The calculations in this book can be done with a $10 pocket calculator. Yes, there are some formulas—heating and cooling is a science, and math is a part of it. There is not, however, math for its own sake. The mathematical ideas and formulas are backed up with at least one practical example that shows how to use them in the real world.

Examples also show how nonmathematical ideas can be applied. The intention is to make the theory of heating and cooling systems accessible and uncomplicated. Examples are given wherever needed to explain an idea.

Also included throughout the book are service tips. These hints relate the topic under discussion to situations often found in the field. They describe what to look for on a service call and how the information in the book applies to the conditions found there. Technicians can apply these ideas directly to jobs to help in problem solving.

The ideas covered in Chapter 1 are heat, energy, and power. Naturally, any discussion of heating and cooling systems should start at the beginning. Physical principles of heat energy are the most fundamental concepts behind the operation of every system. A review of temperature, humidity (psychrometrics), and the implications of warm air buoyancy are covered.

Heat transfer is discussed in Chapter 2. Included in this chapter are descriptions of the theories of heat transfer via conduction, convection, and radiation. Heat transfer formulas involved in air and water flow systems are covered in detail. Examples show how to estimate heat delivery or removal in many situations. Insulation is also reviewed here. This chapter covers available materials and their appropriate use. Advantages and problems that can arise with different materials are described.

Chapter 3 covers building heat losses and gains. Methods for estimating the capacity of heating and cooling systems are described in detail. Included are practical explanations of solar heat gains and the effects of insulation on the building's perimeter. Straightforward, practical methods of estimating heating and cooling system loads are given.

Descriptions of air flow and the many applications of forced air systems are provided in Chapter 4. Components described include fans and ducts (including layouts and sizing), as well as discussions of noise control and flow measurement. Equipment used in most HVAC air distribution systems is covered. Practical methods for selecting components are given for typical situations. Different air distribution systems are described, along with the operating principles of each. Enough of an overview is given to allow the technician to approach these systems with a good background in their design and operation. Duct layout and sizing are covered also. Good duct design is a science, not an art. This book covers the fundamentals and details needed to ensure that these systems work well and deliver the performance needed.

Chapter 5 covers water-based heating and cooling distribution systems. Pipes and fittings are described, along with details explaining the best ways to install them. Detailed explanations of the operation of control valves, heat exchangers, pumps, and other parts are provided. Enough information is included to allow the technician to size and select the proper components. Methods of economically and properly sizing pipes and systems are discussed. Proper water treatment is described, as well as how corrosion, scale, and biological controls can be achieved with commonly available water treatment chemicals. Also covered are water system measurements: pressure, temperature, and flow.

The topic of system startup, testing, and balancing is discussed in Chapter 6. Proper starting of an HVAC system will avoid problems with future operation and help complete the job faster. Basic balancing is discussed in enough detail to help the technician set up a new system. Although most technicians do not specialize in this field, a good understanding of the topic is essential to really know how to properly design a system. Understanding system balancing can help on service calls where the equipment is running but is not performing the way it should. Both air and water systems are covered.

Chapter 7 covers heating systems. Available heating sources (including electric, fossil fuel, heat pumps, and heat recovery units) are described and their differences discussed. Additional topics include forced hot air systems, hydronic systems, and radiant designs. Primary and secondary piping systems are covered, as well as their appropriate applications.

Air conditioning systems are described in Chapter 8. Again, this book starts with the fundamentals: the operation of compression cycle refrigeration systems and their parts are described. Available system types are covered and information on which types work the best in various applications is provided. Included are unitary package equipment, central split units, modular split equipment, and PTACs. Chilled water systems are covered in detail, with explanations of how they work and the functions of their major components. Cooling tower design and operation are explained in detail, along with the service that most of them require.

Refrigerant legislation is also covered. The HVAC technician must be licensed by the Environmental Protection Agency (EPA) and know the laws covering refrigerant release. Details of the regulations are provided to allow the technician to review these important facts. This chapter covers the various "alternate" refrigerant materials that can be substituted for the formerly used CFC compounds. The latest "best practices" of equipment room design are also covered.

Water-source heat pumps are covered in Chapter 9. These systems can be very flexible and suit many applications, but special installation details must be considered for them to work properly. These units are significantly different from standard heating and cooling equipment. System concepts are reviewed and their operation is described. Proper heat pump piping layouts are covered, as well as the other components required to make a complete system. Included are special boiler considerations, cooling towers, and the unique controls water-source heat pump systems require.

Ventilation is discussed in detail in Chapter 10. This topic is becoming very important in the HVAC field. The technician should have a very good understanding of current requirements. Knowledge of proper ventilation techniques is becoming essential to help technicians protect their livelihoods. Many lawsuits are brought and won by dissatisfied or injured building occupants because of poor ventilation systems. Regulations and requirements are described, as well as the hardware used to ensure adequate ventilation for building occupants. Heat exchangers and the use of make-up air are discussed.

Chapter 11 covers humidity control. The needs and uses of humidification and dehumidification systems are described. Some problems with building design and construction can show themselves as humidity problems. The HVAC technician should know how the building's heating and cooling systems can mask these problems or help alleviate them. Components used for air drying and humidification are covered. Both refrigerant and desiccant dehumidifiers are reviewed. Humidifiers (steam injection, evaporative, and atomizing types) are described, and procedures for estimating their capacity are given.

This book provides a comprehensive list of resources and component manufacturers in Appendix A. Telephone numbers are included to allow the technician to locate nearby equipment dealers when installing or servicing systems.

Codes and standards are covered in Appendix B. Although no universal HVAC code exists for all areas, certain standards and requirements are commonly used as the basis for codes in the United States. The HVAC technician should be familiar with these basic requirements when evaluating the proper way to service or install systems.

Appendix C provides SI (metric) conversions. Both individual unit and equation conversions are given. These make it very easy to estimate values in the SI system. Every equation used in the book is given in an SI equivalent.

This book will be a valuable reference for the working technician. Again, it does not contain esoteric theory written in complex equations. It shows, by example, how systems function and what they do. The technician will learn enough to confidently approach any service or installation situation and efficiently do the job.

Chapter

1

Heat

Heat is a form of energy. It can be contained in a substance like other forms of energy (electrical, chemical, or force exerted through a distance) and it has the capacity to do work. Heat can flow from one place or thing to another and is convertible to and from other forms of energy.

Heat energy is a *basic* energy form. It represents at least one product of almost any process. Thermal energy is created by altering and converting other forms of energy. Friction, chemical reactions, and electrical energy produce heat as a by-product. Combustion of fuel and oxygen is a common method of producing heat for HVAC systems.

Units of Measure

Heat energy is measured in various units. The most common measurement used in the United States is the British Thermal Unit (Btu). This represents the amount of heat required to heat 1 lb of water 1°F. Almost all heating and cooling systems sold in the United States are rated by the number of Btu they can deliver or remove in 1 hour.

The most common metric unit of heat energy is the watt-hour (W-hr). This is equal to 1 W of power applied for 1 hour. It is usually used with electric heaters and appliances, but any other energy measurement can be done in watt-hours, as well. A closely related unit of measure to the watt-hour is the kilowatt-hour (kWh). One kilowatt-hour is equal to 1000 W-hr.

Another common metric unit of measure for heat energy is the calorie (cal). One calorie is the amount of heat required to heat 1 g of water 1°C. Finally, to make the metric system even more confusing, there is the kilocalorie (usually written Calorie, with a capital "C," or kcal). This is the amount of heat required to raise the temperature of 1 kg of water 1°C, and is equal to 1000 cal.

Use the formulas in Table 1.1 to convert between English and metric measurements. For example, it is possible to find the number of Calories equivalent to 3000 Btu by using the formula

$$3000 \text{ Btu} \times 2 \text{ Calories/Btu} = 756 \text{ Calories}$$

Similarly, 400 W of electric power run for 5 hours can be converted to Btu with the formula

$$400 \text{ W} \times 5 \text{ hr} \times 3.415 \text{ Btu/W-hr} = 6830 \text{ Btu}$$

TABLE 1.1 U.S. and Metric Units of Energy and Power

To find	Multiply	By
Btu	Calories	3.968
Btu	Foot-pounds	0.001286
Btu	Watt-hours	3.415
Calories	Btu	0.2520
Calories	Foot-pounds	0.0003241
Calories	Watt-hours	0.8606
Foot-pounds	Btu	777.6
Foot-pounds	Calories	3086
Foot-pounds	Watt-hours	2656
Watt-hours	Btu	0.2928
Watt-hours	Calories	1.162
Watt-hours	Foot-pounds	0.0003766
Btu/hour	Hp	2546
Btu/hour	Hp—boiler	33,480
Btu/hour	Tons	12,000
Btu/hour	Watts	3.415
Hp	Btu/hour	0.0003927
Hp	Hp—boiler	13.15
Hp	Tons	4.713
Hp	Watts	745.7
Hp—boiler	Btu/hour	0.00002987
Hp—boiler	Hp	0.07605
Hp—boiler	Tons	0.3584
Hp—boiler	Watts	0.0001020
Tons	Btu/hour	0.00008333
Tons	Hp	0.2122
Tons	Hp—boiler	2.790
Tons	Watts	0.0002846
Watts	Btu hour	0.2928
Watts	Hp	0.001341
Watts	Hp—boiler	9804
Watts	Tons	3514

Multiply the electrical power in watts by the time it is running before finding the equivalent heat energy in Btu. Don't neglect the time that electric energy is being consumed when calculating energy consumption. Multiplying watts by the factor 3.412 gives Btu per hour, not Btu. This will be useful later to find power and heat flow.

Temperature

The amount of heat energy a substance contains is different from its temperature. It is incorrect to believe that a higher temperature object necessarily contains more heat than a cooler one.

The temperature of a substance is determined by the submicroscopic motion of its molecules. Constantly moving atoms and molecules make up all substances. Heat energy causes them to vibrate and move against each other. Higher temperature substances have more energetically moving molecules. Raising the temperature of an object makes the molecules vibrate with more energy. Hotter objects have molecules that, on average, vibrate faster. The total amount of energy in a substance depends on more than the speed of its molecule's vibrations.

Temperature is the one thing that determines in which direction heat flows. Heat always flows from a higher temperature substance to a lower temperature one. This is a fundamental law of nature that cannot be violated.

Systems that seem to violate this rule (air conditioners and refrigeration equipment) actually rely on it. While heat may appear to flow from a low temperature area to a higher one, this is not the case. External energy is sued to create temperatures that move heat in its natural (hot to cold) direction.

It may be useful to think of temperature as analogous to pressure. Consider two air tanks connected to each other with a hose. A small tank filled with high-pressure air will push some of its air into a larger tank filled with lower pressure air. Similarly, heat flows from a small, high-temperature substance into a larger, lower temperature one. A higher temperature object has more heat "pressure," but not necessarily more heat.

Keep in mind that a flow of heat energy from a hotter object to a cooler one does not mean that the hotter object contains more heat energy. It may contain less total heat than the cooler one, but it sends heat to the cooler one because of their temperature differences.

The amount of heat in any object, including a quantity of water or air, depends on two things besides temperature: its specific heat and its mass (weight).

Specific Heat

The specific heat of a substance is a physical property determined by testing and experimentation. The actual definition of specific heat is that amount of heat required (in Btu) to heat 1 lb of material 1°F, compared to water. See Table 1.2 for the specific heats of some common substances. These values are averages that, while they may vary slightly with changes in the substance's temperature, are close enough for normal HVAC work.

Specific heat values correspond to how easily a material can be warmed or cooled. Those with higher specific heat values take longer to warm than those with lower values if the same amount of heat is delivered to each material. More heat energy is required to cause a temperature change in high specific heat substances.

For example, note that water has a specific heat of 1.0 and ice has a specific heat of 0.49. This means that, when applying equal amounts of heat to liquid water and ice, the ice will warm about twice as much as the liquid water. It reaches a higher temperature after the same amount of heat energy addition because of its lower specific heat rating. Objects with higher specific heat ratings store more energy than those with lower specific heats. The higher the specific heat, the more energy the substance can absorb as it heats and release as it cools.

Note that air is not listed in Table 1.2. This is because air is not a simple substance, but a complex mixture of gases. One complicating factor of air is its humidity, or how much moisture it contains in the gaseous (vapor) state.

The formula used to determine the amount of heat contained in an object is

$$\text{Heat} = \text{Mass} \times \text{Specific heat} \times \text{Temperature}$$

This equation means that multiplying the mass (or weight) of a given substance by its specific heat and temperature gives the amount of heat contained in that object.

TABLE 1.2 Specific Heat of Common Substances

Material	Specific heat
Brick	0.22
Concrete	0.156
Glass	0.199
Granite	0.195
Ice	0.487
Machine oil	0.4
Soil	0.44
Distilled water	1
Wood	0.63

Note that this formula allows a substance to have a "negative energy." An object with a temperature below zero produces a negative result. This is fine. By convention, energy in air is usually calculated with the temperature measured in degrees Fahrenheit. Negative energy numbers are allowed; do not be concerned about this.

Most HVAC applications work with the *change* of energy in a substance caused by a change in its temperature. Because of this, the original heat values are subtracted from each other and the answers come out correct in the end.

Standard convention has defined that the energy in water is calculated by using its temperature above 32°F. Therefore, subtract 32 from the water's temperature reading (in degrees Fahrenheit) when estimating its energy.

For example, we have 1 gal of water at 75°F. How much heat does it contain?

1 gal of water = 8.3 lb

Specific heat of water = 1.0 Btu/lb/°F

Temperature = 75°F

Btu in the water = 8.3 lb × 1.0 Btu/lb × (75°F – 32°F) = 357 Btu

Remember to subtract 32 from the temperature before multiplying.

Just one more word regarding the possibility of air, or any other substance, having a "negative energy." The most scientifically accurate way to describe the temperature of a substance is to use its temperature reading in degrees Fahrenheit plus 460. This new figure is its temperature in degrees Rankine. This is its *absolute* temperature, and energy calculations made with it will never be negative. There is no temperature colder than –460°F. It represents the point where the atoms and molecules in a substance have almost entirely stopped moving and vibrating. Because they cannot go more slowly, the temperature of the substance cannot get any lower.

Remember to use standard engineering conventions when determining energy levels in air or water. Use normal degrees Fahrenheit for air and degrees Fahrenheit minus 32 for water.

Mass

Heavier objects can hold more heat than lighter ones (assuming they are made of the same substance). For example, 2 gal of water can hold about twice as much heat as 1 gal. This is important to remember: The amount of heat a substance can contain depends on its mass.

Effects of the mass of an object on its heat-carrying ability are important in HVAC work. The amount of heat transfer material flowing in a system (gallons per minute of water, or cubic feet per minute of air) is as important as its temperature. Proper operation of heating and cooling systems depends both on the heat transfer material's temperature and on a sufficient flow of the material. Too little flow (and therefore mass) and the amount of heat delivered will be insufficient for the job.

Another example: We have 100 gal of water at 45°F. How much heat is in this quantity of water?

$$100 \text{ gal of water} = 830 \text{ lb}$$

$$\text{Temperature} = 45°\text{F}$$

$$\text{Btu} = 830 \text{ lb} \times 1.0 \text{ Btu/lb} \times (45°\text{F} - 32°\text{F}) = 10{,}790 \text{ Btu}$$

Note that this much larger quantity of colder water contains considerably more heat than the warmer water in the previous example. Heat quantity does not depend on temperature alone. The larger mass (weight) of the water in this example allows it to hold more heat energy.

A more practical example: A hydronic heating system begins operating with its water at 70°F and heats it to 180°F. The heating system holds 125 gal. How much heat did the boiler have to add to the water (not including any heat delivered by the system to the building during the heating process)?

$$125 \text{ gal} \times 8.3 \text{ lb/gal} = 1038 \text{ lb}$$

The energy at the start was

$$\text{Btu} = 1038 \text{ lb} \times 1.0 \text{ Btu/lb} \times (70°\text{F} - 32°\text{F}) = 39{,}444 \text{ Btu}$$

The energy after heating was

$$\text{Btu} = 1038 \text{ lb} \times 1.0 \text{ Btu/lb} \times (180°\text{F} - 32°\text{F}) = 153{,}624 \text{ Btu}$$

The difference between the starting and ending energies represents the heat the boiler added to the water and is found by subtracting:

$$153{,}624 \text{ Btu} - 39{,}444 \text{ Btu} = 114{,}180 \text{ Btu}$$

The same answer can be found by subtracting the starting and ending water temperatures in the first place. This eliminates the need to subtract 32 from each temperature.

$$\text{Btu} = 1038 \text{ lb} \times 1.0 \text{ Btu/lb} \times (180°\text{F} - 70°\text{F}) = 114{,}180 \text{ Btu}$$

There is no need to subtract 32 from each temperature when working with temperature changes; just subtract the temperatures. Because the answer is the amount of the water's energy *change*, the difference is the same whether the 32 is first subtracted or not.

Heat in Air (Psychrometrics)

As noted earlier, there is no simple specific heat value for air. The addition of water vapor (humidity) to air in variable amounts changes the amount of heat required to warm it. The effects of heating and cooling moist air is called *psychrometrics*.

Adding or removing humidity not only changes the way air holds and reacts to heat, it changes the amount of energy contained in the air. The more moisture (humidity or water vapor) contained in a quantity of air, the more energy the air contains. This additional energy is called *latent heat*. It is just as real as additional energy derived from an increase in temperature, but it does not increase the air's temperature.

Keep this idea in mind: More humidity in the air means there is more energy in the air at any given temperature. This extra energy, or latent heat, came from the heat required to evaporate the water from a liquid form. This is the water's *heat of vaporization*, and it is present in humid air.

Water requires heat energy to change from a liquid to a vapor state. Approximately 1070 Btu is needed to evaporate 1 lb of water. Energy does not disappear when the water is evaporated. It exists in the vapor as the potential energy absorbed during evaporation. Each pound of water vapor in air contributes its 1070 Btu of evaporation energy to the total energy contained in the air.

The type of heat that is most familiar is called *sensible heat*. Adding sensible heat to air increases its temperature. Sensible heat is added by most heating systems to increase the air's temperature. It is what is most commonly thought of as heat. This form of heat is called *sensible* because it makes the air feel warmer.

The total amount of heat in air is the sum of its latent and sensible components. This total heat is referred to as the air's *enthalpy*. Consider enthalpy to mean the same as *total energy*.

The amount of sensible heat in air compared to the total of the sensible and latent heat combined is the *sensible heat ratio*. Hotter and drier air has a higher sensible heat ratio because more of its total heat is of the sensible type. Cool, damp air has a lower sensible heat ratio because much of its heat is the latent type.

It is entirely possible for one parcel of hot, dry air and another of cool, moist air to have the same total energy. While their sensible heat ratios are different, the total amount of energy in each can be the same.

Table 1.3 shows the energy contained in humid air at various temperatures and humidities. It also gives the air's dew point temperature and the specific volume (in cubic feet of 1 pound of air) at those conditions.

TABLE 1.3 Psychrometric Values*

Temperature (°F)	Relative humidity	Total heat (Btu/lb)	Total heat (Btu/ft³)	Volume (ft³/lb)
50	0	12	0.94	12.8
	10	12.9	1.00	12.9
	20	13.8	1.07	12.9
	30	14.6	1.13	12.9
	40	15.3	1.19	12.9
	50	16.1	1.25	12.9
	60	17	1.32	12.9
	70	17.8	1.37	13
	80	18.5	1.42	13
	90	19.4	1.49	13
	100	20.2	1.55	13
55	0	13.2	1.02	13
	10	14.4	1.11	13
	20	15.2	1.17	13
	30	16.1	1.24	13
	40	17.1	1.31	13.1
	50	18.1	1.38	13.1
	60	19.1	1.46	13.1
	70	20.1	1.53	13.1
	80	21.1	1.61	13.1
	90	22.1	1.69	13.1
	100	23.2	1.76	13.2
60	0	14.2	1.08	13.1
	10	15.6	1.19	13.1
	20	16.6	1.27	13.1
	30	18	1.36	13.2
	40	19.1	1.45	13.2
	50	20.1	1.52	13.2
	60	21.5	1.63	13.2
	70	22.6	1.70	13.3
	80	23.9	1.80	13.3
	90	25.2	1.89	13.3
	100	25.5	1.92	13.3
65	0	15.5	1.17	13.2
	10	17	1.29	13.2
	20	18.2	1.37	13.3
	30	19.8	1.49	13.3
	40	21.2	1.59	13.3
	50	22.6	1.69	13.4
	60	24.2	1.81	13.4
	70	25.6	1.91	13.4
	80	27.1	2.02	13.4
	90	28.5	2.11	13.5
	100	30.1	2.23	13.5
70	0	16.8	1.26	13.3
	10	18.8	1.40	13.4
	20	20.1	1.50	13.4
	30	21.9	1.63	13.4
	40	23.5	1.74	13.5

Temperature (°F)	Relative humidity	Total heat (Btu/lb)	Total heat (Btu/ft^3)	Volume (ft^3/lb)
	50	25.3	1.87	13.5
	60	27	2.00	13.5
	70	28.7	2.11	13.6
	80	30.5	2.24	13.6
	90	32.3	2.36	13.7
	100	34.1	2.49	13.7
75	0	18	1.33	13.5
	10	20	1.48	13.5
	20	22	1.62	13.6
	30	24.6	1.81	13.6
	40	26.1	1.92	13.6
	50	28	2.04	13.7
	60	30.1	2.20	13.7
	70	32.2	2.33	13.8
	80	34.2	2.48	13.8
	90	36.4	2.64	13.8
	100	38.5	2.77	13.9
80	0	19.2	1.41	13.6
	10	21.6	1.58	13.7
	20	23.9	1.74	13.7
	30	26.3	1.92	13.7
	40	28.7	2.08	13.8
	50	31.1	2.25	13.8
	60	33.6	2.42	13.9
	70	36.1	2.60	13.9
	80	38.6	2.76	14
	90	41.1	2.94	14
	100	43.7	3.10	14.1
85	0	20.4	1.49	13.7
	10	23.1	1.67	13.8
	20	26	1.88	13.8
	30	28.7	2.06	13.9
	40	31.7	2.26	14
	50	34.5	2.46	14
	60	37.5	2.66	14.1
	70	40.4	2.87	14.1
	80	43.4	3.06	14.2
	90	46.3	3.24	14.3
	100	49.5	3.46	14.3
90	0	21.4	1.55	13.8
	10	24.9	1.79	13.9
	20	28	2.00	14
	30	31.5	2.23	14.1
	40	34.8	2.47	14.1
	50	38.2	2.69	14.2
	60	41.7	2.92	14.3
	70	45.1	3.15	14.3
	80	48.9	3.40	14.4
	90	52.4	3.61	14.5

TABLE 1.3 (Continued)

Temperature (°F)	Relative humidity	Total heat (Btu/lb)	Total heat (Btu/ft³)	Volume (ft³/lb)
95	0	22.8	1.63	14
	10	26.6	1.89	14.1
	20	30.4	2.16	14.1
	30	34.4	2.42	14.2
	40	38.4	2.69	14.3
	50	42.5	2.95	14.4
	60	46.4	3.20	14.5
	70	50.6	3.49	14.5
	80	54.9	3.76	14.6
100	0	24	1.70	14.1
	10	28.5	2.01	14.2
	20	33.1	2.31	14.3
	30	37.6	2.61	14.4
	40	42.2	2.91	14.5
	50	47.1	3.23	14.6
	60	51.8	3.52	14.7

*The values in the table correspond to the amount of heat contained in 1 lb of air and the volume of 1 ft³ of air.

If the exact value of the temperature or humidity is not on the table, it isn't hard to estimate the energy in the air. Find the table's temperature and humidity values above and below the location of the "problem" air's conditions. The problem air's conditions should be between the chosen values. Then, estimate an enthalpy value between the charted values. It is usually permissible to use a value approximately halfway between the charted values if the problem's conditions are roughly halfway between those shown. Don't spend too much time making these estimates. The value does not need to be extremely precise for most heating and cooling work.

It is important to figure out the amount of energy in air. We need it to estimate the required capacity of HVAC equipment, particularly air conditioning. Air conditioning systems usually have to remove both sensible and latent heat from air. Energy values obtained from Table 1.3 give the total amount of heat in the air the cooling system has to handle.

For example, how much heat must be removed from a room 24 ft long by 18 ft wide by 9 ft high? The air in the room is at 85°F with 60% relative humidity and we wish to cool it to 75°F with 50% relative humidity.

$$\text{Volume of the room} = 24\text{ ft} \times 18\text{ ft} \times 9\text{ ft} = 3888\text{ ft}^3$$

Table 1.3 shows that each cubic foot of air at 85°F and 60% relative humidity contains 2.66 Btu. The total energy is

$$2.66\text{ Btu/ft}^3 \times 3888\text{ ft}^3 = 10{,}342\text{ Btu}$$

Table 1.3 shows that air at 75°F and 50% relative humidity contains 2.04 Btu/ft^3. Energy at the cooler temperature in the same 3888 ft^3 of air is

$$2.04 \text{ Btu/ft}^3 \times 3888 \text{ ft}^3 = 7932 \text{ Btu}$$

The difference is the amount of heat removed from the air:

$$10{,}342 \text{ Btu} - 7932 \text{ Btu} = 2410 \text{ Btu}$$

We can simplify the math by subtracting the differences in the air's Btu contents before multiplying. The equation is

$$3888 \text{ ft}^3 \times (2.66 \text{ Btu/ft}^3 - 2.04 \text{ Btu/ft}^3) = 2410 \text{ Btu}$$

That is the amount of cooling an air conditioning system needs to provide to initially cool the example room. This does not include cooling for additional heat loads and ongoing heat removal needed to keep it cool.

Another important property of air's behavior is the change in relative humidity with temperature. The relative humidity of a parcel of air decreases as the air's temperature increases. It feels drier. The relative humidity increases as the air is cooled. Air can hold greater and greater amounts of moisture as it is heated. Because the mass of water contained in the air parcel does not change, the ratio of the moisture contained in the air compared to its moisture holding capacity goes down. Relative humidity compares the amount of water held in the air to the maximum amount the air could hold (at a given temperature).

Air feels drier when it is warmed even though the amount of water vapor has not changed. Increasing the air's temperature increases its sensible heat, but does not change the mass of water vapor in the air. Changing the mass of water vapor in the air requires physically removing it. An increase in the air's temperature (increasing its sensible heat) will not do that.

Similarly, removing sensible heat from moist air, causing its temperature to decrease, increases the air's relative humidity. Colder air cannot hold as much moisture as warmer air. Reducing the air's temperature causes the amount of moisture contained in it to become larger compared to the amount of water the air could possibly hold. The air becomes damper as it is cooled.

If the air is cooled far enough, it reaches its *dew point* temperature. This is the temperature where the air becomes saturated with water vapor. Its relative humidity is 100% and it cannot hold any more water. If the air is cooled further some of the water will come out of the air and form liquid water. This process is called *condensation*, and it is the opposite of evaporation. Water changes from the vapor state to the liquid state.

During condensation, the water returns the energy that had originally gone into its evaporation. Each pound of water that condenses out from its vapor phase delivers about 1070 Btu of sensible heat energy. Removing water from the air also removes latent heat. This is the heat energy that is changed into sensible heat during condensation, keeping the total amount of energy in the system constant. Cooling moist air below its dew point (as many air conditioning systems do) causes this heat conversion.

Knowing the constant dew point behavior of sensible heated or cooled air can help when using Table 1.3. Once the air's initial dew point is found on the table, the final air's condition will have the same dew point.

An example of sensible air heating will be helpful. A room 12 ft by 16 ft by 9 ft is at 60°F and 50% relative humidity. Find how much heat is needed to warm it to 75°F. All of the heat supplied is sensible.

$$\text{Volume of the room} = 12\text{ ft} \times 16\text{ ft} \times 9\text{ ft} = 1728\text{ ft}^3$$

At 60°F and 50% relative humidity the air has 1.52 Btu/ft^3.

$$\text{Total energy of the air} = 1728\text{ ft}^3 \times 1.52\text{ Btu/ft}^3 = 2627\text{ Btu}$$

The initial dew point of the air is 41.5°F. The closest condition shown on Table 1.3 is where air has a dry bulb temperature of 75°F, a dew point of 41.5°F, and a relative humidity of 30%. At those conditions the air has an energy content of 1.81 Btu/ft^3. Therefore, the total heat energy added can be found:

$$\text{Total energy of the air at the final condition} = 1728\text{ ft}^3 \times 1.81\text{ Btu/ft}^3 = 3128\text{ Btu}$$

$$\text{Energy added to the air} = 3128\text{ Btu} - 2627\text{ Btu} = 501\text{ Btu}$$

Similarly, cooling moist air without condensation of water vapor maintains the air's dew point at a constant temperature. The amount of water in the air at initial conditions is the same as at the end. Subtracting the air's starting enthalpy from its enthalpy at the end gives the energy removed from the air.

If water evaporates into air, and its temperature is approximately equal to the air's wet bulb temperature, evaporation causes the air to cool. Sensible heat energy is removed from the air (causing its temperature to drop) as the air's latent heat is increased by the addition of the evaporated water. Sensible heat is exchanged for latent heat, but the total amount of heat in the air is not changed. Again, the air loses approximately 1070 Btu of sensible heat for each pound of water evaporated into it. This sensible heat reduction causes the air's temperature to decrease.

As water evaporates into the air it cools and its relative humidity increases. The enthalpy, however, remains constant. Note that if we initially had air at 85°F and 20% relative humidity, its enthalpy is 26 Btu/lb (see Table 1.3). Cooling this air through evaporation to 65°F leaves us with the air at approximately 75% relative humidity. The enthalpy has not changed, but the air is cooler and moister with less sensible heat and more latent heat.

Note that if the evaporating water is hotter or colder than the air's wet bulb temperature it can change the total amount of energy in the air. Spraying hot water into air can heat the air, in spite of the evaporative cooling that takes place. Similarly, spraying air with chilled water usually provides much more cooling than is caused by evaporation alone. Most often, however, the effect of water evaporation on air is to cool it. Keep in mind that the total energy of the air does not change because of the water's evaporation.

Standard air

An important psychrometric idea is that of *standard air* conditions. Dry air (with no water vapor, or 0% relative humidity) at normal atmospheric pressure (which is all we have been considering) and at a temperature of 70°F is considered standard air. At these conditions the volume is 13.34 ft^3/lb and the specific heat is 0.240 Btu/lb/°F. Heating and cooling equipment manufacturers measure the performance of equipment with standard air.

Standard air is also useful for calculating the energy requirements of systems. Calculations performed with standard air properties are usually very good approximations of the actual conditions for most HVAC work.

Humidity definitions

It may be helpful to review and define some commonly used terms associated with humidity and the study of air conditions.

Condensation. The process of changing water from its vapor phase to its liquid phase. As water condenses, it releases heat energy. Each pound of water that condenses liberates 1070 Btu of heat. This heat increases the temperature of the surface and the surrounding space it is condensing into.

Dew point. That temperature that causes humidity in air to condense, or cause dew. This is the temperature that causes air to be totally saturated with water, meaning it cannot hold any more. Cooling air below its dew point causes water to condense out of it.

Dry bulb. A temperature reading obtained with the thermometer's sensing bulb kept dry. This is the same as an ordinary temperature reading.

Enthalpy. The total amount of energy in air. It is equal to the sum of the sensible and latent heat.

Evaporation. The process of changing water from its liquid phase to its vapor phase. This process requires that energy be added to the water. Specifically, 1070 Btu is needed to evaporate 1 lb of water. Evaporation can cause surrounding temperatures to decrease because of this sensible heat absorption.

Humidity. The water vapor contained in air. The quantity of water contained in air is expressed in various terms, including grains of water per pound of air.

If this measure is encountered, remember that 1 lb is equal to 6985 grains. Typical humidity levels correspond with 50 to 100 grains/lb of air. Water vapor makes up a small amount of the total mass of a humid air mixture.

Latent heat. The moisture present in air. The reason water vapor's presence is a form of heat is because this vapor absorbed energy when it originally evaporated and will return it if it is condensed.

Relative humidity. The amount of water contained in air at a given temperature compared to how much it could possibly hold at that temperature. This is also referred to as *degree of saturation*. At the dew point the relative humidity is 100% because the air cannot hold any more moisture. Air with no water vapor in it has a relative humidity of 0%.

Saturation. The condition where air cannot hold any additional moisture. Air that is saturated is at its dew point temperature and further cooling causes condensation.

Sensible heat. Heat energy in air manifested by temperature. The more sensible heat in air, the higher its temperature. Do not assume, however, that a quantity of higher temperature air must have more sensible heat than a quantity of lower temperature air. The total amount of energy also depends upon the mass (weight) of the air and the amount of latent heat (water vapor) in it.

Standard air. Dry air at normal atmospheric pressure (14.7 psi) and at 70°F. Its volume is 13.34 ft^3/lb and its specific heat is 0.240 Btu/lb/°F.

Vapor pressure. The actual pressure exerted by the water vapor contained in air. The total pressure exerted by air comes from the various gases in the air, with nitrogen and oxygen making the largest contributions. Water vapor contributes a small amount to the total pressure, but when it is higher in one area than another it moves to the lower pressure area.

Wet bulb. A temperature reading obtained with a thermometer that has its sensing bulb kept wet. It is usually lower than the dry bulb, or ordinary, temperature because the evaporation of water from the bulb's surface causes a cooling effect. The less humidity in the surrounding air, the cooler the wet bulb temperature is compared to the dry bulb temperature. For example, 70°F air at 30% relative humidity has a wet bulb reading of 53°F. If the 70°F air's wet bulb temperature was also 70°F it would indicate that the air was saturated and at 100% relative humidity.

Buoyancy and the stack effect

Most substances expand when heated and contract when cooled. This is also true of air. The density decreases and volume increases as heat is added.

This means that a given mass of air gets larger and lighter as heat is added. This occurs whether the added heat is latent or sensible. Adding heat or humidity causes air to swell, reducing its density. This causes it to become lighter and float on top of cooler, drier air.

This effect leads to the notion that "heat rises." Heat energy, itself, does not move unless driven by a temperature difference. It flows upward only if it is moving from a hotter area to a colder one. Hot air, on the other hand, tends to float above cooler air if given the chance.

Natural draft chimneys depend on this effect. The hot gases inside are much more buoyant than the cooler air around them. They lift to the top and rise out without danger of leaking into the building. While the stack effect is useful and positive with chimneys, it can have negative effects on the performance of HVAC systems.

The change in air density because of temperature differences can have serious implications in buildings over three stories high (or two stories plus a basement). Warmer air migrates to the top and cooler air settles to the bottom. A distinct layering of air occurs. This can affect the pressure in the building.

When heating a building surrounded by cold, outdoor air, the hot air delivered by the heating system accumulates near the upper parts of the building. This accumulation leads to an increase in pressure there and pushes the warm air outside. The air will go through any gaps or spaces available in the building's outside walls or roof (see Fig. 1.1). While this is happening, outside air is trying to get into the building at the bottom where the pressure is lowest. It enters the building to make up for the air being lost at the top.

The entire system sets up a large convective system. Cold outdoor air enters at the bottom, is warmed by the heating system, and rises to the top. It is lost to the outside again because of the higher pressure there.

Building height and the differences in outside and inside temperature directly affect the intensity of the stack effect. Differences between indoor and outdoor temperature are usually the greatest influence on air density differences. Tall buildings in very cold climates are more likely to have more stack effect problems than those in milder areas.

Stack effects can make servicing tall buildings a challenge. If called in for "too cold" calls, find out if the parts of the building with the most comfort problems correspond to areas where cold outside air may be migrating inside. Of course, the stack effect is not the reason that outside air can force its way inside. Strong winds on one side of the building can also force in air.

The cures for stack effect related complaints often have nothing to do with the HVAC system. Sealing leaky doors and windows or gaps in the building's siding is the best way to solve the problem. Some cases, however, may be more difficult. Balancing the air delivery system or adding supplementary heating may be necessary.

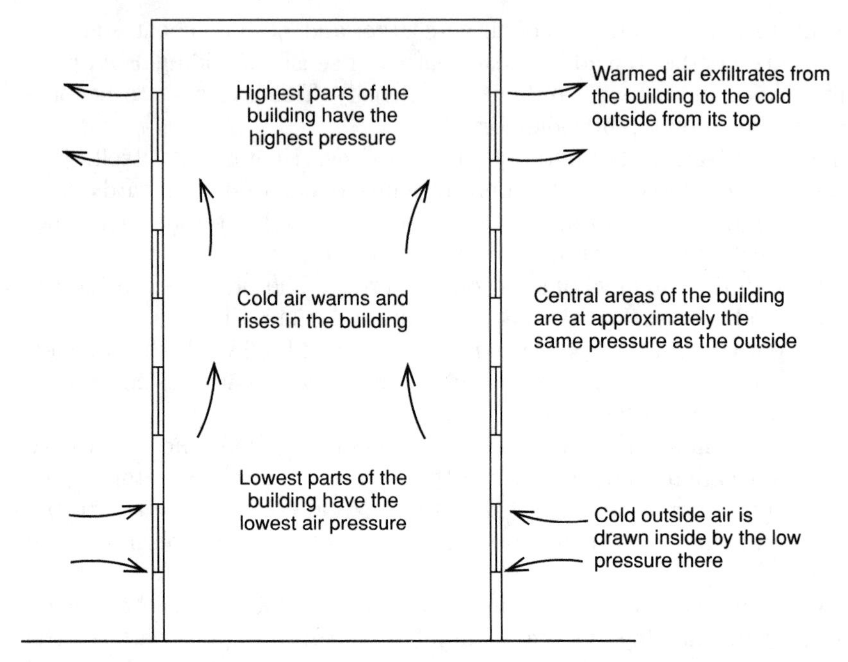

Figure 1.1. Building stack effect and pressure differences compared to outside air.

Finally, when installing HVAC systems in a new or renovated tall building, try to anticipate this effect. Expect cold drafts on lower floors in the winter or "too warm" conditions in upper areas in the summer. Heating and cooling technicians usually arrive at a building long after it is designed and the shell is up. They usually have no say about the choice of materials. Use good design practice as described in this book and problems will be minimized.

Power

Closely related to energy is the idea of power. Power is energy applied over time. One process uses more power than another if it uses more energy over the same amount of time, or the same amount of energy over a shorter period of time. Power involves the flow of energy, and most HVAC applications involve the flow of heat.

Mathematically, power is the energy applied divided by the time it is used:

$$\text{Power} = \frac{\text{Energy}}{\text{Time}}$$

For example, 1000 Btu of heat flow measured over a 1-hour period would be equal to a power of 1000 Btu/hr. What if the time required was 2½ hours?

The power would be

$$\text{Power} = \frac{1000 \text{ Btu}}{2.5 \text{ hr}} = 400 \text{ Btu/hr}$$

We can use the power equation to find the amount of time required to heat or cool something. We need to know the energy required and the power output available from the heating or cooling system. For example, a 55-gal tank of water is at a temperature of 45°F. We want to find out how long it takes to heat it to 70°F. with a 30,000 Btu/hr heater. We know that

$$\text{Power} = \frac{\text{Btu required}}{\text{Time}}$$

Changing the equation around to solve for Time gives us

$$\text{Time} = \frac{\text{Btu required}}{\text{Power}}$$

In our example we need to heat a quantity of water. Fifty-five gallons of water at 8.3 lb/gal weighs

$$\text{Weight} = 55 \text{ gal} \times 8.3 \text{ lb/gal} = 457 \text{ lb}$$

$$\text{Btu required} = 457 \text{ lb} \times 1.0 \text{ Btu/lb} \times (70°\text{F} - 45°\text{F}) = 11{,}425 \text{ Btu}$$

The time required would be

$$\text{Time} = \frac{11{,}425 \text{ Btu}}{30{,}000 \text{ Btu/hr}} = 0.381 \text{ hr}$$

That is the answer we were looking for. We can convert it to the number of minutes needed by multiplying 0.381 hours by 60 min/hr to get 23 minutes. It takes the heater 23 minutes to warm the water.

There are many commonly used units for defining power. Table 1.1 contains most of those needed when working with HVAC systems. Special mention should be made of the power unit *ton*. This is commonly used for describing air conditioning system capacity. One ton is equal to 12,000 Btu/hr. A 5-ton air conditioner would have a capacity of

$$\text{Capacity} = 5 \times 12{,}000 \text{ Btu/hr} = 60{,}000 \text{ Btu/hr}$$

Another common power unit is MBH. This is an abbreviation for 1000 Btu/hr. One ton is equal to 12 MBH.

The most common metric unit of power measure is the watt. We often use this measurement when dealing with electric appliances and heaters. Electrically, the power (in watts) is usually equal to the number of volts a system operates at multiplied by the number of amperes flowing through it. Electrical circuits have an additional mathematical term called the *power factor* that further defines a circuit's power consumption. The electrical power equation is

$$\text{Power} = \frac{\text{Volts} \times \text{Amperes}}{\text{Power factor}}$$

For most electrical circuits the power factor is very close to 1.0, so it usually is ignored for HVAC load and power calculations. A circuit's power factor must be considered only when trying to closely define the watt rating of motors and other magnetically driven electrical equipment.

Converting back and forth between watts and Btu/hr is easy. Use the following equations:

$$\text{Btu/hr} = \text{Watts} \times 3.412$$

$$\text{Watts} = \frac{\text{Btu/hr}}{3.412}$$

Find the number of Btu per hour available from a 1200-W electric heater:

$$\text{Btu/hr} = 1200 \text{ W} \times 3.412 = 4094 \text{ Btu/hr}$$

Another example: Find the time it takes to heat 55 gal of water at 45°F to 70°F with a 1000-W heater. From the earlier example, we found that 55 gal of water weighed 457 lb and required 11,425 Btu to increase its temperature from 45°F to 70°F. The Btu per hour available from the electric heater is

$$\text{Btu/hr} = 1000 \text{ W} \times 3.412 = 3412 \text{ Btu/hr}$$

$$\text{Time} = \frac{11{,}425 \text{ Btu}}{3412 \text{ Btu/hr}} = 3.35 \text{ hr}$$

Find the required electric heater capacity in Btu per hour and watts to warm a room 10 ft by 12 ft by 8 ft. It starts at 50°F and must be warmed to 70°F in 15 minutes (0.25 hr). The room starts with a relative humidity of 50% and will be at 25% relative humidity when heated to 70°F.

First, the room contains 10 ft × 12 ft × 8 ft = 960 ft^3. Using Table 1.3, the air in the room initially contains 1.25 Btu/ft^3. The total amount of heat in the room's air is

$$\text{Btu} = 960 \text{ ft}^3 \times 1.25 \text{ Btu/ft}^3 = 1200 \text{ Btu}$$

Conditions in the room after the air is heated can be found in Table 1.3 by figuring halfway between the Btu per cubic foot of 20% and 30% relative humidity (1.50 and 1.63 Btu/ft^3, respectively). The air, therefore, contains about 1.57 Btu/ft^3 of heat. Its total heat after the heater has run is

$$\text{Btu} = 960 \text{ ft}^3 \times 1.57 \text{ Btu/ft}^3 = 1507 \text{ Btu}$$

The amount of energy the heater delivered to the room is

$$\text{Btu increase} = 1507 \text{ Btu} - 1200 \text{ Btu} = 307 \text{ Btu}$$

Finally, we know that Power = Btu required/Time, so

$$\text{Power} = \frac{307 \text{ Btu}}{0.25 \text{ hr}} = 1228 \text{ Btu/hr}$$

We can find the electric heater's wattage by dividing our 1228 Btu/hr by 3.412 to arrive at

$$\frac{1228 \text{ Btu/hr}}{3.412 \text{ W/Btu/hr}} = 360 \text{ W}$$

Note that our answer is a reasonable approximation. Although not included in our calculations, the mass of air remaining in the room at the end of its heating was probably less than at the start. As the air heated it expanded. To avoid increasing the pressure in the room some of it would escape to the outside. The volume of the room did not change, so after heating, all of the air would not fit without increasing its pressure. Remember, initially the air required 12.9 ft^3/lb, but Table 1.3 shows that after heating it needed 13.4 ft^3. This air loss is usually not very important for HVAC estimating.

We can repeat the above calculations using the properties of standard air. This shows how little difference using standard air's properties really makes.

Room volume	960 ft^3
Room air's initial energy	16.1 Btu/lb × 960 ft^3/13.34 ft^3/lb = 1159 Btu
Room air's final energy	21 × 960 ft^3/13.34 ft^3/lb = 1511 Btu
Energy added to the air	1511 Btu – 1159 Btu = 352 Btu
Power required	352 Btu/0.25 hr = 1408 Btu/hr
Heater wattage required	1408 Btu/hr/3.412 = 413 W

Using the original weight of air for our calculations, we arrived at an answer that would cause us to install a slightly smaller heater. The heater chosen with the standard air properties is a little larger. Which is correct? Neither.

The exact amount of heat, allowing for the air's expansion in the room as it warms, requires a very complex calculation involving integral calculus. It is not necessary to do this kind of calculation. Standard air properties usually provide an acceptable answer. The solution found using only the air's initial and final properties (the first way the above example was done) selected a heater only slightly smaller than the standard air's estimates. Using either method produced an answer within a few percent of the most accurate calculus-derived answer. Both methods are perfectly acceptable for heating and cooling work.

Chapter 2

Heat Transfer

Heating and cooling systems depend entirely on the transfer of heat. The previous chapter discussed the ideas of what heat is, the forms it can take, and some of its effects. This chapter discusses how heat moves from one place or thing to another, and how to calculate the amount of heat transferred.

Heat transfer takes place by three modes: conduction, convection, and radiation. In a building these transfer methods work together, moving heat from higher temperature areas to those of lower temperature. They can be used as part of a high-quality HVAC system that provides comfort and economical operation.

Conduction

Heat transfer by conduction requires direct contact between the substances carrying heat. Molecules of nonmetallic materials (gases, insulation, and most fluids) conduct heat by vibrating within the material. These molecules send energy to neighboring particles when their vibration causes them to hit and move them. Energy transfers from the impact of faster molecules into slower ones. Like a cue ball breaking up a rack of balls in a billiard game, faster moving molecules deliver energy to slower ones.

Higher temperature molecules move faster than the cooler ones. Hotter molecules deliver energy to cooler ones when they strike them. Energy transfers from these molecular impacts. Hotter molecules move more slowly and cooler ones move faster after they are hit. This increases the temperature of the slower moving molecules.

Metallic solids, or any material that conducts electricity, transfer heat the same way they conduct electricity. Electrons move from one part of the microscopic crystal to another carrying energy (as heat) with them.

To predict how heat conducts through materials, a few basic laws have been developed. It's important to learn them and understand how heat transfer takes place so you will understand the operating principles of heating and cooling systems.

Heat conduction through an object depends on three things: the size of the object, a property called its *thermal conductivity*, and the temperature difference driving the heat. Formulas describe the effects of these factors.

The basic formula for heat conduction through an object is

$$q = k \times \left(\frac{A}{L}\right) \times T_d$$

where q = heat flow, Btu/hr
k = heat conductivity, Btu/hr/ft/°F
A = area through which the heat is flowing, ft^2
L = length of material the heat is flowing through, ft
T_d = high temperature – low temperature, °F

This formula shows that the heat transferred through an object increases when any of the following increase:

- The value of the object's thermal conductivity
- The area available for the heat to flow through
- The difference in the temperature from the object's hot side to the cold side

On the other hand, increasing the length of the path that the heat must flow through decreases heat transfer.

The thermal conductivity of most substances varies with changes in temperature, but an average value can be used for most materials in HVAC applications. Table 2.1 gives the thermal conductivities for some common substances.

TABLE 2.1 Thermal Conductivities and R-Values

Building materials	Thermal	Resistance R-value	
Material	conductivity	Per inch	Nominal value
Gypsum board	1.13	0.88	
Cement plaster	5.00	0.20	
Gypsum plaster	1.49	0.67	
Plywood	0.80	1.25	
Hardwoods	1.15	0.87	
Softwoods	0.87	1.15	

Building materials	Thermal	Resistance R-value	
Material	conductivity	Per inch	Nominal value
Hardboard siding	1.49	0.67	
Brick, common	5.00	0.20	
Brick, face	9.09	0.11	
Aluminum siding			
Hollow backed			0.61
⅜ in. insulation board backed			1.82
⅜ in. foil-backed insulation board			2.96
Hollow clay tile			
3 in.			0.80
4 in.			1.11
6 in.			1.52
8 in.			1.85
10 in.			2.22
12 in.			2.50
Concrete block			
Common, structural			
4 in.			0.71
8 in.			1.11
12 in.			1.28
Cinder			
3 in.			0.86
4 in.			1.11
8 in.			1.72
12 in.			1.89
Lightweight			
3 in.			1.27
4 in.			1.50
8 in.			2.00
12 in.			2.27
Stone	1.25	0.80	
Concrete, structural	13	0.077	
High density		0.19–0.59	
Lightweight aggregate		0.86–1.43	
Perlite aggregate		0.86–2.00	
Asbestos cement roof shingles			0.21
Asphalt roll roofing			0.15
Asphalt roof shingles			0.44
Built-up roof			0.33
Fiberboard–acoustic tile	0.40	2.50	

Insulating materials	R-value	
Material	Conductivity	Per inch
Mineral or glass fiber batts	0.31	3.20
Glass fiberboard	0.25	4.00
Expanded polystyrene	0.25	4.00
Cellular polyurethane	0.16	6.25
Cellular polyisocyanurate	0.14	7.20
Wood-cement slabs	0.53	1.90
Cellulose (paper) loose fill	0.29	3.50

TABLE 2.1 (Continued)

Insulating materials	R-value	
Material	Conductivity	Per inch
Sawdust or wood shavings	0.45	2.22
Mineral fiber loose fill	0.40	2.50
Field-sprayed polyurethane	0.17	75
Calcium silicate	0.48	2.08
Cellular glass board and block		
Low temperature	0.30	3.33
High temperature	0.45	2.22
Polyurethane foam	0.18	5.56

Excerpt from *ASHRAE Handbook: 1985 Fundamentals*, reprinted with permission.

Using the formula, we can find the heat flow through a 6-in. thickness of structural concrete and the same thickness of clay tile. The size of the concrete slab and tile structure are 22 ft by 28 ft. The temperature on one side is 30°F and 75°F on the other side. The k for concrete is 13 Btu/hr/ft/°F from Table 2.1. The area is 12 ft × 18 ft = 216 ft^2. The length of the concrete heat is flowing through its thickness is 6 in. (0.5 ft).

$$q = 13 \text{ Btu/hr/ft/°F} \times \left(\frac{216 \text{ ft}^2}{0.5 \text{ ft}}\right) \times (75°\text{F} - 30°\text{F}) = 252{,}720 \text{ Btu/1hr}$$

Determining the heat flow through the clay tile requires considering the tile's *R-value*, or resistance to heat. This is a method of rating the heat transfer resistance of building materials, insulation, equipment ducts, and pipes. Heat flow through a material is related to its R-value by the formula

$$q = A \times \frac{T_d}{R}$$

where $R = L/k$

L = length of the heat flow path, or thickness of the material, inches

k = thermal conductivity, Btu/hr/ft^2/°F

This means that the thicker the material (meaning the longer the heat path) or the lower its heat conductivity, the higher the R-value.

The example 6-in. thickness of clay tile has an R-value of 1.52, so the heat flow through it is

$$q = 216 \text{ ft}^2 \times \frac{(75°\text{F} - 30°\text{F})}{1.52} = 6395 \text{ Btu/hr}$$

If we settle on a simple, common material thickness (say 2 in.) for estimating heat losses, then each material has an R-value per inch equal to

$$\frac{R}{\text{inch}} = \frac{1}{k}$$

Table 2.1 also provides R/inch values for most of the materials listed. For example, an insulating material has an R-value of 6. Find the heat flowing through a 6-ft by 12-ft area. The temperature on one side is 20°F and the other side is 130°F. The formula is

$$q = 6\text{ ft} \times 12\text{ ft} \times \frac{(130°\text{F} - 20°\text{F})}{6} = 1320\text{ Btu/hr}$$

R-values are useful for finding the effects of combined materials. It is easy to estimate the amount of heat that flows through layers of stacked materials if the R-value of each material is known.

If heat is forced to flow through several materials in turn, it is easy to find the total heat flow using R/inch values. First, multiply each material's R/inch value by its thickness in inches. This gives each material's individual R-value. Next, add the R-values of the materials together. This gives the total R-value. Finally, use the heat flow equation for R-values (using the total R-value) and the answer is the total heat flow.

$$R_{total} = R_1 + R_2 + R_3 + \ldots$$

Note that the individual R-values are shown numbered. Their sum is the total R value.

$$\text{Btu/hr} = \frac{A \times T}{R_{total}}$$

For example, find the heat flow through 4 in. of concrete backed up with 2 in. of polystyrene foam (see Figure 2.1). The total area is 16 ft^2. The temperature on one face is 0°F and 75°F on the other.

R/inch for concrete	0.077
R-value for the concrete	4 in. × 0.077 = 0.308
R/inch for polystyrene foam	4.0
R-value for the polystyrene foam	2 in. × 4.0 = 8.0
R_{total}	0.308 + 8.0 = 8.308
q	16 ft^2 × (75°F – 0°F)/8.3 = 145 Btu/hr

Another example (see Figure 2.2): Find the heat flow through a 500-ft^2 roof. It consists of 2 in. of glass fiber blanket insulation and 4 in. of structural concrete. The bottom surface is 70°F and the top is 20°F. First you must find the R-values of each material:

R/inch for glass fiber	3.2
R-value for the glass fiber	2 in. × 3.2 = 6.4
R/inch for concrete	0.077
R-value for the concrete	4 in. × 0.077 = 0.308

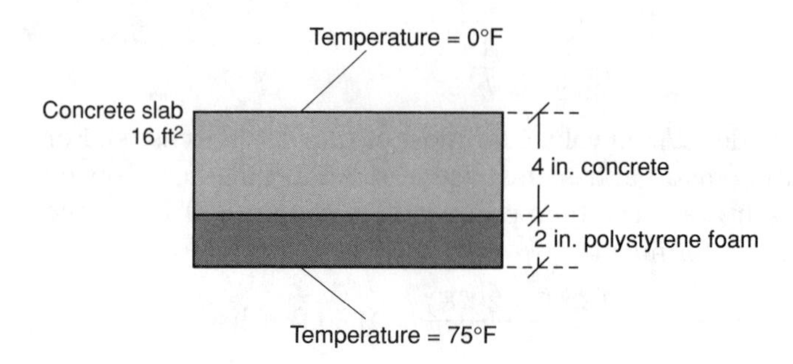

Figure 2.1. Heat conduction through concrete and foam insulation.

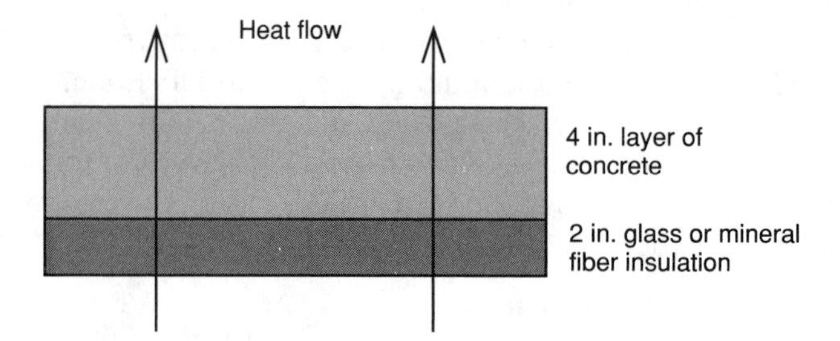

Figure 2.2. Example roof construction with heat flow.

Apply the heat flow equation.

$$q = 500 \text{ ft}^2 \times \frac{(70°\text{F} - 20°\text{F})}{(6.4 + 0.308)} = 3727 \text{ Btu/hr}$$

R-values and R/inch values are powerful and useful for HVAC calculations. They provide an easy way to add the effects of materials that make up a heat conduction path.

The internal temperature of a material varies as it is conducting. Material in contact with the hotter surface assumes a temperature close to the surface's. Similarly, material at the colder surface reaches a temperature nearly equal to it. Within the material the temperature varies from the hottest to coldest. The way the temperature changes within a material is its *temperature profile*. This profile is usually a straight line. The internal temperature varies smoothly and evenly within a material (see Figure 2.3).

For example, halfway through the material, the temperature would be midway between the high and low surface temperatures. Cylindrical materials, like the insulation used to cover pipes, have the greatest temperature change close to the inside surface and the smallest near the outside surface. Heat continues to flow within the material, driven by the differences in its internal temperature.

A material's temperature profile can be important. Condensed moisture can damage insulating materials. Condensation can occur within the insulation if it is covering a component colder than the surrounding air. Moist air penetrating into the insulation cools and forms liquid water inside. The insulation quickly becomes saturated with water.

The previous discussion and formulas on heat conduction assumed that the conduction was taking place in a *steady-state* condition. There were no variations in the temperature over time; the temperatures driving heat transfer were constant. While this is very often true, and can represent the worst case of many heat transfer situations, the condition where temperatures vary with time should be considered.

As temperatures change over time, the process becomes more complicated than simple heat conduction. The actual heating or cooling of the insulating material must be considered. This brings us to the need to consider the insulating material's mass (weight) and specific heat.

More massive materials, or those with higher densities, take longer to change their temperature than those with lower mass or density. The specific heat of a material must be considered when the speed of the tempera-

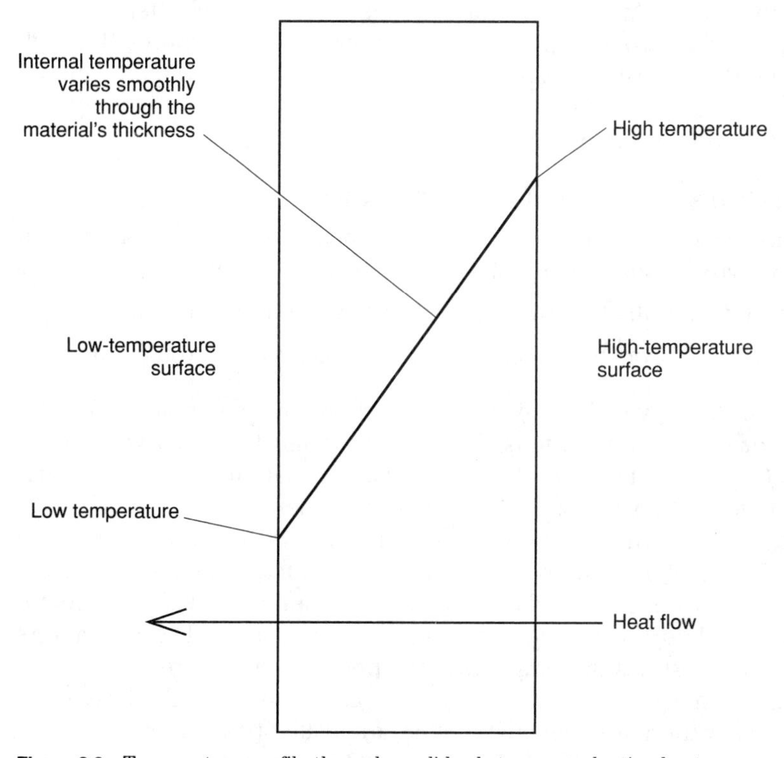

Figure 2.3. Temperature profile through a solid substance conducting heat.

ture change is important. High specific heat materials take longer to heat or cool than those with lower specific heats.

Together, the combined effects of mass (or density), specific heat, and thermal conductivity make certain materials very good at resisting temperature changes. Masonry materials (concrete, stone, and brick) are excellent for storing and slowly releasing heat. Buildings constructed of these materials often stay cool late into a warm day. Before it has a chance to warm the inside air, the building materials absorb and store incoming heat. Later, as the material cools in the evening the building stays warm from the released heat.

When calculating heating loads it is best not to depend on the surrounding building materials to help limit heat losses. It is more conservative to use peak loads as if they were the normal conditions. Detailed engineering calculations take these variations in heating and cooling loads into account, but the HVAC service technician need not be concerned with them.

Insulation

Heating, ventilating, and air conditioning applications make use of various types of insulation. Specific materials are generally used for different applications. Materials are chosen depending upon the temperatures they will handle and their moisture resistance and physical strength requirements.

There are many important functions for insulation in the heating and cooling field. Some common uses for insulation include

- Saving energy by reducing heat losses or gains
- Stopping condensation on cold surfaces (uncontrolled condensation can lead to corrosion and other damage of adjacent materials)
- Preventing injury by keeping exposed surfaces at safe temperatures
- Reducing noise from vibration and sound transmission through ducts

Many local energy codes specify using certain minimum quantities of insulation for various applications. The technician must find out what codes apply to the project being worked on. The technician must be certain that an installation or repair satisfies all code requirements.

Usually, HVAC insulation with a specified minimum fire and smoke rating must be used. Building codes often call for pipe, duct, and equipment insulation that is used in occupied areas to have fire spread and smoke generation ratings of no greater than 25 and 50, respectively. The insulation's manufacturer certifies these ratings in the product's literature.

Insulation's most important property, the ability to resist the passage of heat with a low thermal conductivity value, is a result of its internal structure. These materials usually consist of a matrix of fibers or a mass filled with small pores that makes heat flow difficult. The small connections between the fibers, and the thin material surrounding the internal pores, require heat to take a long path through a small area. It is common for the heat path to be far longer than the distance measured across the material itself. The combination of a long heat path and the small area available for the heat to travel through makes the thermal conductivity very low. Most insulation also resists the passage of air to limit heat transfer caused by the flow of warm or cool air.

Porous insulating materials that have no connections between their internal pores are *closed cell* materials. Porous insulations with interconnected internal cells are *open cell* materials (see Figure 2.4). An example of an open cell material is an ordinary cellulose sponge. It is a fairly efficient insulator, but soaks up water very well because of its interconnected pores. If it becomes saturated with water, its insulating properties are lost. The water acts as a very efficient heat conductor.

Most HVAC insulation must be moisture resistant. In cooling applications it is common to have the insulation and the material being insulated well below the dew point of the ambient air. This leads to condensation wherever moist air is in contact with a cold material.

As noted previously, the dew point can occur inside the insulation (see Figure 2.5). Use insulation with an integral vapor barrier on its outside surface, or install a separate vapor barrier on the warm side of open cell insulation. Alternately, you can use closed cell insulation impervious to moisture. This type of insulation is resistant to moisture penetration even if the vapor barrier's surface is cut or damaged.

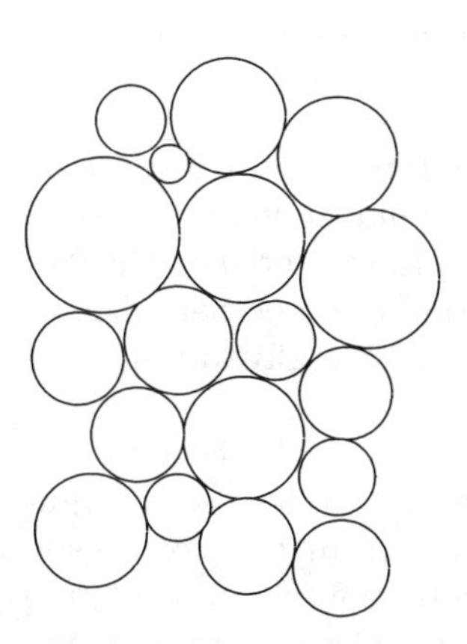

Figure 2.4. Microscopic representation of closed and open cell insulation. Open cell's porous fibers permit fluid to flow through the material.

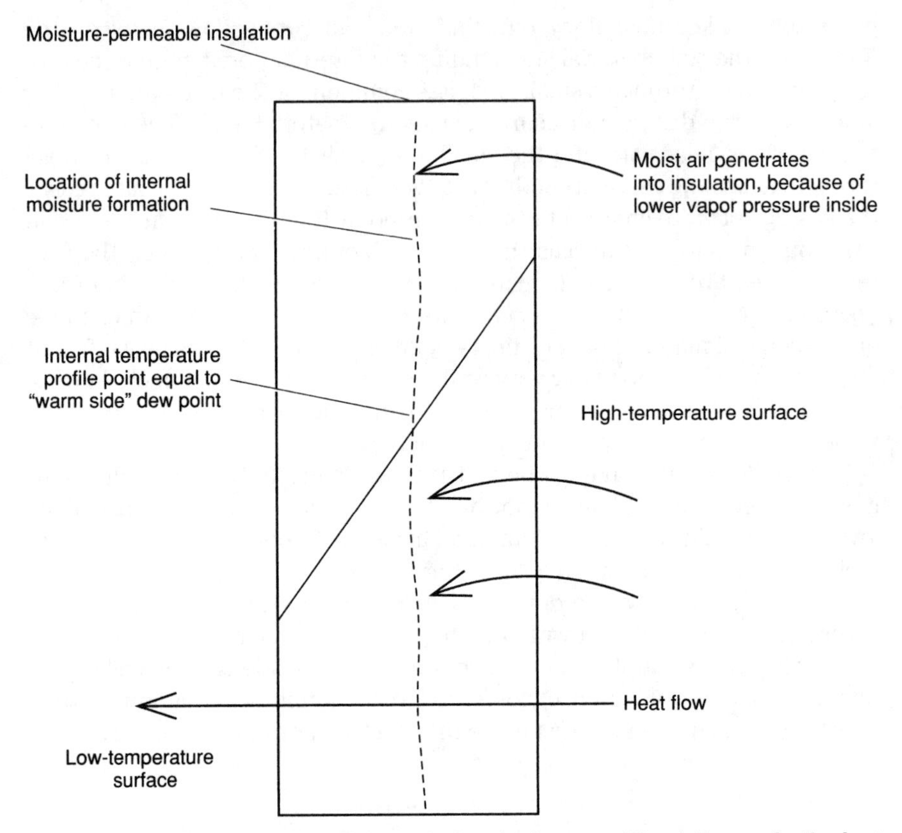

Figure 2.5. Temperature profile and dew point location in open cell insulation conducting heat.

Proper installation and workmanship are critical for a vapor barrier to be effective. Small gaps or holes in the barrier allow humid air and moisture to enter. This can degrade the insulating properties of the material and lead to further damage, allowing even more air and moisture to enter. Seal all gaps and holes in vapor barriers. Caulk and tape the joints liberally to ensure an airtight seal all around the insulation.

In an environment where the surrounding air is at 100% relative humidity, it can be impossible to prevent condensation. Under these conditions the surface of the insulation becomes wet from the condensed water vapor. Plan to provide extra moisture protection when insulating cold surfaces in very humid areas. Closed cell or moisture-proof insulation is mandatory in these applications. Cold water pipes or refrigerant tubing installed in boiler rooms with poor ventilation may have these conditions.

Materials

A small variety of materials are used for insulation. The choice depends on the temperature of the application. High temperature applications, up to 2000°F (used for boiler breaching, chimney insulation, steam pipe insulation, and some high-temperature water service), use inorganic materials. These include mineral fibers (rock wool), calcium silicate, special fiberglass, and cellular glass.

Moderate temperature applications, up to 170°F to 200°F, include organic materials. Common insulating materials are expanded polystyrene, polyisocyanurate, polyurethane, rubber, and cork. Glass fiber insulation is often used for these applications, as well.

Some insulations are manufactured with a thin, metallic foil vapor barrier. Often this is an aluminum foil sheet bonded directly to the underlying material. Boards of polyisocyanurate and some glass fiber boards and blankets are available with this barrier. These are very effective at keeping moisture out and resisting heat transfer via radiation.

Glass fiber pipe insulation is often supplied with a nonporous paper or composite vapor barrier. An integral adhesive seal is often bonded to the part of the barrier that overlaps onto the adjacent cover. These form a fairly effective vapor barrier, but may not completely prevent moisture generation.

The thermal performance of inorganic insulations is not quite as good as most organic materials. This, and their moderately higher cost, make them impractical for use at lower temperatures. Also, some inorganic insulations are not as moisture resistant as some organic insulations and can't resist damage from condensation.

Table 2.1 includes data on common insulating materials. Included are the forms available, typical thermal conductivities, and the material's R/inch values.

The flexible, closed cell organic pipe insulation commonly used for suction lines on split system air conditioners requires special consideration. This material, usually a form of expanded polyurethane, is an excellent insulator with good moisture resistance. Some of these products are not suitable for outdoor applications, however. Ultraviolet light in sunlight can degrade some of these materials, causing them to become brittle and crack. Moisture penetrates inside and the deterioration worsens. Sometimes the insulation becomes a source of water leaks into a building as it carries water in from the outside. This is especially common where the insulated tubing penetrates a roof. Be certain to specify insulation with a high degree of ultraviolet resistance for outdoor use. Better yet, install weathertight metal or plastic jackets over any insulation used outdoors.

Applications

Heating and cooling systems require insulation on pipes, ducts, and equipment. We shall consider these applications in the following paragraphs.

Pipes. Pipe insulation is usually supplied in lengths cut in half cylinders or as full tubes slit along one side. They usually come with tape to seal the slit after the insulation is installed. Common pipe insulating materials include glass fiber and closed cell flexible polyurethane and polyethylene. Manufactured fittings are available with plastic vapor barriers for elbows, valves, and other fittings. These make a neat, professional installation easy to achieve. It is important to seal all insulation and vapor barrier joints to prevent condensation or extra heat loss.

An additional vapor barrier can be brushed on to insulation surrounding pipes and equipment. These coatings are also effective as weather protection for pipes installed outdoors. They form a flexible, tough skin that is impervious to water and can help extend the life of the insulation. If used outdoors, however, they may have to be reapplied periodically to prevent cracks from forming due to sunlight and temperature expansion.

On pipes carrying hot fluids, the insulation should be thick enough to prevent excessive heat loss and keep the exposed surface temperature low enough to prevent burns. For pipes carrying steam or hot water, glass fiber insulation should be

- 1½ in. thick for pipes up to 1 in.
- 2 in. thick for 1¼– to 2-in. pipes
- 3 in. thick for pipes up to 6 in.
- 3½ in. thick for pipes larger than 6 in.

Small pipes carrying cold fluid, up to 1 in. in size, should have at least ½ in. of insulation. Pipes over 1 in. in size should have at least 1 in. of insulation. Remember that code requirements may be different. Install at least the minimum amount of insulation that codes require for every application. More insulation can often reduce operating costs, so use the codes only as minimum amount guidelines.

Ducts. It is often necessary to insulate metal ducts. Insulation prevents excessive heat losses or gains and limits the possibility of condensation forming on the duct's surfaces. Use insulated ducts where air supplies or returns are installed outdoors, in unconditioned spaces, or where air conditioning supply ducts have to positively prevent exterior condensation.

It is common for interior air conditioning supply ducts to sweat when the system is first started and the surrounding air is warm and humid. Proper insulation of the ducts can prevent moisture from forming and limit the noise transmitted from the duct.

Ducts used to exhaust inside air from a space do not require insulation when run inside a building. There is no temperature difference between the inside and outside surfaces, so condensation should not be a problem.

Exhaust ducts discharging moist air from a building might need insulation if they run outside. If they must carry the exhaust air for long distances before the discharge point, insulation prevents moist air from condensing in the duct. If condensation does occur, the water can run back into the building and cause damage.

Insulation for metal ducts is made from a variety of materials. Rigid glass fiberboards, often with a factory applied foil vapor barrier, are commonly used. These are effective for limiting heat losses (or gains) and noise transmission. Noise is reduced as it travels through the duct's walls and while it propagates inside the duct. These insulation boards should be glued, clipped, pinned, or banded to the duct's outside surface.

Glass fiber blanket insulation is also available for ducts. This is usually supplied with a foil vapor barrier. Blanket insulation is often pinned to the outside of a metal duct. The vapor barrier should face the warm side, usually away from the outside surfaces of air conditioning ducts.

Organic fiber materials can also be used for duct insulation. They give excellent service when protected from moisture with an adequate vapor barrier.

The required R-value of duct insulation is usually dependent upon the expected difference in internal and external temperatures. A good rule to follow to find the required R-value for duct insulation is to use the formula

$$\text{R-value} = \frac{T}{15}$$

Again, local codes may require different amounts of insulation. If the above formula is acceptable, then no insulation is required if the temperature difference is less than 15°F. For example, a heating duct run outdoors carries 120°F air while it is –5°F outside. The required insulation R-value would be

$$\text{R-value} = \frac{(120°\text{F} - -5°\text{F})}{15} = \frac{125}{15} = 8.3$$

Avoid excessive heat losses by using more insulation on unusually long duct runs. Consider any exterior duct longer than 30 ft as a candidate for additional insulation.

Install thermal insulation on the outside surfaces of the duct. Duct liners, primarily used for noise control, are installed on the inside. Some of these products are not intended to provide thermal performance. Their primary use is to eliminate noise carried within the duct's interior. They absorb sound to prevent it from being carried along the length of the duct, limiting the noise transmitted through the duct's walls.

Ducts installed outdoors must be weatherproofed. Cover the duct (whether bare metal or its insulation) with a mastic cement or an EPDM (synthetic rubber) sheet. These work very well if applied properly. All edges must be overlapped so that water will drain away from the joints without penetrating inside (see Figure 2.6). Each rubber-to-rubber splice should have a minimum of a 3-in. overlap.

Equipment. Rigid insulation blocks, boards, and sheets are usually used to insulate heated or cooled equipment surfaces. High-temperature applications use glass fiber, calcium silicate, or cellular glass materials. Low-temperature applications can use glass fiber or closed cell organic materials (often polyurethane) that will prevent moisture intrusion.

Insulating irregular surfaces can be time-consuming. The materials must be cut to fit each part of the surface and attached with clips or adhesives (for moderate temperature applications) to hold them in place. Coat the insulation with cement reinforced with plastic or glass mesh to seal the insulation and provide a degree of rigidity and strength. This coating can also provide vapor barrier protection to prevent condensation within the underlying insulation. Again, this is only necessary if the equipment operates at temperatures below the surrounding air's dew point. Sheet metal is sometimes used over the softer mastic to prevent damage from physical abuse.

Equipment installed on concrete or steel surfaces often requires insulation at its mounting points. Wood or insulated concrete is often used. Extend insulation and vapor barriers along steel supports if the equipment operates at temperatures cold enough to allow condensation.

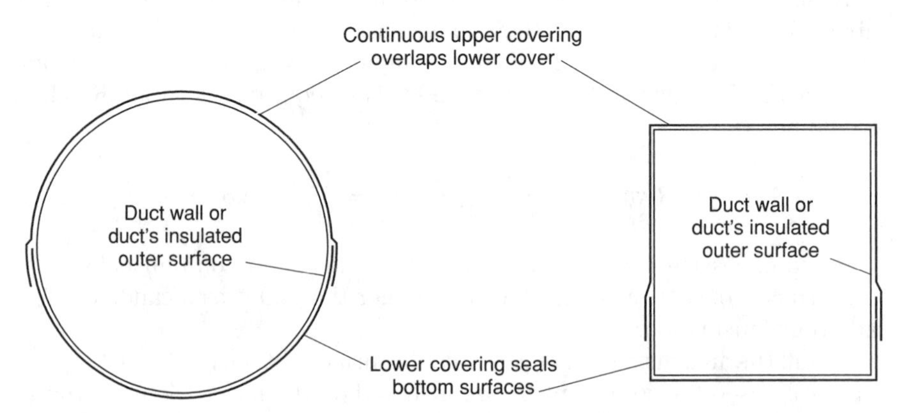

Figure 2.6. Proper duct waterproofing laps to prevent leaks. Always allow water to drain away from the joints.

Convection

Convection is heat transfer caused by air or water in motion. Convection can be either natural or forced. Natural convection occurs because of the natural buoyancy of regions of hotter fluid compared with cooler areas. Hotter fluid is less dense than cool fluid and rises away from a source of heat. Cooler fluid flows in to take the hotter fluid's place. The result is that heat is transferred from the hot surface to the fluid. The same effect occurs with cooling, except that the convective flow runs away from the cold surface.

Forced convection depends on the fluid being forcibly driven by the wind, a fan, or a pump. The additional fluid motion of forced convection allows more heat to be transferred than with natural convection.

Convection is at the heart of many HVAC systems. It is the mechanism that allows coils to take heat out of the fluid passing inside them and distribute it to the air moving over them. All heat exchangers use convection, as do cooling towers. It is an important heat transfer mechanism.

Convective heat transfer depends on two effects. First, heat is conducted through the fluid as described previously. Fluids have their own thermal conductivities and R-values, and they conduct heat like other materials. Second, the fluid flows around and to some extent mixes within itself. Fluid not in contact with the surface mixes with the fluid that is and carries some of the heat to it. This "unused" fluid carries more heat as the mixing effect brings it into contact with heated or cooled surfaces. Fluid mixing contributes a great deal to the total heat transfer taking place during convection.

The most important idea for the technician regarding convection is the large effect the speed of the moving fluid has on the amount of heat transferred. Very slow fluid speeds dramatically reduce the amount of heat transfer that can take place. Fluid forms an almost immobile layer on the surface of the object it is passing over. This layer of fluid is the *boundary layer* (see Figure 2.7).

Friction between the moving fluid and the surface it is touching causes the boundary layer. There is always a boundary layer, no matter what the speed of the fluid.

To illustrate a nonheat transfer effect of boundary layers, consider the dust that accumulates on fan blades. Dust settles on fan blades even when they are constantly moving through the air. This is because, in the boundary layer of air "attached" to the fan blades, there is no air movement. Dust particles can lie on the fan and the microscopically thin layer of immobile air does not push them off. The boundary layer also affects heat transfer.

The faster the fluid moves over the surface (or the surface moves through the fluid), the thinner the boundary layer becomes. Also, the part of a surface closest to the leading edge of the moving fluid flow has a thinner boundary layer than the surface farther away. A thicker boundary layer reduces heat transfer more than a thinner one.

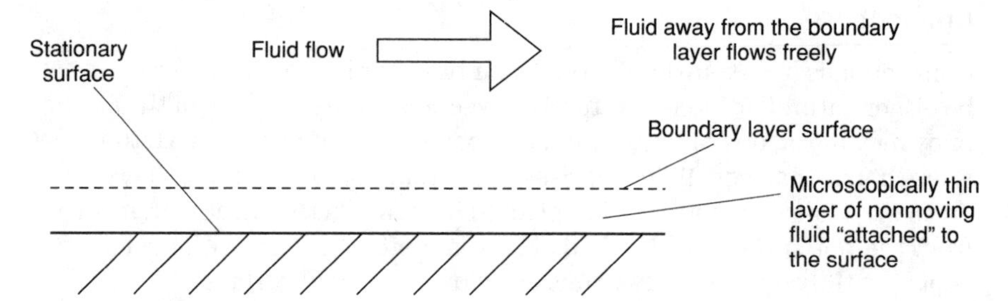

Figure 2.7. Moving fluid boundary layer over a stationary material.

At very low speeds, the moving fluid moves over the boundary layer as *laminar* flow (see Figure 2.8). This means that the material forms distinct layers as it passes through and around objects. Very little internal mixing takes place between the layers. Laminar flow, and the thick boundary layer that forms when it occurs, severely limits the amount of heat the fluid can absorb.

For example, water passing slowly over a heated metal plate would have a boundary layer of stationary water attached to the plate. This acts like a layer of insulation, keeping heat from transferring efficiently. Heat must be conducted into the stationary layer of water immediately next to the boundary layer. This layer will warm and send its heat on to the next slowly moving layer.

All of the thin layers of water repeat this process. Because there is no physical mixing of the layers of water, the metal plate's heat is lost entirely through conduction. It is still considered convective heat transfer, however, because the moving fluid carries heat away from the plate.

Many natural convective systems have at least some laminar flow. Baseboard fin tube radiators, for example, operate almost entirely with laminar flow of air between the fins.

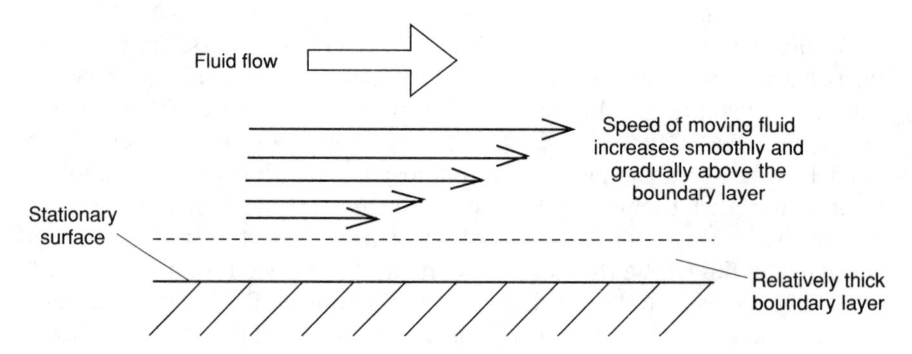

Figure 2.8. Laminar flow profile and boundary layer.

If the fluid moves fast enough, it becomes turbulent (Figure 2.9). Turbulent flow is much more effective for transferring heat. This type of flow thins the boundary layer (minimizing the layer of nonmoving fluid) and causes mixing within the fluid itself. In turbulent flow the fluid swirls and mixes on a microscopic level. Mixing in this way, it shears off much of the boundary layer and constantly introduces cool fluid to the heated surfaces and heated fluid to cooler areas.

Figuring out whether a flow will be laminar or turbulent is a complex calculation. The point at which a fluid flow becomes turbulent depends on several factors. These include the fluid's viscosity, density, velocity, and the distance it has flowed over a surface. The turbulence of flow in a pipeline also depends on the diameter of the pipe. The best way to ensure turbulent flow through a system is to maintain velocities at reasonable levels. Insufficient flows, those where fluid moves very slowly, may cause laminar flow and poor system performance.

For most HVAC service or design situations it is not necessary for the technician to figure out whether fluid flow is laminar or turbulent. This book does not provide the formulas for performing the calculations because it is not necessary for technicians to know them. Design engineers must consider how the fluid behaves as it flows through the equipment. It is their job to ensure that the equipment works as intended. It is not necessary to do this when servicing equipment. Keep the air and water flows through equipment at "normal" levels for their capacity.

If the technician must design a heat exchanger, avoid systems that provide extremely low air or water velocities. Most manufacturers' heat transfer components, except natural convective heaters, are designed to work with turbulent fluid flows. Maintain minimum flow rates to ensure that sufficient heat transfer can take place. Consult with the manufacturer or his representative when in doubt.

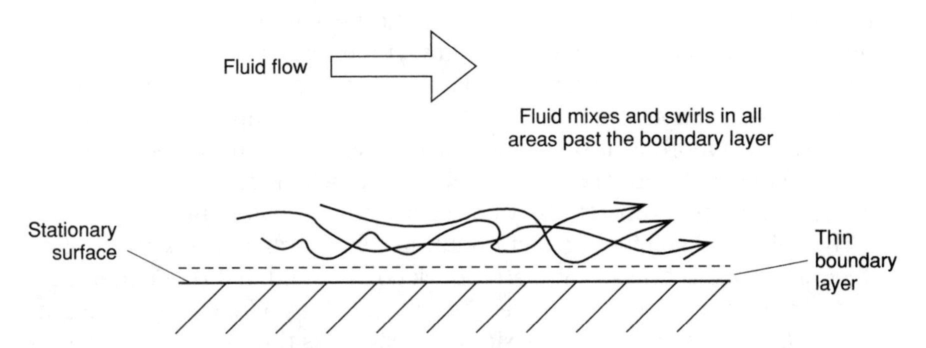

Figure 2.9. Turbulent flow profile and boundary layer.

Radiation

The third method of heat transfer is through radiation of energy. Physical contact between the objects transferring heat is not required. No fluid or substance carries the energy transferred with radiant heating. Electromagnetic radiation transfers the heat energy. This is *infrared radiation*. It is similar to light, except that infrared radiation has less energy and longer wavelengths than visible light. In other respects they behave the same.

All objects radiate heat energy as infrared radiation. The object's temperature decides the energy level of the radiation emitted. Higher temperature sources have more energetic waves with shorter wavelengths. Cooler temperature sources have less energy and the wavelengths of the radiation are longer.

The total amount of heat transferred by radiation depends on more than the temperature difference between the objects. The relationship between the temperatures of the objects and the amount of transferred energy is more complex than with conductive or convective heat transfer. The energy transferred between two objects via infrared radiation is related to the following:

- The difference between the object's temperatures, raised to the fourth power
- The color, shininess, or reflectivity of the objects (collectively, these properties contribute to an object's *emissivity*)
- The angle between the objects, the distance between them, and their relative sizes, all relating to how well one object can "view" the other

Considering these effects, an analysis of infrared radiation and heat transfer is complicated. A few general observations are appropriate for the HVAC technician.

Heat transfer is related to the temperature difference between the objects raised to the fourth power. This represents a tremendous temperature effect. Conductive heat transfer relates directly to temperature differences: double the temperature difference and the heat flow doubles; triple the temperature difference and the heat flow triples; and so on.

With radiative heat transfer, doubling the temperature difference results in 16 times the amount of heat flow, and tripling the temperature difference gives 81 times more heat flow. This extreme sensitivity to temperature difference contributes to the high efficiency of radiant heat transfer. Small changes in temperature greatly vary the amount of heat delivered.

The color and shininess effect on infrared heat transfer is a result of an object's surface acting as a radiator. Dark colored and dull (compared with shiny or glossy) objects are more efficient radiators. While it is not important to understand all of the physics that make this the case, a basic understanding is helpful.

This effect occurs because light, shiny objects are good reflectors that absorb very little of the radiation that falls on them. To prevent an object from absorbing and accumulating an unlimited amount of energy, it has to radiate energy away as efficiently as it can absorb it. It follows that objects that are very good reflectors are very poor radiators, and vice versa.

The best color for radiant heating efficiency is black. Dull, nongloss black finishes provide more radiant heating ability compared to lighter colors. The worst surface finish for a radiator is a shiny metallic one. A shiny metallic surface can be more than 10 times worse at radiating heat than a dull black surface. While many clients object to black radiators, do not coat them with shiny metallic paints. Flat finishes are better.

Finally, the way the heat source "views" the objects being heated is significant. How much radiative heat transfer occurs depends on the angle the surfaces make to each other, how far apart they are, and their sizes. The better the view, the better the heat transfer.

Radiative heat transfer drops off as the distance between the objects increases. Doubling the distance between objects transferring heat reduces the amount of heat to ¼ of its previous value. This relationship is the *inverse square law*. Remember that, for best efficiency, the radiator should be in full view of and close to the people or objects to be warmed.

Airflow Formulas

Heat is easily and reliably transferred by moving air. There are specific formulas used to estimate the amount of heat delivered.

Heat flow, as we are considering it, is really power delivery. There is an amount of heat delivered over a time, usually expressed as Btu per hour. Because temperature differences drive heat flow, the temperature of the air bringing energy to (or removing the energy from) the space must be different from the space's temperature. Air used for heating a space must be hotter than the desired room temperature. When cooling, the delivered air must be colder than the desired room's temperature.

Use the properties of standard air for heat flow formulas. As discussed in Chapter 1, standard air is dry, has a pressure of 14.7 psia, a temperature of 70°F, a density of 13.34 ft^3/lb, and a specific heat of 0.240 Btu/lb°F. We can use the formulas derived from the specific heat discussion in Chapter 1 to show how much heat air delivers. First, the basic formula:

$$\text{Heat} = \text{Mass} \times \text{Specific heat} \times \text{Temperature difference}$$

Find the amount of heat added to or taken away from an object by multiplying its mass by both its specific heat and the temperature difference (T_d).

$$T_d = \text{Higher temperature} - \text{Lower temperature}$$

For standard air, we know that specific heat is 0.240 and its specific volume is 13.34 ft^3/lb. Substituting into the heat equation, we find

$$\text{Heat} = \frac{\text{ft}^3}{13.34} \times 0.240 \times T_d$$

Note that if we divide both sides of the equation by hours, we have

$$\text{Btu/hr} = \left(\frac{\text{ft}^3}{\text{hr}}\right) \times \left(\frac{0.240}{13.34}\right) \times T_d$$

Because almost all air delivery systems are rated in terms of cubic feet per minute (cfm), we adjust the formula one more time:

$$\text{Btu/hr} = \frac{\text{cfm} \times 60 \text{ min}}{\text{hr}} \times \left(\frac{0.240}{13.34}\right) \times T_d$$

Simplifying, the above formula reads

$$\text{Btu/hr} = 1.08 \times \text{cfm} \times T_d$$

Note that the heat delivered (or removed) in the equation is completely sensible. This formula does not account for latent heat. In almost all heating applications the requirement is for sensible heating only. Both sensible and latent heats are removed, and must be considered, in most air conditioning applications.

An example of a heating application: Air is delivered to a room at a rate of 1000 cfm. The delivered air is 125°F and the room is kept at 70°F. How much heat is delivered to the space?

$$\text{Sensible Btu/hr} = 1.08 \times 1000 \text{ cfm} \times (125°\text{F} - 70°\text{F}) = 59{,}400 \text{ Btu/hr}$$

If the room also happened to be losing this much heat continuously, its temperature would not change. If the room were losing less heat it would warm toward the 125°F delivered air temperature.

The previous equation can be rearranged to solve for its other terms. In most applications the problem is to find the required cubic feet per minute or the air temperature to satisfy a requirement. If we know the heat flow required (Btu/hr), the cubic feet per minute can be found:

$$\text{cfm} = \frac{\text{Btu/hr}}{(1.08 \times T_d)}$$

Earlier we considered a snow covered roof that lost 2800 Btu/hr of heat from the space below it. Assuming we have a warm air source of 120°F, and we maintain the 70°F interior temperature, how many cubic feet per minute are required to offset the roof's heat loss?

$$\text{cfm} = \frac{2800 \text{ Btu/hr}}{(1.08 \times (120°\text{F} - 70°\text{F}))} = 52 \text{ cfm}$$

The temperature required (T_r) for the supply air can be found if the required heat delivery (Btu/hr) and cubic feet per minute are known. The formula is

$$T_r = \frac{T + \text{Btu/hr}}{(1.08 \times \text{cfm})}$$

If, in our roof example, we had 100 cfm of air available, what temperature would we have to heat it to maintain the 70°F ambient temperature?

$$T_r = 70°\text{F} + \frac{(2800 \text{ Btu/hr}}{(1.08 \times 100 \text{ cfm}))}$$

$$T_r = 70 + \frac{2800}{108}$$

$$T_r = 70 + 25.9 = 95.9°\text{F}$$

Note that in heating applications it is not usually safe or practical to allow the delivered air temperature to go significantly above 125°F. While some special cases do call for extremely warm air (like industrial product drying or hardening), almost all comfort heating systems should limit the delivered air temperature to 125°F. Adjust the volume of air delivered at this temperature to deliver the proper amount of heat.

All of the previous examples have considered heating applications. The formula is also applicable to cooling (sensible heat removal only). For example: What quantity of sensible heat (in Btu/hr) can be removed from a building with a 4500 cfm supply of 50°F air? The room is maintained at 75°F.

$$\text{Sensible Btu/hr} = 1.08 \times 4500 \text{ cfm} \times (75°\text{F} - 50°\text{F}) = 121{,}500 \text{ Btu/hr}$$

It is common for commercial comfort air conditioning systems to deliver air at approximately 50°F to 55°F. Air delivered at this temperature dehumidifies 75°F room air to 40% to 50% relative humidity and provides a high degree of comfort. These assumptions work if the heat load in the space being air conditioned is predominantly sensible. Most commercial air conditioners' ratings are based on removing between 70% and 75% sensible heat. The rest of their rated capacity is allocated to removing latent heat (humidity). This is typical of most space and comfort cooling loads.

We can use the previous assumptions, 50°F to 55°F discharge air temperature and predominantly sensible cooling loads, to develop a simple rule for estimating air conditioning cubic feet per minute: Allow 400 cfm of air flow for each ton of air conditioning capacity required. One ton of heat flow equals 12,000 Btu/hr.

As is true in heating systems, there are special cases where the temperature of the delivered air may be varied. This can provide more cooling capacity or lower humidity. However, almost all standard systems provide approximately 50°F air and deliver an air flow of about 400 cfm/ton of cooling capacity.

An example of cubic feet per minute requirements for air conditioning calculations: An air conditioner must keep a room at 75°F. The room has a "normal" (predominantly sensible) heat gain of 34,000 Btu/hr. Find the required cubic feet per minute.

$$\text{Tons capacity} = \frac{34{,}000\ \text{Btu/hr}}{12{,}000\ \text{Btu/hr/ton}} = 2.83\ \text{tons}$$

$$\text{cfm required} = 2.83\ \text{ton} \times 400\ \text{cfm/ton} = 1132\ \text{cfm}$$

If a package air conditioner were being chosen for this application, a 3-ton air conditioner delivering approximately 1200 cfm would be correct. If 50°F to 55°F air from a central system was delivered to this space, the previous calculation would provide the air flow required.

Total heat transfer (including both latent and sensible) can take place by the flow of air. While it is only necessary to know the temperature differences when calculating sensible heat transfer, total heat transfer estimates must use differences in enthalpy. The formula is

$$\text{Total Btu/hr} = 4.5 \times \text{cfm} \times \text{Enthalpy difference}$$

This formula provides the total heat transferred by a flow of air. For example, an air conditioning system is used to precool 350 cfm of outdoor air delivered into a room. The outdoor air is at 90°F and 70% relative humidity. It is cooled to 50°F and 90% relative humidity. Find the total, sensible, and latent heat removed from the air. Find the enthalpies from the data listed in Table 1.3.

$$\text{Outside air enthalpy} = 45.1\ \text{Btu/lb}$$

$$\text{Cooled air enthalpy} = 14.4\ \text{Btu/lb}$$

$$\begin{aligned}\text{Total Btu/hr} &= 4.5 \times 350\ \text{cfm} \times (45.1\ \text{Btu/lb} - 19.4\ \text{Btu/lb})\\ &= 4.5 \times 350 \times 25.7 = 40{,}478\ \text{Btu/hr}\end{aligned}$$

$$\begin{aligned}\text{Sensible Btu/hr} &= 1.08 \times 350\ \text{cfm} \times (90°\text{F} - 50°\text{F})\\ &= 1.08 \times 350 \times 40 = 15{,}220\ \text{Btu/hr}\end{aligned}$$

$$\text{Latent Btu/hr} = \text{Total} - \text{Sensible} = 40{,}478 - 15{,}120 = 25{,}358\ \text{Btu/hr}$$

It should be clear from this example that, when doing air conditioning calculations, the total heat load of the air must be considered. Calculating cooling loads based on sensible heat alone can provide too small an estimate.

Another air formula sometimes required is calculating the number of air changes per hour (ACH) for a specific space. Some ventilation requirements specify a minimum number of ACH. Air changes per hour refer to the number of times a room- or other space-sized volume of air is delivered in 1 hour. The formula is

$$\text{ACH} = \frac{\text{cfm supply} \times 60}{\text{Space volume}}$$

For example, find the ACH if a 30-ft by 18-ft by 9-ft room receives 350 cfm.

$$\text{Room volume} = 30 \text{ ft} \times 18 \text{ ft} \times 9 \text{ ft} = 4860 \text{ ft}^3$$
$$\text{ACH} = 350 \times 60/4860 = 4.3$$

This formula can be adjusted to find the cubic feet per minute required if the room size and the number of ACH are known:

$$\text{cfm} = \frac{\text{ACH} \times \text{Space volume}}{60}$$

Find the cubic feet per minute required to provide 3 ACH in a 40-ft by 36-ft by 12-ft room.

$$\text{cfm} = \frac{3 \times (40 \times 36 \times 12)}{60} = 864 \text{ cfm}$$

Water Flow Formulas

Water is an excellent vehicle for transferring heat. It is more efficient than air because the heat capacity and density of water are much higher than the same properties of air. More heat can be transferred with smaller hydronic equipment than with forced air systems. Smaller pipes and pumps deliver the same Btu per hour as larger ducts and fans.

As with air, the heat capacity available from a given flow of water is easily derived from the original formula:

$$\text{Heat} = \text{Mass} \times \text{Specific heat} \times T_d$$

Again, the term T_d refers to the difference in temperature of the water as it changes its heat content. The specific heat of water is 1.0 and the mass is 8.3 lb/gal. Therefore

$$\text{Heat} = \text{gal} \times T_d \times 8.3 \text{ lb/gal}$$

Dividing both sides by hours, we get

$$\text{Btu/hr} = \text{gal/hr} \times T_d \times 8.3 \text{ lb/gal}$$

Assuming we want to express the water flow in the standard gallons per minute (gpm), we get

$$\text{Btu/hr} = \text{gpm} \times T_d \times 8.3 \text{ lb/gal} \times 60 \text{ min/hr}$$

Cleaning up the terms, the final equation is

$$\text{Btu/hr} = \text{gpm} \times T_d \times 498$$

Find the heat, in Btu/hr, delivered by a boiler to a 50-gpm flow of water. The water's temperature increases from 115°F to 137°F as it goes through the boiler's heat exchanger.

$$\text{Btu/hr} = 50 \text{ gpm} \times (137°\text{F} - 115°\text{F}) \times 498 = 547{,}800 \text{ Btu/hr}$$

This equation can be rearranged to allow us to find the gallons per minute water flow or the temperature change if the other terms are known. The equations are

$$\text{gpm} = \frac{\text{Btu/hr}}{(T_d \times 498)}$$

$$T_d = \frac{\text{Btu/hr}}{(\text{gpm} \times 498)}$$

Find the gallons per minute required to transfer 120,000 Btu/hr from a boiler with a 20°F temperature rise:

$$\text{gpm} = \frac{120{,}000 \text{ Btu/hr}}{(10°\text{F} \times 498)} = 24.1 \text{ gpm}$$

These formulas work equally well for heating and cooling. Find the temperature drop of a 60-gpm water flow through a 30-ton capacity chiller:

$$\text{Btu/hr} = \text{tons} \times 12{,}000 \text{ Btu/hr/ton} = 30 \times 12{,}000 = 360{,}000 \text{ Btu/hr}$$

$$T_d = \frac{360{,}000 \text{ Btu/hr}}{(60 \text{ gpm} \times 498)} = 12.0°\text{F}$$

Keep in mind that the term T_d is the lower temperature value subtracted from the higher. This always gives a positive heat flow. It is usually obvious which way heat is flowing: Heat delivered to the water causes its temperature to increase. Removing heat causes the water to become colder.

Some conventions always subtract the temperature of the water entering the system from its temperature leaving the system. In those cases a positive answer shows that the water was heated, while a negative answer shows that cooling took place. These results can be confusing. In this book I always subtract the higher temperature from the lower, giving a positive answer. But it is important to remember the direction of heat flow. Heaters, boilers, and coils in air conditioning systems deliver heat to the water. Chillers, radiators, and hydronic heating system coils take heat away from it.

The amount of heat delivered to the space by the water system changes with the difference between the air's temperature and the average water temperature. The higher the difference, the more heat that can transfer. The exact amount of heat transferred depends on the details of the heat exchanger, but the amount of heat transfer capability with changing temperature differences is constant.

For example, assume that an average water temperature of 190°F and an air temperature of 72°F are used as "standards." Changing either value away from those standards changes the heat transfer in predictable ways. This can be useful if you are installing a fin tube (or other heat exchanger) rated to deliver a particular amount of heat at one water temperature and the hydronic system provides another. A larger or smaller heater may be used to compensate for the difference in temperature.

Table 2.2 shows the changes in heat transfer ability for a water-to-air heat exchanger as the average water temperature changes. Average water temperature is found by adding the inlet and outlet temperatures and dividing the sum by two. Find the average water temperature of the hot water heating unit and use it to determine the relative amount of heat output from Table 2.2.

TABLE 2.2 Heating Capacities for Hot Water Coils at Various Temperatures*

Average water temperature (°F)	Heating capacity (%)
190	100
185	96
180	92
175	87
170	83
165	79
160	75
155	70
150	66
145	62
140	58
135	53
130	49
125	45
120	41
115	36
110	32

*Assume air temperature = 72°F. Capacity is compared to an average water temperature of 190°F.

For example, assume that a hydronic system was designed to provide 180°F water that is to be cooled by the system's fin tube radiators to 165°F. Therefore, the fin tube's incoming water is 180°F and the outgoing water is 165°F. The fin tubes are rated by their manufacturer to deliver 120,000 Btu/hr of heat at an average water temperature of 180°F. How much heat will they deliver with the system's lower water temperature?

$$\text{Average system temperature} = \frac{(180°\text{F} + 165°\text{F})}{2} = 173°\text{F}$$

The capacity of fin tubes at 173°F (from Table 2.2) is about 85%. The capacity at 180°F is 92%. To find the output with the lower temperature water, use the formula

$$\text{Output Btu/hr} = \frac{\text{Rated Btu/hr} \times \text{Operating \%}}{\text{Rated \%}}$$

The example gives these values:

$$\text{Output Btu/hr} = \frac{120{,}000 \times 85}{92} = 220{,}870 \text{ Btu/hr}$$

The lower water temperature has decreased the output capacity of the fin tubes by over 9000 Btu/hr. It may be necessary to install more fin tube units to compensate for this reduction.

Excessive temperature changes (more than 25°F) would imply a flow restriction. Very small temperature changes (less than 5°F) may show that there is too much flow for proper system operation or that the heating or cooling device is not operating properly.

Most hydronic heating systems use water temperatures of at least 130°F to provide heat. These temperatures are very common with radiant heating systems. Convective heating systems usually are designed to use hotter water, often close to 200°F, to provide full heat output. Lower temperature systems run a bit more efficiently than high-temperature systems, but the difference is not significant unless fuel costs are extremely high.

It is possible to check the temperature change of water flowing through a boiler, chiller, or coil to get an idea of its flow. Most designs call for about a 10°F to 20°F temperature change as the water flows through these components. Most chillers and chilled water coils operate at the lower end of this range, with water temperature changes near 10°F. Boilers and hot water exchangers often operate with higher temperature changes close to 20°F. Although this is not always the case, temperature changes very different from these may imply flow problems.

Chapter

3

Building Heat Losses and Gains

Heating and cooling systems are designed to control the heat in buildings. Building occupants remain comfortable as the systems add or remove heat from the space to counteract the natural flows of air to and from the outside. Heating and cooling systems must work well under extreme conditions and still be economical during normal conditions. To meet these goals, it is important to select the system that has the required capacity to do the job without being oversized.

Some misinformation exists regarding system sizing. Many systems are chosen based on the number of square feet or cubic feet space being served. This is usually a very poor way to estimate the required capacity of a system. Rules of thumb like these are generally useful for only one type of building or one set of conditions. If the building's construction differs from that used when the rule was developed, the system will be a poor match.

Another widespread misconception is that it is always "safe" to install a much larger heating or cooling system than is really required. The idea is that it is better to have too much than not enough. Large systems, so the theory goes, will run for less time to fit the smaller need.

Systems that are very oversized for their application generally operate poorly. A system that is too large can cause drastic swings in temperature as it cycles on and off. System efficiency suffers and the occupants become uncomfortable and dissatisfied.

Installing air conditioning systems that are grossly oversized can contribute to occupants' illnesses, as well as structural damage to the building. Air conditioners that run for short periods of time because of oversizing do not properly control humidity. This can lead to structural damage and the uncontrolled growth of mold and mildew. These can contribute to sickness and serious disease for the building's occupants.

Properly estimating heating and cooling loads is not difficult. Correctly sized systems will, most likely, be smaller than those chosen with a rule of thumb approach. Smaller systems are less expensive to purchase and install, leading to more satisfied clients and more sales for the contractor. The systems operate efficiently and thus are more economical to run.

Energy codes adopted throughout the United States might require the designer to follow specific procedures or make certain assumptions when estimating heating or cooling system capacities. Verify any special code-related considerations before using any other methods. If there are no code conflicts, the procedures outlined in this chapter provide a method that gives accurate results.

Note that the methods provided here are sufficiently accurate for most heating and air conditioning systems. More exact methods can be found in the American Society of Heating, Ventilation, and Air Conditioning Engineer's (ASHRAE) *Fundamentals Handbook*. This reference also contains weather data, heating degree days, and other information that is useful for estimating heating and cooling loads.

Although the text and examples show calculations for estimating whole building heating and cooling loads, similar methods are used for estimating the requirements for a single room or area. Ignore adjacent areas that are maintained at the same temperature as the space being estimated. They don't contribute to the heating or cooling load. Only surfaces exposed to significantly different temperatures need to be considered.

Heat Losses

Buildings lose heat to the outside almost entirely at their perimeters (outside surfaces) (see Figure 3.1). Heat conducts through the floor slab, foundation, walls, windows, ceiling, and roof, driven by the temperature difference between outside and inside. Air infiltration due to the stack effect or general negative pressure inside the building causes a net loss of heat on a cold day. Colder outdoor air enters through the building's skin and is warmed to the interior temperature. Air exiting from the building carries heat to the outside with it. Fresh air ventilation also causes heat losses. Outdoor air brought to the inside for improved indoor air quality, combustion, or to make up for air exhausted to the outside contributes to a large building heating load.

It is usually easy to deal with ventilation heat losses by heating the outside air before it enters the building space. Most ventilation systems can be designed to deliver air to a few specific locations where small, dedicated heaters can temper it. This prevents cold drafts in the building. A heat exchanger installed to limit the heat losses associated with ventilation can

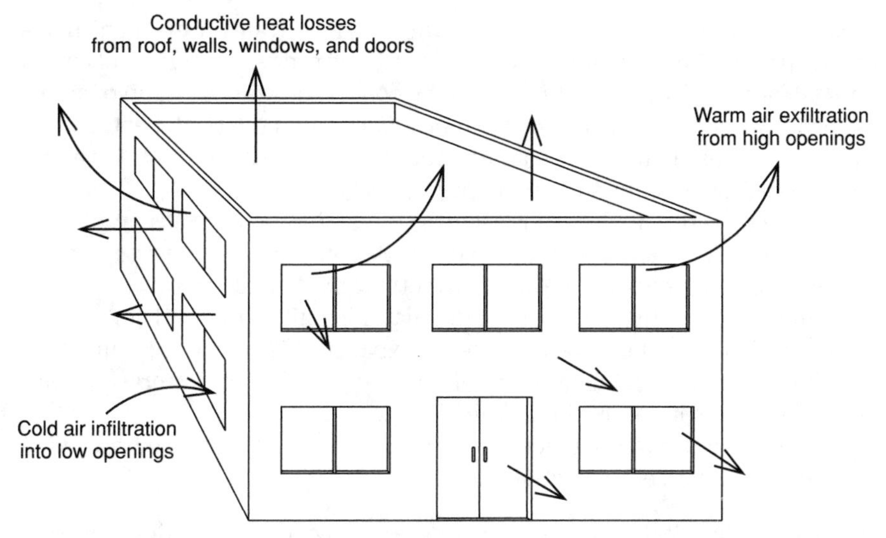

Figure 3.1. Building heat losses. All losses occur at the outside surface of the building.

save significant amounts of energy. See Chapter 10, "Ventilation," for information on proper energy conservation practices.

Estimating building heat losses and gains is one of the most common problems facing the heating system designer or service technician. Heating loads are caused by conduction through the building's siding and roof structure and heat losses and gains caused by infiltration. A service call that describes an area or building as "always too cold" may be because the system does not have enough capacity. This type of call can also be the result of changes in the building's use that might have made a previously adequate heating system unacceptable. If the heating system is operating properly but still can't keep up, calculate the required capacity for the job. It is possible that the system has become too small for the load it is handling.

It is proper to install a heating system with a reasonable (up to 30%) amount of overcapacity. This amount of reserve is not excessive and won't lead to operational problems. Because most heating systems must run at their full capacities for only a few hours a year (sometimes not at all during a mild winter), they are usually a little oversized for routine operation. A reasonable amount of reserve capacity allows the system to handle changes in outdoor temperatures and system thermostat settings with rapid response.

Grossly oversizing the heater, with an excess capacity beyond 50%, can cause problems. The heater cycles off and on repeatedly as it quickly warms the space to the required temperature. During the comparatively long "off" time the space might feel drafty as the space loses heat and cold air infiltrates inside.

Because building heat loss involves only two heat transfer mechanisms—conduction and convection—the calculation of proper system capacity is straightforward. There is little need to be concerned with the time of greatest heating need. The design can be chosen to match the peak heating load the system must handle. We can assume that the greatest conductive and convective losses occur simultaneously.

Calculate conductive heat losses for the building's roof, walls, and along the perimeter where it contacts the earth. Account for windows and doors, which cause both conductive and convective heat losses.

Estimate convective heat losses for natural infiltration and for power-driven make-up air (because of powered exhausts). Most codes require that buildings receive a certain amount of outdoor air. This air requires system-supplied heat as it enters the building's space.

Total heating system capacity requirements are estimated by following the steps summarized below:

- Estimate the total conductive heat losses. Divide the building's outside surfaces into different categories with similar heat flow resistances. Group all walls of similar construction together. Do the same with windows, doors, and the roof. Estimate the area of each surface and calculate the heat loss from each.
- Calculate convective heat losses. Estimate the amount of infiltration and the energy needed to warm the incoming air.
- The total heat loss is found by summing the conductive and convective losses. Select a heating system that provides this heating capacity with up to 30% reserve.

Conductive losses

Thermal conduction through the "skin" of a building follows the formulas for conductive heat flow. The formula, developed in Chapter 2, is

$$\text{Btu/hr} = \frac{\text{Area} \times T_d}{\text{R}}$$

Walls and roofs. The effective R-value and the area of the outside surface determine the net heat loss due to conduction. Usually the outside surface is not uniform, with the walls made up of many components (gypsum board, concrete block, insulation, and brick, for example). Windows and doors may also contribute significantly to conductive heat flow.

Like walls, roofs are usually constructed of several components. Gypsum board or lay-in ceiling tiles often make up the inside surface of the roof. There might be space above the ceiling that is part of the heated space. Some attics are ventilated to the outside, while others are not. Insulation is often used, and the roof assembly itself might have multiple components. The deck (wood, steel, or concrete) and the membrane (rubber, shingle, or multiple-layer built-up) might be used in a number of combinations.

Estimating the total heat flow requires breaking the wall and roof structures into their component parts and adding their individual heat losses together. Because the heat conduction formula requires knowing the R-value, square foot area, and temperature difference in order to calculate heat flow, these factors are considered separately. There are a few shortcuts that make the conductive heat loss calculations simple but accurate enough to determine the required capacity of a heating system.

First, consider the R-value and surface area of the building components. The idea of a material's total R/inch value was covered in Chapter 2. This is very useful for estimating heat losses through complex assemblies such as walls and ceilings. Heat flowing through a wall must pass through each material in turn as it travels from inside to outside. It is easy to calculate the total heat loss for a wall or roof assembly if the R-values of the individual components are known.

Multiply the R/inch value of each material in the wall or roof structure by its thickness in inches. This gives the material's R-value as installed. After calculating each part's R-value, add them to give the total R-value for the assembly. Use this value in the heat flow formula.

Of course, if a material's total R-value is already known or specified, there is no need to use an R/inch value and thickness calculation. For example, most insulations are rated with the total R-value. Use the published R-value directly in the calculation.

We begin analyzing an example commercial two-story building (see Figure 3.2). For this example, find the total R-value of the exterior walls with the following construction:

- ½ in. gypsum board
- 3½ in., R-11 glass fiber insulation on wood framing
- 8 in. hollow-core concrete block
- 4 in. common-face brick

From Table 2.1, the R/inch of gypsum board is found to be 0.88. Multiplying the gypsum board's R-value gives us

$$0.5 \text{ in.} \times 0.88 = 0.44$$

We are told that the insulation's R-value is 11, 8-in. concrete block has an R-value of 1.11, and common 4-in. face brick has an R-value of $4 \times 0.11 = 0.44$. The total R-value is thus

$$0.44 + 11 + 1.11 + 0.44 = 12.99, \text{ or } 13$$

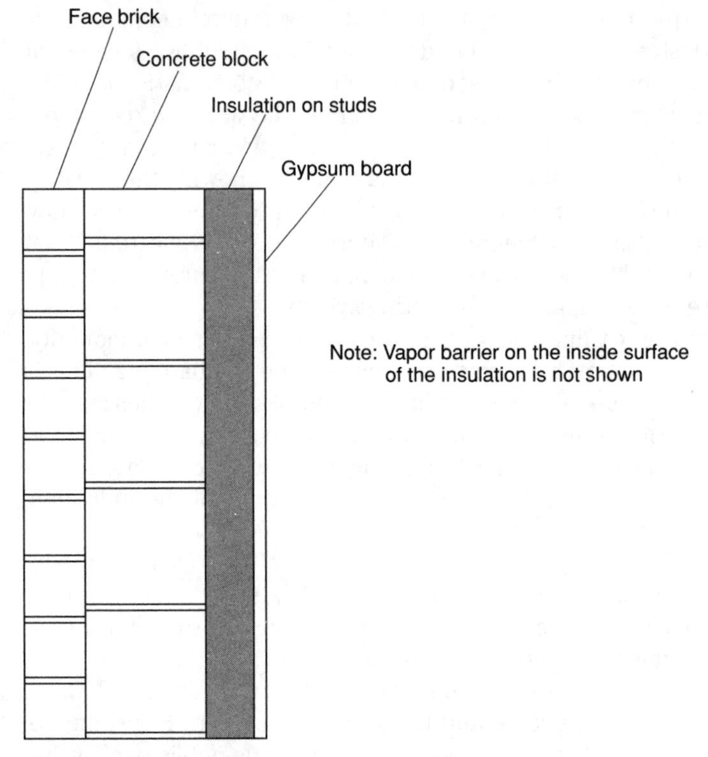

Figure 3.2. Section through example building's wall.

Note that the effect of an air film or boundary layer that resists heat flow on the inside and outside surfaces was not considered. Some references consider air film effects and include them in their calculations. Not doing so simplifies the problem and makes the solution a bit more conservative, leading to selection of a slightly higher capacity heating system.

Use the following for the example building's roof construction:

- ½-in.-thick suspended ceiling (unheated, unventilated space above)
- 2 in. glass fiber batting
- Steel roof deck
- ½ in. plywood roof decking
- Built-up three-ply roof

Find the total R-value by adding the individual parts' R-values. The R for ceiling tiles is 2.5 per in. (so the ½-in. thickness has an R-value of 1.25). The R/inch of glass fiber batting is 3.2, giving an R-value of 6.4 for 2 in. of insulation. Steel has no insulating properties, giving the roof deck an effective

R-value of zero. The ½-in. plywood roof deck, with an R/inch of 1.25, represents an R-value of 0.63. Finally, the roof itself will contribute an R-value of 0.33 to the construction. The total roof assembly R-value is

$$1.25 + 6.4 + 0 + 0.63 + 0.33 = 8.6$$

Use the total square foot area of the walls and roof in the heat flow formula. First, measure the perimeter of the building. This is the sum of the lengths of all exterior walls. With rectangular buildings it is particularly easy to figure out the perimeter length. Add the width and the depth of the building and double the sum.

If, for example, our example building has a rectangular shape measuring 48 ft by 76 ft, its perimeter length is equal to

$$\text{Perimeter length} = 2 \times (48 \text{ ft} + 76 \text{ ft}) = 248 \text{ ft}$$

Find the wall area by multiplying the perimeter length by the height of the building. A reasonable estimate is 11 ft per floor for office and residential buildings, so our example building, because it has two stories, would have an exterior wall area of

$$\text{Total wall area} = 2 \times 11 \text{ ft} \times 248 \text{ ft} = 5456 \text{ ft}^2$$

Modify the procedure if the building has some areas with different numbers of stories. Find the perimeter for each part, multiply by the number of floors in each part, and add the products together. The goal is to get a reasonable estimate of the total wall area for the entire building.

Calculate roof areas by multiplying the building's width by its length. It might be necessary to separately estimate some parts of the building and add them together if the building has a complex shape. Be sure you get an accurate total size estimate.

Our example building's roof area is easy to estimate. Multiply the building's width by its depth:

$$\text{Roof area} = 48 \text{ ft} \times 76 \text{ ft} = 3648 \text{ ft}^2$$

In the event that the building has a sloped roof, use the "flat roof" area of the ceiling as the heat loss area (see Figure 3.3). The sloped or pitched square footage is larger. However, unless the roof is severely pitched, the heat loss area can be calculated using the equivalent flat roof area.

For example, the width and depth of a building with a sloped roof is 40 ft by 60 ft. Use 40 ft × 60 ft = 2400 ft^2 as the heat loss area. The actual roof area is larger, but this calculation will adequately account for the heat loss.

An unheated attic space, if well ventilated to the outside, assumes the outdoor temperature. We can ignore its presence for our calculations and use the total R-value from the ceiling to the attic.

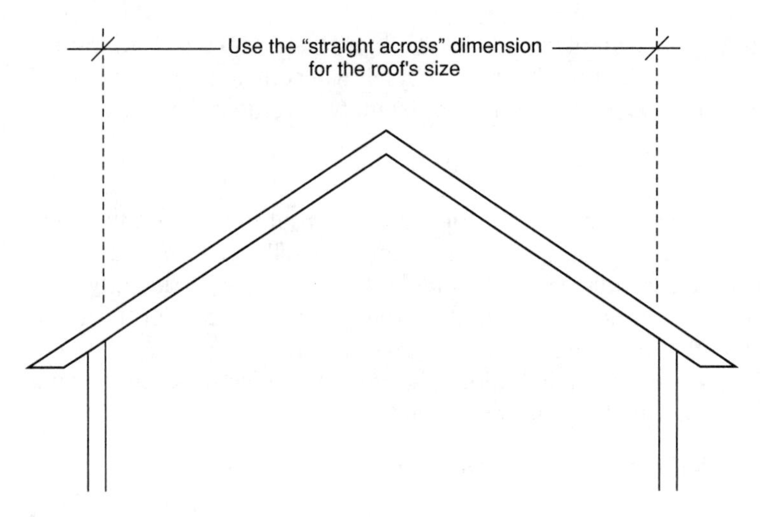

Figure 3.3. Roof area to use for heat loss estimates (up to a 4:12 pitch). Steeper roofs should be estimated with their actual surface area.

Normal office and residential buildings have window areas equal to roughly 30% of the total wall area. This is a reasonable estimate for most buildings. Some buildings, however, have much more. It is not uncommon for some buildings to have more than 75% window area. Manufacturing and warehouse buildings usually have considerably less, sometimes approaching none at all.

Try to estimate the amount of wall space taken up with windows. It is not necessary to measure every window. If a building has many "same size" windows, measure one and multiply its area by the total number of windows used. Again, minor variations in window size are not usually important, but try to get a good estimate of the total window area.

If our example building had 30% of its wall area made up with windows, the window and wall areas would be

$$\text{Window area} = 5456\ \text{ft}^3 \times 0.3 = 1637\ \text{ft}^2$$
$$\text{Wall area} = 5456\ \text{ft} - 1637\ \text{ft} = 3819\ \text{ft}^2$$

Windows. Windows vary widely in thermal performance, with some triple-glazed "low e" designs having high effective R-values. Windows with R-values higher than four are considered to be very efficient. Older, single-pane windows have much lower R-values. If you are calculating heat losses for a construction project where new windows are used, refer to the window's specifications or contact the manufacturer for the effective R-value. If there are existing windows and no information is available, use the following typical R-values. These are generally conservative but not far from the actual values.

Single pane	R = 0.9
Single pane, low e	R = 1.1
Double pane, ¼ in. air space	R = 1.7
Double pane, ½ in. air space	R = 2.0
High-efficiency double pane with 1 in. air space	R = 2.5
Double pane, ½ in. air space, low e	R = 2.6
Triple pane, small air space	R = 2.6
Triple pane, ½ in. air space	R = 3.2

Normal office and residential buildings have window areas equal to roughly 30% of the total wall area. This is a reasonable estimate for most buildings. Some buildings, however, have much more. It is not uncommon for some buildings to have more than 75% window area. Manufacturing and warehouse buildings usually have considerably less, sometimes approaching none at all.

Try to estimate the amount of wall space taken up with windows. It is not necessary to measure every window. If a building has many "same size" windows, measure one and multiply its area by the total number of windows used. Again, minor variations in window size are not usually important, but try to get an accurate estimate of the total window area.

We have already established that our example building has a window area of 1637 ft^2. To continue the analysis, assume that they are double-pane units with ½-in. air spaces and an R-value of 2.0.

Doors. The R-value for doors depends upon the unit's construction. Some insulated steel doors have thermal breaks built into them that make them very efficient. Doors usually make up a small amount of the total outside area of most buildings, so their exact R-value makes little difference to the total heat loss. They are significant, however, if doors make up more than 10% of the outside wall area. Use care when estimating an appropriate general R-value.

Table 2.1 gives an R/inch value for wood of 0.87 to 1.25. Because doors are a composite of hard and soft woods, use an average R-value of 1.0 for their thermal resistance. Most doors are about 1¾ in. thick, so assume that solid wood doors have an R-value of 1.75. Use an R-value of 5.2 for insulated steel doors with internal thermal breaks and an R-value of 2.5 for doors without thermal breaks. Hollow wood doors typically have R-values of 2.5 to 3, with the lower numbers corresponding to doors that include window panels. Treat glass doors like single-pane windows. Consider entrances with air locks (one entrance leading to a vestibule with another set of doors into the building) equivalent to double-pane windows. This is conservative and results in good estimates.

We can assume that our example building has four, 3-ft by 7-ft doors. Two are glass and two are insulated steel units. That equals 42 ft^2 of equivalent single-pane window and 42 ft^2 of R = 5.2 surface.

Foundations. Conductive heat losses also occur along the edge of the building where it meets the ground. Use the following to estimate heat losses:

- Slab-on-grade buildings (with no basement or crawl space), built with no insulation at the slab's edge: Heat loss equals 0.8 Btu/ft of perimeter edge per degree F outside to inside temperature difference.
- Edge insulated at least 3 ft below grade (as most energy codes today stipulate): Heat loss equals 0.6 Btu/ft of perimeter edge per degree F (see Figure 3.4).

Continuing with our example, remember that the building has a perimeter of 248 ft. Its heat loss through an insulated slab edge would be

$$248 \times 0.6 = \frac{149 \text{ Btu}}{\text{degree F difference in temperature}}$$

Buildings with basements have heat losses through their below-grade walls. Uninsulated, heated basements with masonry walls lose about 1.1 Btu/ft of wall length per degree F difference from inside to outside. Use this estimate for basements warmed to at least 50°F during the heating season. Insulating the upper 3 to 4 ft of subgrade wall with R-11 insulation cuts the heat loss almost in half. It equals 0.6 Btu/ft of perimeter length per degree F of temperature difference. See Figure 3.5 for proper basement wall insulation.

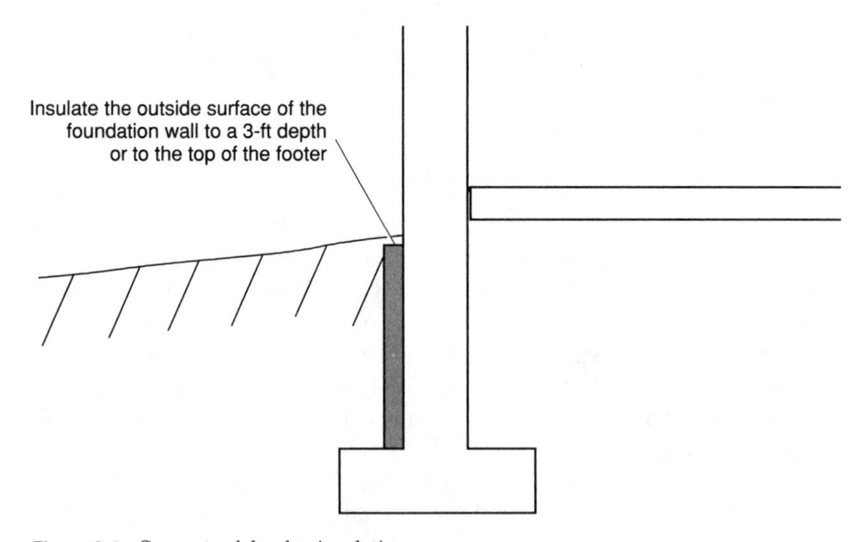

Figure 3.4. Concrete slab edge insulation.

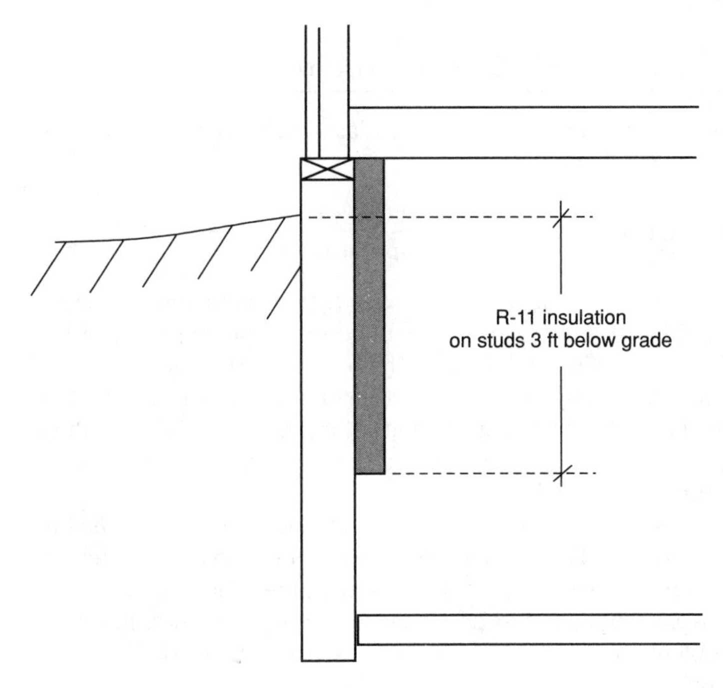

Figure 3.5. Basement wall insulation.

For example, if our building had a basement without insulated walls, its heat loss to the surrounding earth would be

$$248 \times 1.1 = \frac{273 \text{ Btu}}{\text{degree F temperature difference}}$$

Note that a partially insulated basement wall has approximately the same loss as an equivalent building with an insulated slab edge. An uninsulated basement wall has roughly double the loss. The building with the insulated slab edge would probably be less comfortable for the occupants, however, because all of the heat loss occurs near the building's perimeter. Air in the uninsulated basement would convect at the comparatively cold edges. This spreads out the heat loss to a larger area and helps keep the floor warmer near the outside edges.

Heated basement walls that are above grade must be considered the same as other perimeter walls. Find their R-values and calculate the exposed area to arrive at their contribution to total heat loss.

Find the heat loss of an uninsulated high-density concrete perimeter wall 2 ft high. It is 8 in. thick and the basement's area is 30 ft by 55 ft.

Perimeter	$2 \times (30 \text{ ft} + 55 \text{ ft}) = 170 \text{ ft}$
Exposed wall area	$170 \text{ ft} \times 2 \text{ ft} = 340 \text{ ft}^2$

R-value	8 in. × 0.4 R/in. = 3.2
Heat loss 340/3.2	106 Btu/degree F temperature difference

If this above-grade perimeter wall were insulated with 3½ in. of glass fiber (R-11), its heat loss would be

$$340/(11 + 3.2) = \frac{23.9 \text{ Btu}}{\text{degree F temperature difference}}$$

An unheated, ventilated crawl space under a floor contributes significantly to the building's total heat loss. Insulation is usually applied to the floor of the building to minimize the losses that occur there. Calculate these losses in the same way as any other conductive loss through a surface. Use the floor and insulation R-values (or estimate them from the R/inch values and the material's thicknesses). The area is the number of square feet of floor area built above the crawl space.

Some crawl spaces have insulation applied at the side walls with none under the floor (see Figure 3.6). Insulation should completely cover the exposed side walls and extend at least 3 ft below grade. Vents in the crawl space's side walls are supposed to be closed in the winter when moisture will not be a problem condensing on the warm house. This works for crawl spaces up to about 2 ft high.

Heat loss estimates for these buildings are the same as for buildings with insulated floors. Use the floor's area and an R-value equal to double that applied to the slab edge to estimate the heat loss from the crawl space.

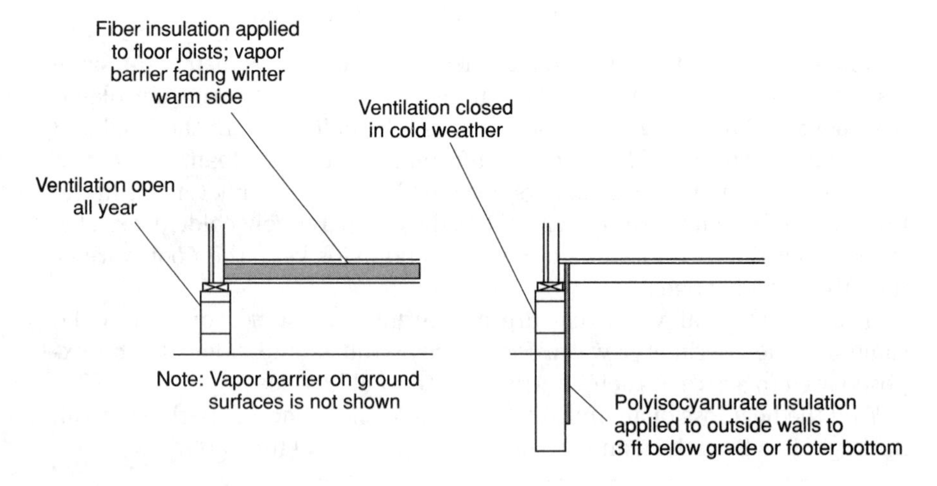

Figure 3.6. Floor and edge applied crawl space insulation. Install a heavy poly vapor barrier on the ground's surface, as well.

For example, consider a 30-ft by 55-ft building with a ventilated crawl space. The floor is 1 in. hardwood with R-19 insulation applied between the joists. The heat loss per degree of temperature difference would be

Area	30 ft × 55 ft = 1650 ft^2
R-value for hardwood	0.87/in.
R-value for the floor	0.87/in. × 1 in. = 0.87
R-value for the insulation	19
Total R-value	19 + 0.87 = 19.87, or 20
Total heat loss	1650/20 = 83 Btu/degree F temperature difference

If this building had 1 in. of polyisocyanurate insulation board installed on the crawl space edges, what is the heat loss?

$$\text{R-value for 1 in. of polyisocyanurate} = 7.2$$
$$2 \times 7.2 = 14.4$$

The heat loss is

$$\frac{1650}{14.4} = 115 \text{ Btu/degree F temperature difference}$$

Temperature estimates. The final number needed to calculate conductive heat loss in the formula is the temperature difference between the inside and outside. Keep in mind that the heating system must provide comfort during any expected weather. Heating systems must handle the most extreme cold that can be expected. In addition, the system should still have enough reserve capacity to keep it operating well even after years of what may be poorly maintained use.

It is important to choose the design outdoor temperature carefully. A reasonable estimate is to use the record low temperature for the area the building is in and subtract an additional 5°F to 10°F from that. Much of the country experiences temperatures well below zero during any given winter. Call the closest National Weather Service office and find out what the record low is for the area.

Continuing with our example building, assume the record low temperature in the area is –15°F. We select a design outdoor temperature of –25°F. For comfort heating, we can assume that the indoor temperature will be 72°F, so the difference in temperature is 97°F (72°F to –25°F).

Note that some energy codes mandate that the designer use specific indoor and outdoor design temperatures to prevent system oversizing. Check the local codes carefully to learn if designs must be limited to the values specified in them.

We now have all of the information needed to estimate conductive heat losses for our example building. To recap:

Roof R-value	8.6
Roof area	3648 ft^2

Wall R-value	13
Wall area	3819 ft^2
Window R-value	2
Window area	1637 ft^2
Glass door R-value	0.9
Glass door area	42 ft^2
Steel door R-value	5.2
Steel door area	42 ft^2
Foundation losses	149 Btu/degree F temperature difference
Temperature difference	97°F

The total conductive heat loss from all of the sources can now be estimated. The formulas are

Btu/hr	Area × T_d/R
Roof Btu/hr	3648 × 97/8.6 = 41,146 Btu/hr
Wall Btu/hr	3819 × 97/13 = 28,496 Btu/hr
Window Btu/hr	1637 × 97/2 = 79,395 Btu/hr
Glass door Btu/hr	42 × 97/0.9 = 4527 Btu/hr
Steel door Btu/hr	42 × 97/5.2 = 783 Btu/hr
Foundation Btu/hr	149 × 97 = 14,453 Btu/hr
Total conductive heat loss	= 41,146 + 28,496 + 79,395 + 4527 + 783 + 14,453 = 168,800 Btu/hr

Infiltration and ventilation losses

The other mechanism that causes buildings to lose heat is convection. Air infiltrates into the building envelope due to the stack effect, wind, and to make up for air discharged through ventilation exhaust system and chimneys. Homes and commercial buildings with fossil fuel heating systems (without dedicated outdoor air supplies) require fresh air for combustion and to develop a proper draft in the chimney flue.

The simplest method of estimating infiltration losses is through calculation of the volume of air delivered by an estimated number of air changes per hour. All of the air coming into the building is outside air, and all of it must be warmed to the indoor temperature. General guidelines for the number of air changes per hour (ACH) are

Building air	ACH
Very energy efficient, tight buildings	0.5
Newer buildings, ordinary construction	1.0
Older buildings	1.5
Extremely leaky buildings	2 to 4

Find the heat loss caused by infiltration for our example building. Assume it is fairly new and not extremely energy efficient. It is 48 ft wide, 76 ft deep, and 22 ft high. The design outdoor temperature is –25°F and the indoor design temperature is 72°F.

$$\text{ACH} = \frac{\text{cfm supply} \times 60}{\text{Space volume}}$$

Rearranging to find cubic feet per minute, the formula is

$$\text{cfm} = \frac{\text{ACH} \times \text{Space volume}}{60}$$

$$\text{Space volume} = 48 \text{ ft} \times 76 \text{ ft} \times 22 \text{ ft} = 80{,}256 \text{ ft}^3$$

$$\text{Infiltration cfm} = \frac{1.0 \times 80{,}256 \text{ ft}^3}{60 \text{ min}} = 1338 \text{ cfm}$$

$$\text{Heat loss} = 1.08 \times 1338 \times (72°\text{F} - -25°\text{F}) = 140{,}169 \text{ Btu/hr}$$

Exhaust air that does not have a dedicated make-up air supply (either a spot unheated supply or one with a dedicated heater) causes additional heat losses. The heat loss formula is the same as used for infiltration, except that the cubic feet per minute value should be the total of all exhaust air discharges in the building. This is added to the heat loss calculated for natural infiltration.

If some heated or spot ventilation air is supplied, the amount of infiltration is smaller. Use the difference between the exhaust and the ventilation supply to figure out how much outside air must be heated by the main system. If the building is positively pressurized by the ventilation system, little or no infiltration occurs. All of the outside air must be tempered by the ventilation system.

The total heat loss for the example building is the sum of its conductive and ventilation losses. These are

Conductive heat loss	168,800 Btu/hr
Ventilation	140,169 Btu/hr
Total heat loss	308,969 Btu/hr

The heating system should be sized to accommodate this need plus a reasonable amount of oversizing. A heating plant able to deliver between 350,000 and 400,000 Btu/hr would be appropriate for this building. A larger system would cost more and perform less satisfactorily than one of the proper size.

Heat Gains

Buildings gain heat from a variety of sources (see Figure 3.7). Major contributors are external sources, but internal heat gains can also be significant. Account for these, too, when designing an air conditioning system. A summary of major building heat gain sources includes

- Conductive heat gains through the building envelope. This includes walls, windows, and the roof. Solar heating of outside surfaces can cause a large part of this heat gain. On sunny days outside surface temperatures are often raised well above the "in the shade" temperatures of the ambient air.

- Solar heating caused by the sun's visible and infrared radiation transmitted through the building's windows. Shades, blinds, and reflective films reduce this.
- Infiltration of warm, moist outdoor air into the cooler and drier interior space. Over half the heat gain from infiltration is often due to the latent heat contained in the air.
- Interior lighting and equipment that adds sensible heat.
- Cooking equipment that can increase both the sensible and latent cooling loads.
- Sensible and latent loads caused by people. These vary with activity levels.

The designer should also be aware of the dynamic nature of many cooling loads (particularly external ones). Heating systems are expected to handle the most extreme conditions at any given time. It is proper to assume that their greatest infiltration loads occur when conductive heat losses are the

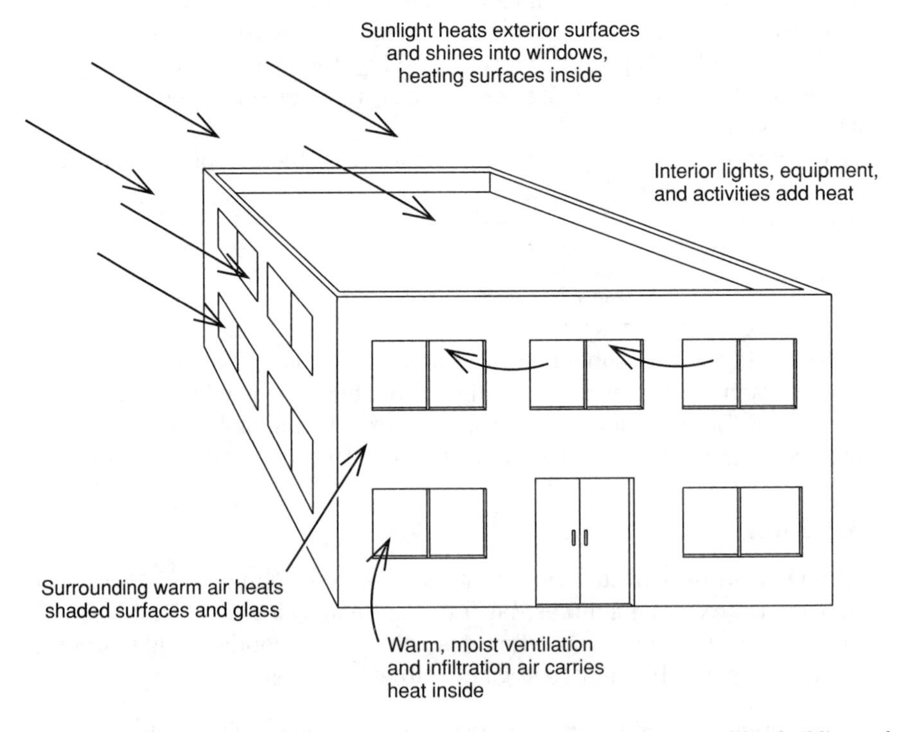

Figure 3.7. Building heat gains. Significant heat sources usually exist both within a building and from its outside surfaces.

highest. Further, the demands on heating systems can continue for long periods (more than 12 hours). This is not usually the case with cooling loads.

Cooling loads change significantly throughout a day. A large external heat source—solar heating—varies throughout a day. Solar heating contributes to the direct heating caused by radiation transmitted into the space through windows and to the heating of the building's outside surface. External solar heating affects the amount of heat conducted into the space. The variable temperature of the surface causes the amount of heat conducted inside to change with time.

Not only does the cooling load change over time, but the orientation of the surfaces that the sun shines onto (or through) is important. Vertical walls and windows facing south, southwest, and west contribute most to solar heating. Surfaces facing east also receive significant amounts of solar radiation and heating, but it occurs early in the day before the day's highest air temperatures occur. North walls and windows receive very little solar heat. Horizontal surfaces (such as flat roofs) are generally not affected by the changing position of the sun unless an adjacent object shields them for part of the day.

Internal contributions to the cooling load can also vary over time. Lighting used during the day and turned off at night does not add as much heat as lights that are left on all the time. Lighting requires several hours to heat its surroundings (primarily by radiation) and cause a cooling load. A time lag occurs between the time the lights are turned on and their contribution to heating the environment.

Similarly, heating caused by people varies with time. Sedentary workers contribute very little to latent heat loads. People's sensible cooling loads are time delayed because much of it is caused by radiation. The radiated heat must warm the surrounding surfaces before it contributes to a cooling load. Convection heating from the skin accounts for a minor heat contribution.

Infiltration loads can also be more complex for cooling estimates than for heating. Doors might be the largest contributor to cooling load infiltration. Simple estimates of ACH, as are done with heating system infiltration losses, are not usually appropriate for cooling systems. A reasonable estimate of the amount of traffic and how often outside doors are used should be done to calculate infiltration heat gains. As discussed earlier, infiltration heat gain calculations must use enthalpy changes to account for the humidity removal required.

Finally, sizing an air conditioning system must be done more carefully than for a heating system. While a reasonably oversized heating system won't cause problems, the same assumption might not be true for an air conditioning system. Most cooling systems must reduce both sensible and latent heat (cool and dehumidify the air). While sensible cooling can take place quickly, latent heat removal requires continuous cooling. Turning off an air conditioning system because the thermostat is satisfied might still allow the humidity to remain high. The key to controlling humidity with an air conditioning system is to have it run almost constantly. On-time cycles should be long and off-time cycles comparatively short. This maintains dry indoor air conditions.

Commercial air conditioning systems with capacity controls (staged compressors, hot gas bypass systems, or unloaders) can handle nonpeak load conditions well. They reduce their Btu capacity and continue operating throughout low demand periods. Small commercial and residential systems, however, rarely have any type of capacity control capability. As soon as the thermostat is satisfied they shut down. During this time the humidity can increase. To avoid this, install an undersized system compared with the expected peak loads.

Residential air conditioners normally must operate almost continuously day and night. Night operation allows them to "catch up" with the day's cooling load. The inside temperature increases during the day and returns to the thermostat's set point at night.

Most residential applications do not "set forward" the thermostat's setting (increase the temperature setting) at night. The house and contents store a significant amount of cooling capacity to use during the following day. This helps prevent wide swings in inside temperature when cooling loads reach their peak in the early afternoon. A moderately undersized air conditioner serves the customer well. It maintains low humidity conditions in the house and provides comfort in the hottest weather.

Air Conditioning System Estimates

The considerations outlined previously illustrate that estimating the capacity of an air conditioning system is more involved than any simple "tons per square foot" rule might suggest. To calculate the proper system capacity, estimate the contributions of each part of the building's cooling requirements. Certain assumptions may be made, however, to arrive at an acceptably accurate estimate of cooling loads.

The greatest cooling loads usually occur in the afternoon, between 2:00 P.M. and 6:00 P.M. local time. This period usually coincides with the local high ambient air temperature and the time of greatest solar exposure. In addition, heat gains from the morning have accumulated and are, in some part, still contributing to the load in the afternoon. Finally, solar heating of the building walls takes some time to migrate through and warm the inside surfaces. This effect depends on the building's materials of construction (with heavy masonry having the longest transmission time lag). However, most buildings of ordinary construction reach their conductive peaks in the early to mid afternoon.

Internal transient heat sources such as lights turned on in the morning and the sensible heat from people are also significant by the afternoon.

Considering these factors, it is reasonable to estimate the cooling loads for the afternoon and use these estimates for system sizing.

It is important to consider how certain building features (particularly windows) are oriented. South, southwest, and west exposures admit significantly more heat than north or east. Wall surfaces oriented in these directions receive significant solar heating, contributing to more conductive heat flow. Therefore, separate the building's exterior walls and windows into two categories: those facing south, southwest, and west, and all other orientations.

To summarize, the total load on a building air conditioning system is the sum of the following heat sources:

- Conduction through the roof and walls. The outdoor temperature used for calculating cooling depends on the wall's orientation. North and east facing walls will be at the ambient air's temperature, others become hotter because of solar heating. The roof's temperature also becomes much higher than the air's temperature.
- Conduction through glass. The outdoor temperature to use for calculations should be equal to (or slightly above) the outdoor air temperature.
- Solar heat gains through south, southwest, and west facing windows.
- Any heat gains from adjoining walls where the neighboring building is not air conditioned.
- Heat gains from air infiltration (including latent and sensible).
- Internal heat gains from lights and equipment.
- Heat gains from people.
- Heat gains from the air conditioner's fans and duct losses.

Specific formulas and methods for each of these heat gains are described in the following sections. Calculate these separately and add the results together to estimate the total cooling system capacity needed.

Conductive gains

The conductive cooling load contribution of roofs, walls, and windows is calculated with the formula

$$\text{Btu/hr} = \frac{\text{Area} \times T_d}{\text{R}}$$

Cooling system conductive heat gains are estimated much like heating system estimates. Wall areas are calculated based upon the height of the building multiplied by the length of its side. Deduct window areas from the wall areas. Roof areas are also done exactly as before.

Again, figure R-values for walls and the roof by adding the individual component's R-values together. The sum is the total R-value to use for the heat gain estimate. Use the suggested window R-values given previously to determine heat gain estimates.

For walls facing north and east, use the anticipated July highest temperature as the outdoor design temperature. Frequently, energy codes restrict indoor design temperatures to no lower than 78°F. This is a reasonable figure to use in any case, as it realistically represents indoor conditions in many buildings.

Walls facing south, southwest, and west should use an outside surface temperature of 110°F to 130°F for conduction estimates. The exact temperature depends on the severity of summer sunshine. For hotter climates a higher number should be used. In almost all cases, however, use an outdoor wall surface temperature of at least 110°F for walls exposed to sunlight during the afternoon. Also, darker colors tend to absorb greater amounts of solar energy and become hotter. Use a higher outside surface temperature for darker walls.

Roofs become hotter than walls during the day because the sun constantly shines on them. For dark roofs, a surface temperature of 160°F is reasonable in most areas of the country. Expect white roofs not exposed to smog or excessive dust to reach 140°F on a hot, sunny day. Please note that heat gain from the roof can be the greatest contributor to building cooling requirements. This is particularly true for a one- or two-story building with a large floor area.

Figure out the amount of heat gained from the roof by using the difference between the roof's surface temperature and the interior space temperature. Heat gains take place the same way losses do in cold weather, with all of the heat flowing through each material. Estimate the total R-value to determine the heat gain.

A ventilated attic below a roof reduces the amount of heat gained by the building. Assume an attic will reach a temperature of 130°F. Then, use the conduction formula to estimate the heat gained from the attic through the ceiling.

Calculate the conductive heat transfer formula using the total attic through ceiling R-value. Add the individual R-values of the materials encountered as heat flows from the attic through the building's ceiling and into the space. The temperature difference to use is the indoor design temperature subtracted from the 130°F attic interior temperature.

Windows absorb very little solar energy. Expect those facing north and east to be at the outdoor July high air temperature in the afternoon. Those facing south, southwest, or west run about 5°F hotter than the outdoor air temperature. Use these temperatures when estimating conductive losses through windows. Again, use the conductive heat formula, with the area and R-value estimated in the same way as for the heat loss estimates done earlier.

Heat gains from interior walls facing a non-air conditioned space are entirely conductive and are found from the formula

$$\text{Btu/hr} = \frac{\text{Area} \times T_d}{\text{R}}$$

Calculate the area and R-values as described earlier. Use the outdoor July high temperature as the design temperature for the non-air conditioned surface unless it is significantly different.

Solar gains

South, southwest, and west facing windows admit a significant amount of heat into a building as solar energy. The formula for estimating the admitted solar heat is

$$\text{Btu/hr} = 125 \times \text{Area} \times \text{SC}$$

where SC is the window's shading coefficient.

The shading coefficient is a number that compares the amount of heat admitted by a particular window to that transmitted by a ⅛-in.-thick pane of clear glass. Window shading coefficients vary with the type of glass and any blinds or draperies used to shield them. Typical shading coefficients are

0.9	Single pane of glass (normal thickness)
0.85	Double-pane glass
0.5	Interior blinds (slat type)
0.4	Interior shades
0.3–0.6	Reflective coated glass (consult manufacturers' specifications)

Multiply the individual shading coefficients together if more than one measure is used to reduce transmitted solar energy. This gives a total coefficient for the window. For example, if 15 ft^2 of double-pane windows use interior shades, how much heat is added to the building from solar radiation?

$$\text{Total SC} = 0.85 \times 0.4 = 0.34$$

$$\text{Btu/hr} = 125 \times 15 \times 0.34 = 638 \text{ Btu/hr}$$

Note that windows facing north or east don't contribute to peak solar heat gains; they only have conductive heat contributions.

Infiltration and ventilation gains

Infiltrating air from the outside contributes both latent and sensible heat to the space. The air conditioning system must remove both. The formula for estimating total heat gain from infiltration and ventilation air is

$$\text{Total Btu/hr} = 4.5 \times \text{cfm} \times \text{Enthalpy difference}$$

Conditions of 78°F and 50% relative humidity correspond to an enthalpy of 29.6 Btu/lb and 2.17 Btu/ft^3 of air. This is a reasonable figure to use for most comfort cooling conditions. Enthalpy of outside air at design conditions can be found using Table 1.3. Find out from the local National Weather Service Office what the peak relative humidity will be at the time of the expected July high temperature. Locate the enthalpy value corresponding to the expected conditions on Table 1.3 and use it in the formula.

Most codes require a minimum amount of fresh air supply to a space while it is occupied. Ventilation systems often deliver 5 to 20 cfm of outdoor air for each person in the space. If no energy recovery heat exchangers are used, the air conditioning system must cool all of the ventilation air. This can be a major load.

The building should be positively pressurized by the fresh air supplied by the ventilation system. Air leaks around windows and doors will be from inside to the outside, so no other heat gains should occur from them. Also, plenty of air is available for bathroom exhaust fans, so their operation won't cause infiltration.

If the building includes large exhaust systems, however, it may have more air mechanically discharged than the ventilation system is providing. If so, figure out the difference between the total exhaust and the supplied ventilation as the air volume requiring conditioning. If there is more exhaust than ventilation, infiltration is their difference. As long as the amount of ventilation supply air exceeds the exhaust, infiltration should not be a problem. The ventilation supply air will, however, require cooling.

An example: A building houses 30 people and receives 20 cfm of outdoor air per person. The outdoor air's peak July temperature is 95°F with a relative humidity of 60%. How much total and sensible heat is added to the building from the ventilation air?

$$\text{Total ventilation cfm} = 30 \times 20 = 600 \text{ cfm}$$

$$\text{Enthalpy of the outdoor air} = 46.4 \text{ Btu/lb}$$

$$\begin{aligned}\text{Total heat added} &= 4.5 \times 600 \text{ cfm} \times (46.4 \text{ Btu/lb} - 29.6 \text{ Btu/lb})\\ &= 45{,}360 \text{ Btu/hr}\end{aligned}$$

$$\text{Sensible heat added} = 1.08 \times 600 \text{ cfm} \times (95\,°\text{F} - 78°\text{F}) = 11{,}016 \text{ Btu/hr}$$

$$\begin{aligned}\text{Latent heat added} &= \text{Total heat} - \text{Sensible heat} = 45{,}360 - 11{,}016\\ &= 34{,}344 \text{ Btu/hr}\end{aligned}$$

Note that the sensible heat is a small part of the total heat admitted with ventilation air.

Not all buildings receive the amounts of fresh air that the newer codes mandate. If you are working on an older building, it will be necessary to estimate infiltration. Calculating the amount of air infiltrating a building requires some knowledge of the building's operation. Normal double-leaf outside doors without air locks admit about 100 ft^3 of air every time a person enters or leaves. An air lock entrance effectively cuts that volume in half. Multiply the appropriate cubic foot figure (either 100 or 50) by the number of times the building occupants or the architect estimates each outside door is opened in 1 hour. Divide this answer by 60 to get the cubic feet per minute of outside air admitted by the door. Understandably, this provides a rough approximation, but it can reveal significant cooling loads.

Wind driven infiltration can also contribute large heat gains. It is safe to estimate 0.2 ACH for buildings that receive minimal ventilation air. Note that bathroom exhausts contribute to this, but the typical number of these small exhaust systems need not be added to this infiltration estimate.

Lighting and equipment gains

Lights and electrically operated equipment in the space contribute to the cooling system's heat load. If the lighting or equipment is used all of the time, the heat load is

$$\text{Btu/hr} = \text{Watts} \times 3.41$$

Note that watts = volts × amperes and that 1 kW = 1000 W. Equipment nameplates list the voltage and either the watts or amps that a piece of equipment uses.

Fluorescent lights include ballasts that contribute heat to the space. Assume that the light fixture's total power is equal to the total of the lamp's wattage rating plus 20%. Multiply the bulb's wattage by 1.2 to get the total watts. Standard four-tube, 4-ft fluorescent fixtures contribute approximately 180 W each. Two-tube, 4-ft light fixtures consume about 90 W each.

Lights and equipment used for 9 or 10 hours (and turned off at night) do not contribute fully to space cooling loads. Their surfaces require time to warm, and then the surrounding objects are heated by their radiant energy. This creates a time lag between their starting and the full heat loading. In these cases multiply the total watts by 0.7 to estimate their cooling load contribution.

An example: An office has fifteen 4-ft fluorescent fixtures. Each fixture has four lamps. Five fixtures run 24 hours a day and the other 10 are turned on at 7:30 A.M. and off at 5:00 P.M. What is the total heat contribution from the lights?

For the lights that stay on all night:

$$\text{Btu/hr} = 5 \text{ fixtures} \times 180 \text{ W} \times 3.41 = 3069 \text{ Btu/hr}$$

For the lights that are turned on in the morning

$$\text{Btu/hr} = 0.7 \times 10 \text{ fixtures} \times 180 \text{ W} \times 3.41 = 4297 \text{ Btu/hr}$$

$$\text{Total Btu/hr} = 3069 + 4297 = 7366 \text{ Btu/hr}$$

Gains from people

People contribute to air conditioning loads by adding both sensible and latent heat to the space. The amount of heat produced is dependent upon their activity level. Also, the sensible heat contribution requires a comparatively long time to become significant. Latent heat contributions take place immediately.

People engaged in light office work produce a total heat load of approximately 500 Btu/hr per person. This total is roughly half sensible and half latent heat, for a gain of approximately 250 Btu/hr for each. Light exertion, including walking and routine physical activity, increases the heat load to approximately 300 Btu/hr sensible and 300 to 350 Btu/hr of latent heat. Finally, heavy physical activity produces 500 Btu/hr of sensible heat and 1000 Btu/hr of latent heat.

For example, find the sensible and latent cooling system loads, the sensible heat ratio, and the tons of air conditioning capacity required by 30 persons. Ten are performing light office work and the rest are doing light assembly work.

First, consider the office workers:

$$10 \text{ persons} \times 250 \text{ Btu/hr} = 2500 \text{ Btu/hr sensible heat}$$

The latent heat load from these persons is also 2500 Btu/hr. Now consider the 20 persons engaged in moderate activity:

Sensible heat gain	20 × 300 Btu/hr = 6000 Btu/hr
Latent heat gain	20 × 325 Btu/hr = 6500 Btu/hr
Total sensible heat gain	2500 Btu/hr + 6000 Btu/hr = 8500 Btu/hr
Total latent heat gain	2500 Btu/hr + 6500 Btu/hr = 9000 Btu/hr
Total heat gain	8500 Btu/hr + 9000 Btu/hr = 17,500 Btu/hr
Sensible heat ratio sensible heat/total heat	8500 Btu/hr/17,500 Btu/hr = 0.486
Cooling tons	17,500 Btu/hr/12,000 Btu/ton = 1.46 tons of air conditioning capacity

Calculations of people's heat contributions to a space are usually general estimates. Most people's activity levels vary widely during a given period and their heat contributions fluctuate with those changes. In situations where the total number of people and their activity level is known (for example, a movie theater, gym, or restaurant) it is easy to calculate accurate heat contributions. Other cases, such as production work, can be more difficult to define. The heat gain estimates above are conservative, but are accurate enough for most HVAC work.

Fan heat gains

Another air conditioning system load to be considered is the air delivery fan. Include the fan and motor contributions when they are inside the conditioned air stream or in a cooled space. Approximately half the heat contribution is from the motor and half is from the fan's impeller. Losses in the drive (pulleys and bearings) are included with the motor's losses.

Sensible heat gains from the motor and fan are found with the formula

$$\text{Btu/hr} = \text{Motor hp} \times 2040 \text{ Btu/hr/hp}$$

A 1-hp motor delivers approximately 2040 Btu/hr to the air. This causes a temperature rise in the air as it passes through the fan and around the motor. The temperature increase is found with the formula

$$\text{Temperature rise} = \frac{\text{Btu/hr}}{(1.08 \times \text{cfm})}$$

Substituting this equation for motor horsepower, the final formula is

$$\text{Temperature rise} = 1890 \times \text{Motor hp/cfm}$$

For example, find the heat contribution and temperature rise of 2200 cfm of air passing through a fan powered by a ½ hp motor:

$$\text{Heat contribution} = 0.5 \times 2040 \text{ Btu/hr/hp} = 1020 \text{ Btu/hr}$$

$$\text{Temperature rise} = \frac{1890 \times 0.5}{2200 \text{ cfm}} = 0.43°\text{F}$$

These equations assume that both the motor and fan are in the air stream or cooled space to contribute heat to the air. Many large air handlers keep the motor out of the air flow and have fan rooms that are isolated from the conditioned air. In these cases only about half the temperature increase would occur. Only the impeller causes air heating that takes place in the fan, and this accounts for roughly one-half of the heat gain.

Chapter

4

Airflow

Air movement is used for almost every heating, ventilation, and air conditioning application. Air delivers or removes heat and provides an acceptable environment by providing ventilation. Understanding how Airflows and the components used for its delivery is essential to understanding HVAC systems.

Air is a fluid, and all fluids flow from high pressure areas to lower pressure ones. Just as temperature always drives heat flow, pressure always drives fluid flow. The greater the pressure difference from one area to another, the greater the flow.

Atmospheric air is always under pressure. On average, the atmosphere at sea level exerts a pressure of 14.7 pounds per square inch (psi) against all surfaces. This 14.7 psi is equal to the total weight of a 1-in.^{2} column of the atmosphere stacked up from the earth's surface to outer space. Because of this, a pressure of 14.7 psi is sometimes called *1 atmosphere* (atm).

Pressure can be measured in more than one way. Besides expressing it as a force acting over an area (as pounds per square in.), it can also relate to the height of a column of liquid. Water and mercury are fluids often used to express pressure. The weight exerted by a column of these liquids causes pressure. Because the properties of these liquids are well known, they have become standards used to express an amount of pressure.

A 1-in.^{2} column of mercury 29.92 in. high exerts a weight of 14.7 lb at the bottom. Similarly, a column of water 33.9 ft high with a 1-in.^{2} area also weighs 14.7 lb. Expressed as the height of a column of water, 1 atm (14.7 psi) is equal to 33.9 ft (or 407 in.) of water. A column of mercury 29.92 in. tall also exerts a pressure of 1 atm. The relationship of the height of a column of water or mercury to pressure allows any pressure value to be expressed as the height of a column of one of these liquids.

The expression of pressure as the height of a fluid is properly called *head*. While the terms pressure and head are often used interchangeably, technically head is the height of a column of fluid supported by pressure. For our purposes, however, there is no need to be concerned with this distinction.

Table 4.1 shows pressure and head equivalents. Here are some examples of pressure and head conversions:

Find the number of inches of water equivalent to 0.5 psi:

$$0.5 \text{ psi} \times 27.7 \text{ in. water/psi} = 13.8 \text{ in. water}$$

What pressure in psi is equal to 18 in. mercury?

$$18 \text{ in. mercury} \times 0.602 \text{ psi/in. mercury} = 10.8 \text{ psi}$$

TABLE 4.1 Head, Pressure, and Vacuum Equivalents

To find	Multiply	By
ft water	in. water	0.0833
	in. mercury	1.13
	KPa	0.335
	psi	2.31
in. water	ft water	12.0
	in. mercury	13.6
	KPa	4.01
	psi	27.7
in. mercury	ft water	0.882
	in. water	0.0735
	KPa	0.295
	psi	2.04
KPa	ft water	2.99
	in. water	0.249
	in. mercury	3.39
	psi	6.90
psi	ft water	0.433
	in. water	0.0361
	in. mercury	0.491
	KPa	0.145

Water (ft)	Water (in.)	Pressure (in. mercury)	Vacuum (in. mercury)	psia	psig
−33.93	−407.2	−29.92	29.92	0	−14.70
−31.76	−381.1	−28	28	0.94	−13.76
−30.62	−367.5	−27	27	1.43	−13.27
−29.49	−353.8	−26	26	1.93	−12.77
−28.35	−340.2	−25	25	2.42	−12.28

Water (ft)	Water (in.)	Pressure (in. mercury)	Vacuum (in. mercury)	psia	psig
–27.22	–326.6	–24	24	2.91	–11.79
–26.08	–313.0	–23	23	3.40	–11.30
–24.95	–299.4	–22	22	3.89	–10.81
–23.82	–285.8	–21	21	4.38	–10.32
–22.68	–272.2	–20	20	4.87	–9.83
–21.55	–258.6	–19	19	5.37	–9.33
–20.41	–245.0	–18	18	5.86	–8.84
–19.28	–231.4	–17	17	6.35	–8.35
–18.15	–217.7	–16	16	6.84	–7.86
–17.01	–204.1	–15	15	7.33	–7.37
–15.88	–190.5	–14	14	7.82	–6.88
–14.74	–176.9	–13	13	8.31	–6.39
–13.61	–163.3	–12	12	8.80	–5.90
–12.48	–149.7	–11	11	9.30	–5.40
–11.34	–136.1	–10	10	9.79	–4.91
–10.21	–122.5	–9	9	10.28	–4.42
–9.07	–108.9	–8	8	10.77	–3.93
–7.94	–95.27	–7	7	11.26	–3.44
–6.80	–81.66	–6	6	11.75	–2.95
–5.67	–68.05	–5	5	12.24	–2.46
–4.54	–54.44	–4	4	12.73	–1.97
–3.40	–40.83	–3	3	13.23	–1.47
–2.27	–27.22	–2	2	13.72	–0.983
–1.13	–13.61	–1	1	14.21	–0.491
0	0	0	0	14.70	0
0.231	2.770	0.204		14.80	0.100
0.462	5.540	0.407		14.90	0.200
0.693	8.310	0.611		15.00	0.300
0.923	11.08	0.814		15.10	0.400
1.154	13.85	1.018		15.20	0.500
1.385	16.62	1.221		15.30	0.600
1.616	19.39	1.425		15.40	0.700
1.847	22.16	1.628		15.50	0.800
2.078	24.93	1.832		15.60	0.900
2.308	27.70	2.035		15.70	1.000
4.617	55.40	4.071		16.70	2.00
6.925	83.10	6.106		17.70	3.00
9.233	110.8	8.141		18.70	4.00
11.54	138.5	10.18		19.70	5.00
13.85	166.2	12.21		20.70	6.00
16.16	193.9	14.25		21.70	7.00
18.47	221.6	16.28		22.70	8.00
20.78	249.3	18.32		23.70	9.00
23.08	277.0	20.35		24.70	10.0
46.17	554.0	40.71		34.70	20.0
69.25	831.0	61.06		44.70	30.0
92.33	1108	81.41		54.70	40.0
115.4	1385	101.8		64.70	50.0
138.5	1662	122.1		74.70	60.0
161.6	1939	142.5		84.70	70.0
184.7	2216	162.8		94.70	80.0

TABLE 4.1 (Continued)

Water (ft)	Water (in.)	Pressure (in. mercury)	psia	psig
207.8	2493	183.2	104.7	90.0
230.8	2770	203.5	114.7	100
253.9	3047	223.9	124.7	110
277.0	3324	244.2	134.7	120
300.1	3601	264.6	144.7	130
323.2	3878	285.0	154.7	140
346.3	4155	305.3	164.7	150
369.3	4432	325.7	174.7	160

Pressures within a system can be further defined as *gauge* or *absolute* readings. Gauge pressure refers to the reading obtained on an ordinary pressure gauge read within the atmosphere. It is the system's pressure above the ambient atmosphere's normal 14.7 psi. When specifying gauge pressure, the abbreviation is psig. The "g" refers to gauge.

Absolute pressure refers to the pressure within a system compared to a total vacuum. The atmosphere is at 14.7 psi compared with a vacuum, so its pressure (14.7 psi) is an absolute reading. Absolute pressure readings are usually written as psia. It is easy to convert gauge pressure to absolute pressure and back. Use the relationships

$$\text{psia} = \text{psig} + 14.7$$

$$\text{psig} = \text{psia} - 14.7$$

Vacuum readings are generally given in inches of mercury, but with zero referring to atmospheric pressure and 29.9 meaning a full vacuum. The readings run this way (with lower absolute pressures having higher readings) because they express the intensity of the vacuum compared to the surrounding atmospheric pressure. Higher vacuum readings correspond to more pressure applied by the atmosphere.

An example: What is the absolute pressure in psia of a 12-in. vacuum?

$$12 \text{ in. mercury} \times 0.602 \text{ psi/in. mercury} = 7.22 \text{ psi}$$

This is a "negative" pressure compared with the atmosphere, and should be subtracted from the atmosphere's absolute pressure:

$$14.7 \text{ psia} - 7.22 \text{ psi} = 7.48 \text{ psia equivalent to a 12-in. mercury vacuum}$$

This idea of atmospheric pressure leads to the consideration of Airflow. Air always flows from high to low pressure. Because the atmosphere is at 14.7 psia, the removal of some air (or an object occupying space in air) causes air to flow. It moves into the space that was formerly occupied. Removing the object "pulls" the air behind it. A partial vacuum is formed for an instant as the object is moved, then air moves in to occupy that vacuum.

Air movement also occurs if an object pushes into it. The object causes an instantaneous increase in pressure at its leading edge. Air molecules move from this high-pressure area to somewhere else where things aren't so crowded. Again, it is pressure that moves the air.

Confined air moving inside a duct, however, causes a force against the duct walls. Air, pressurized by the fan, tries to expand and pushes against the duct walls. Return air systems, connected to the fan's inlet, experience a similar pressure as a partial vacuum. The pressure that causes this force is the air's *static pressure*. Air speed does not change static pressure. They are not related. Conditions may exist where the air is moving slowly but its static pressure is high.

Moving air has a form of pressure caused by its movement known as *velocity pressure*. Velocity pressure exerts a force against objects in the air's path. Velocity pressure from a strong wind can blow people down.

Velocity pressure is related to the speed of the air squared. Doubling the air's speed causes the velocity pressure to quadruple. Tripling the velocity of the air causes its velocity pressure to increase nine times.

Velocity pressure exerts a force only against objects in the path of the air's movement. No velocity pressure is exerted against an object parallel to the flow of the air.

The reverse can also be true. Air can have a high speed and velocity pressure while it also has a low static pressure. Static pressure depends on the restriction of air's flow instead of its speed. Velocity pressure, on the other hand, is always related to the air's speed.

Finally, the *total pressure* of confined, moving air is the sum of the velocity pressure and the static pressure. This total pressure can vary widely as Airflows from one point to another. Changes in the air's speed, duct dimension changes, and turbulence can all cause the total pressure to change.

The velocity pressure of moving air is found with the equation

$$\text{Pressure} = \left(\frac{\text{Speed}}{4005}\right)^2$$

The pressure is expressed as inches of water and speed is expressed as feet per minute (fpm).

Find the velocity pressure of air moving at 850 fpm:

$$\text{Pressure} = \left(\frac{850 \text{ fpm}}{4005}\right)^2 = 0.212^2 = 0.045 \text{ in. water}$$

Find the velocity pressure in pounds per square inch of air resisting the movement of a car. It is traveling at 60 mph, or 88 fps.

First, find its speed in fpm:

$$\text{Speed} = 88 \text{ fps} \times 60 \text{ sec/min} = 5280 \text{ fpm}$$

$$\text{Pressure} = \left(\frac{5280 \text{ fpm}}{4005}\right)^2 = 1.32^2 = 1.74 \text{ in. water}$$

$$\text{psi} = 1.74 \text{ in. water} \times 0.0361 \text{ psi/in. water} = 0.0628 \text{ psi}$$

The previous examples illustrate that velocity pressures are generally low for ordinary air speeds. While the automobile in the above example would experience a reasonable force because of its large area, velocity pressures alone in most HVAC systems generally do not require unusual considerations for strength of ducts or other air handling equipment. Internal static pressure is usually higher than velocity pressure. If not fabricated and supported properly, fans can pressurize ducts enough to damage them. Only very high air speeds (not usually used for HVAC systems) require special duct construction to handle higher velocity pressures.

Fans

Forced air heating systems, and almost all air conditioning systems, rely on the mechanical movement of air from one place to another. Fans are used for this. They use an electric motor spinning an impeller on a shaft to push air from a lower pressure area to a higher pressure area. The fan adds power to the air because it is being forced against the "normal" direction of flow (from higher pressure to lower).

Several things occur as a fan supplies power to a stream of air. The air's pressure increases (generally both its static and velocity pressure) and its temperature is raised. Air temperature might increase from 1°F to 5°F as it passes through a fan.

Fan laws

Several basic laws govern the way fans transfer energy to air. The energy (or power) going in, measured by the horsepower of the fan's motor, is converted into energy added to the air. The air's energy increase is determined by its flow rate and pressure change.

Defining the amount of energy delivered from the fan to the air is the flow in cubic feet per minute (cfm) and the air's static pressure increase. The pressure increase is measured by subtracting the static air pressure at the fan's inlet from that at the outlet. These two factors, Airflow and static pressure increase, fully describe the amount of usable energy delivered by the fan. Fan heating is not included.

Within a given fan type and size (diameter and width) the Airflow, the air's pressure increase, and the fan motor's horsepower requirements all vary with the fan's impeller speed. These relationships (known as *fan laws*) are useful when figuring out the effects of changing the fan's revolutions per minute. The fan laws predict how varying the speed changes the volume of air delivered, its static pressure increase, and the horsepower load on the motor.

Note that the fan laws require that other conditions remain the same. The fan must be blowing air through the same duct system (or other resistance), and there can be no physical changes to the fan's size. Only the fan's speed must change. The fan law relating delivered cubic feet per minute to revolutions per minute is

$$\text{New cfm} = \text{Old cfm} \times \left(\frac{\text{New rpm}}{\text{Old rpm}} \right)$$

Doubling the revolutions per minute doubles the volume of air the fan is delivering. If the air is confined in a duct, the speed of the air inside the duct is doubled.

The fan law for static pressure is

$$\text{New static pressure} = \text{Old static pressure} \times \left(\frac{\text{New rpm}}{\text{Old rpm}} \right)^2$$

Doubling the fan's speed raises the air's static pressure higher by a factor of four. The air's pressure increases much faster than its speed when the fan is run faster.

Fan motor horsepower also has a fan law:

$$\text{New hp} = \text{Old hp} \times \left(\frac{\text{New rpm}}{\text{Old rpm}} \right)^3$$

Doubling the fan's speed requires eight times the horsepower from the motor. Increasing the fan to 50% more than its original speed causes the motor to more than triple its power requirement. Keep this in mind when adjusting a fan's speed. A relatively small speed increase causes a large increase in the motor's power draw.

Adjust fan speeds (by changing or adjusting the drive sheaves) with care. Place a clamp-on ammeter on the motor leads when adjusting a fan's speed to watch the motor's power draw. A motor can easily be forced to deliver more power than it should. This can cause it to overheat, leading to an early burnout or failure.

Use the ammeter to ensure that the current drawn by the motor does not exceed its nameplate rating. The motor horsepower fan law shows that power increases very fast with changes in fan speed. If it is not possible to get sufficient air speed or volume without overloading the motor, replace it with a higher horsepower unit. Be certain that the electrical circuits (wiring, starters, and fuses) can handle the additional load.

Fan manufacturers provide characteristic *performance curves* for their products. An example is shown in Figure 4.1. These show how the fan's ef-

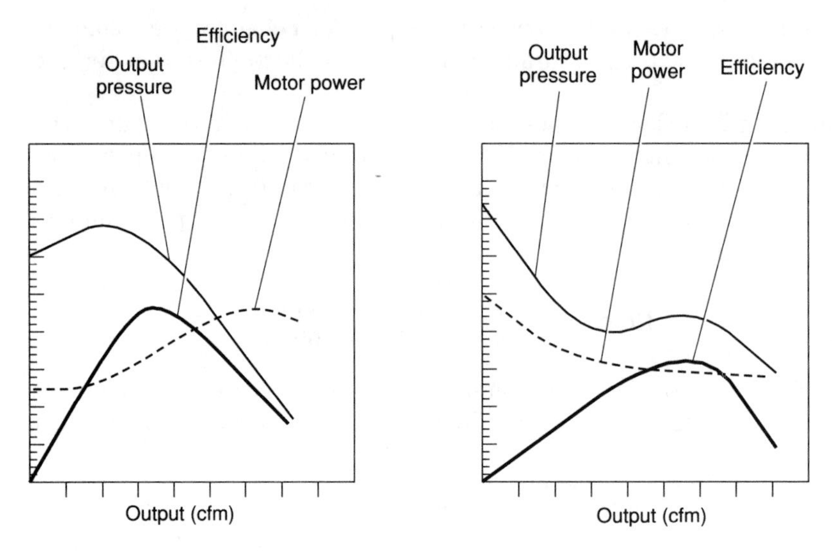

Figure 4.1. Typical axial and centrifugal fan curves.

ficiency, horsepower requirements, and pressure capabilities vary over differing flow conditions. They are derived from standardized test procedures defined by the Air Moving and Conditioning Association (AMCA) and the American Society of Heating, Refrigeration, and Air Conditioning Engineers (ASHRAE).

Fan types

Fans are classified into two broad categories: centrifugal and axial. Centrifugal units are further classified by the type of blades used and their orientation. Axial fans are categorized into propeller, tubeaxial, and vaneaxial versions. Each type of fan, and the subtypes within them, have specific characteristics that apply to the entire group and make them suitable for different situations. These generalizations are useful when predicting how a fan will function in different applications.

Note that the fan laws relate changes in fan speed to volume, pressure, and power requirements within a given fan type. Different fan types have very different characteristics when compared to each other.

Centrifugal fans. Centrifugal fans are available in a huge array of types and capacities. There are miniature units for ceiling-mounted room exhausters that move less than 500 cfm at approximately 0.5 in. water static pressure. On the other extreme there are large industrial and commercial fans that handle over 250,000 cfm and static pressures over 20 in. water. Centrifugal fans provide the designer with the widest range of choices of any type.

They are the most commonly used fans for HVAC applications. Everything from wall-mounted air conditioners and fan coil units to large central station air handlers use centrifugal fans. They are simple and inexpensive to manufacture. Also, most HVAC design engineers know and understand their characteristics.

These fans use an impeller made up of a series of blades mounted between two hubs (making an assembly called a *squirrel cage*). Air flows through the wheel radially from its inside diameter to the outside and gains speed from two effects: (1) the *centrifugal* force caused by the air's movement from the wheel's inside to its outside, and (2) from the blade tips pushing the air as it flows away from them.

The impeller is usually enclosed in a scroll-shaped housing that guides and concentrates the air to a single outlet. These fan housings have inlets on one or both sides that allow air to enter the impeller's inside diameter. The housing's scroll-shaped design changes some of the high-speed air's velocity pressure into static pressure by slowing it down uniformly toward the outlet.

Centrifugal fans change the direction of the airflow. Air enters the fan along the axis of the impeller and leaves it at a right angle to the direction it entered.

A key design element of centrifugal fans is the blade design. These fall into three categories depending on how the blades are "tilted" compared to the rotation direction of the impeller. These are radial, forward curved (or inclined), and backward inclined. Backward inclined units are further broken down into airfoil and curved blade categories. Figure 4.2 shows the major types.

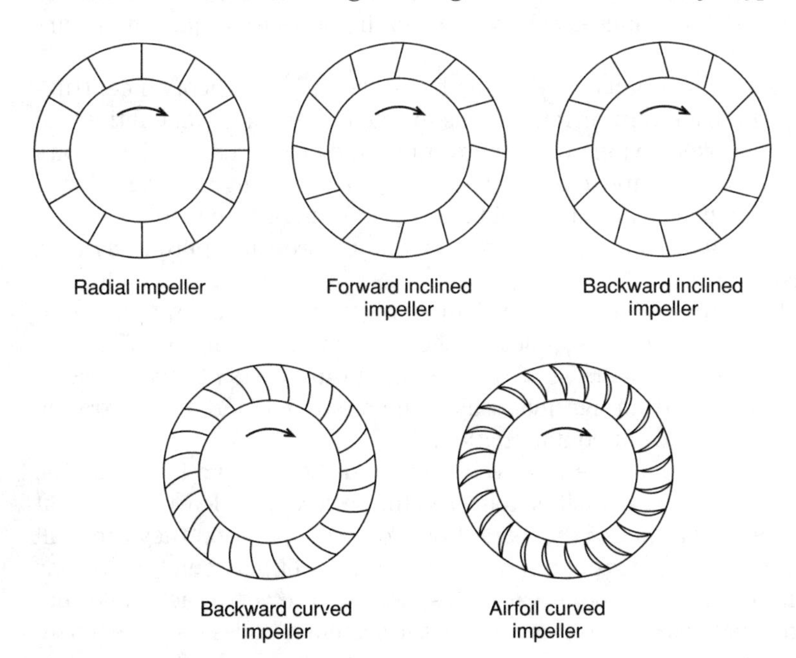

Figure 4.2. Centrifugal fan impeller configurations.

Radial fans have blades that run straight from the inside radius of the hub to the outside. They are straight and perpendicular to the direction of rotation. Forward curved blades are pitched in the same direction as the impeller's rotation. Backward inclined blades tilt in a direction opposite to the fan's rotation.

Radial blade fans are usually not used for HVAC systems. These are commonly used for dust handling and applications where abrasion of the impeller can occur because of materials suspended in the airstream. They are inexpensive to manufacture, rugged, easy to repair, comparatively noisy, and inefficient. Because the space between the blades increases from the inside to the outside there is little tendency to clog. Clearances between the fan housing and the impeller are not critical. Power requirements rise as the amount of airflow increases, despite changes in static pressure.

Forward inclined fans are also inexpensive to manufacture. They use many narrow, flat blades of stamped metal evenly spaced between the impeller hubs. These fans increase the air's output speed higher than the blade tip's rotation speed. This high-velocity air discharge permits them to be physically smaller than other centrifugal units.

Forward inclined fans are often used for low-pressure HVAC applications where cost of construction is an important consideration. Residential furnaces, package rooftop units, and some central air handling units use these fans. They are of moderate efficiency and, if run within their optimal design limits, comparatively quiet. Horsepower requirements increase as the amount of flow is increased, even when the pressure requirements are reduced.

Backward inclined fans provide the best total efficiencies of all centrifugal units. The backward airfoil fan has excellent pressure capabilities and can deliver air over a wide range of flow rates and pressures. The blades are not made of flat or simply curved sheet metal. They are manufactured with a cross section like an airplane's wing to provide a minimum amount of flow restriction. Airfoil blades cost more to manufacture (and purchase) than simple flat blades.

These fans are used in larger air handling systems that operate continuously. Their greater cost is offset by the large improvement in efficiency, lower power requirements, and reduced operating costs. They are easily clogged by dust and foreign material in the airstream, and good filtration must be used to prevent contamination.

Blades on these fans are relatively wide and their profile increases the static pressure of the air as it is delivered to the scroll. Air leaves the fan at a lower speed than that of the blade tips. For best operation they are built with very close tolerances and require an excellent fit between the housing and the impeller. They operate at high speed and are very quiet while running. Removing restrictions to airflow in the system reduces the fan's horsepower requirements.

Backward inclined (or backward curved) fans are slightly less efficient than airfoil units. They can, however, provide the superior static pressure deliveries of airfoil types. Because there is no airfoil profile to the blades, they are less prone to clogging or erosion. Backward inclined fans are also less expensive to manufacture than airfoil fans. These are sometimes used in HVAC applications where forward curved fans do not provide sufficient static pressure capabilities.

Specialized fans include roof-mounted exhausters. These are usually centrifugal units intended for ducted ventilation or fume exhaust. They include fans that handle high temperatures and corrosive gases in their airstream. Exhausters are available with special coatings (including epoxy compounds) that completely cover all of their internal surfaces. This prevents corrosion and damage from the air or gases flowing through them. Some fans are fabricated entirely out of epoxy-coated glass fibers or other corrosion resistant material to survive extreme applications.

There are several available options for installing motors on exhaust fans. Often, for the sake of economy, motors are installed suspended inside the fan's impeller and exposed to the airflow. This helps cool the motor. With exhausters, this might not be feasible if the fan is handling very high temperatures or corrosive atmospheres. Available motor mount options for fume exhausters include placing it out of the airflow with special sealed belt drives.

Also available are separate ventilation ducts used to provide air for the motor. These fans use separate motor cooling systems so they can be used where the air being handled by the fan is corrosive or hot. An outside source of clean, cooler air is ducted to the motor to keep it cool.

Commercial kitchen exhaust fans are often used as roof-mounted exhausters. Called *upblast fans*, they direct their discharge Airflow upward away from the roof. The hot, greasy contents of kitchen exhaust gases can damage the roof. In addition, building codes usually specify that a kitchen hood's exhaust stream must be directed away from nearby structures (including the roof). This prevents a fire on the range or within the hood and duct system from being blown onto the building.

Upblast centrifugal exhausters are available in sizes that will move over 25,000 cfm at total pressures greater than 2 in. water. Smaller models are available that provide lower volumes at more moderate static pressures.

Axial fans. Axial fans do not change the direction of air flowing through them. Air enters at one end, passes straight through the impeller, and discharges out the exit. As noted earlier, these fans include propeller, tubeaxial, and vaneaxial types.

Propeller fans are low-efficiency devices. They aren't usually installed in a duct system because they provide very little static pressure. They can, however, move a large volume of air in a free delivery installation. Propeller fans are available in capacities over 75,000 cfm but can handle a maximum of 1 in. water static pressure. Usually a venturi ring is installed close to the blade tips to provide increased airflow and help to reduce noise (see Figure 4.3). Nevertheless, propeller fans are the loudest axial fans.

Air delivered from a propeller fan swirls as it leaves the fan. This additional movement, not contributing to its forward motion, reduces the static pressure capabilities of these fans.

The hub holding the fan's blades has a small diameter (compared to the total diameter of the fan's blade tips), and two to four blades are mounted on it. High-speed fans use narrow blades and low-speed fans use fairly wide blades. They are often used for spot cooling people or equipment, or for building ventilation, smoke removal, or make-up air when no duct system is required.

Tubeaxial fans use a modified propeller-type impeller as shown in Figure 4.4. The hub diameter might be as large as half the impeller diameter, with four to eight blades installed on it. The impeller is installed in a cylindrical housing, which gives the fan its *tubeaxial* name. Tubeaxial impeller tips fit close to the housing's inside diameter. This improves their efficiency.

These fans can provide medium pressure (over 1.5 in. water) and high flow rates, over 70,000 cfm. If the air discharge is excessively restricted they might become unstable and begin to *surge*. Surge is caused by excessively high-pressure outlet air. Its pressure is high enough to enable it to force itself upstream, back past the fan blades against the direction the fan is running. This condition can cause the blades to crack and severely stresses the bearings.

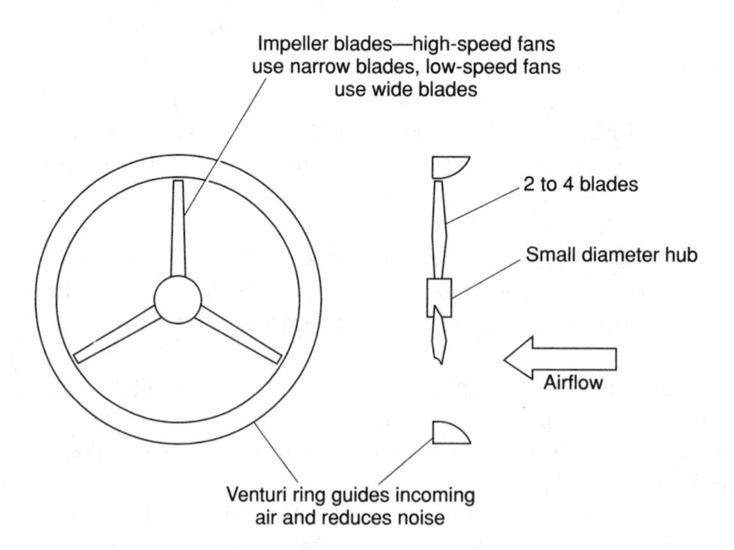

Figure 4.3. Propeller fan detail and section.

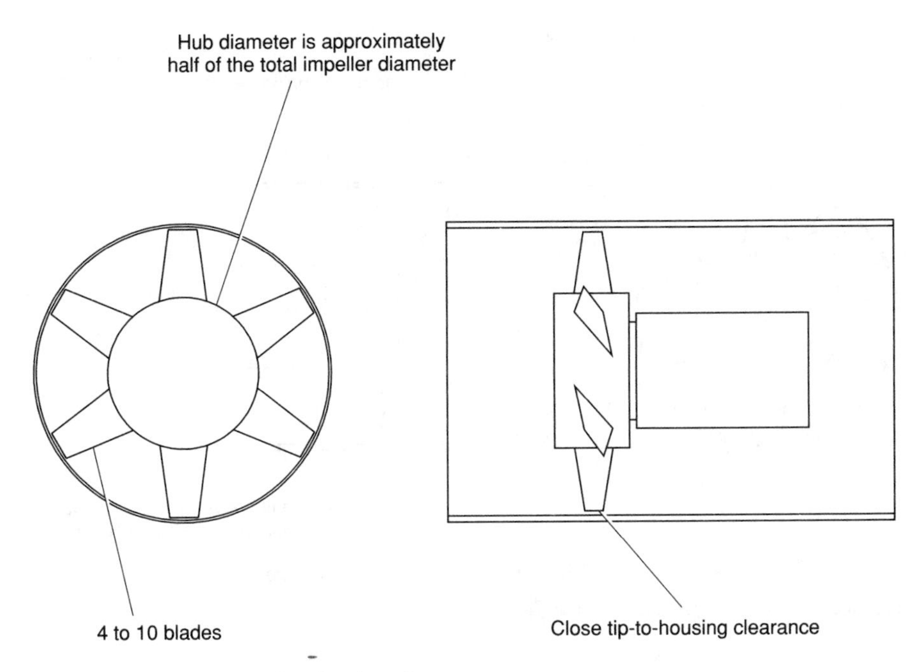

Figure 4.4. Tubeaxial fan detail and section.

Avoid surge conditions to prevent vibration and possible damage to the fan. Keep the fan's discharge open enough to allow at least half the free air delivery to flow in the system.

Tubeaxial fans are commonly used for industrial uses or for low-pressure HVAC applications. Because the discharge air swirls, air distribution to downstream branches close to the fan outlet might not be satisfactory. Their straight-through Airflow design might make them useful, however, where a centrifugal fan cannot fit.

Vaneaxial fans (Figure 4.5) are the most efficient axial fans. These have large diameter hubs, usually more than 50% of the total diameter, with up to 24 narrow blades attached. Higher efficiency is achieved by the blades' airfoil profile. Installed in the fan's housing are curved, stationary vanes, both upstream and downstream of the impeller. The upstream vanes guide air into the impeller at the proper angle, and the downstream vanes straighten the airflow, removing swirl caused by the impeller.

These fans are excellent for high-pressure, medium-to high-volume applications. They are commonly available in capacities over 200,000 cfm and 12 in. water static pressure. Like tubeaxial fans, they should not be excessively restricted. The flow should be at least half its free discharge volume to prevent excessive pressure buildup and surge. The lack of swirl in the output airflow allows excellent duct distribution downstream of the fan.

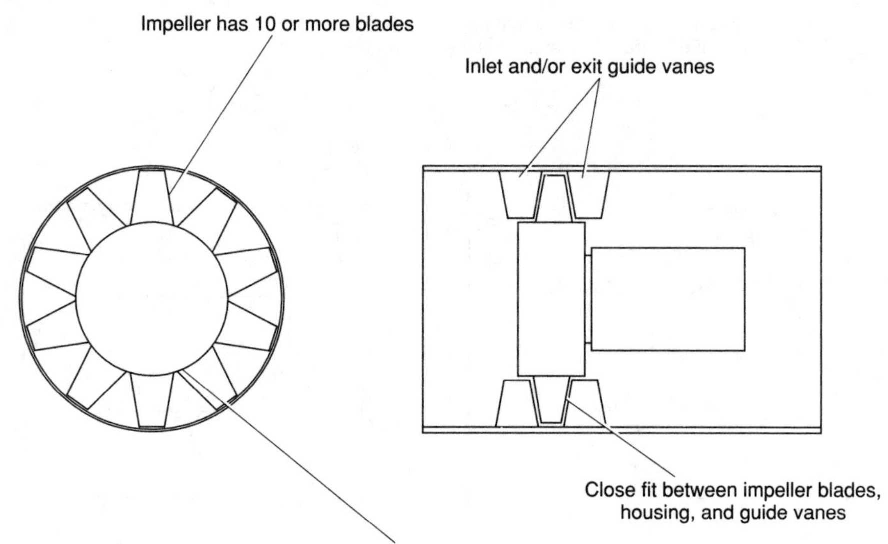

Figure 4.5. Vaneaxial fan detail and section.

Small vaneaxial fans might fit where physically larger centrifugal fans cannot. As discussed earlier, the right-angle change in airflow in a centrifugal fan might not be acceptable in some applications. A vaneaxial fan would be an excellent choice instead.

Ducts

Ducts are the conduits, or pipelines, for delivering air. Proper airflow through ducts is essential to deliver satisfactory heating and cooling system performance.

Ducts and duct systems are available in different materials and types. Accessories help control and guide the flow of air and provide proper air distribution.

Materials and configurations

Either sheet metal or rigid glass fiberboards are usually used for ducts. Proper duct construction is essential to conform with local and national codes governing air handling. The National Fire Protection Agency (NFPA) publishes Standard 90A and Standard 90B governing requirements. These cover commercial and residential supply and return duct systems. Also, the Sheet Metal and Air Conditioning Contractors' National Association (SMACNA) publishes requirements for sheet metal and fibrous materials, and their installation. These standards are often used as a basis for codes around the country.

Underwriters Laboratories' (UL) Standard 181 rates materials used for ducts for proper fire and smoke generation. UL specifies that a Class 0 duct should allow no smoke or flame development. A Class 1 duct is limited to a 25 rating for flame spread and a 50 rating for smoke development. Metal ducts are Class 0 types, while almost all nonmetallic rigid duct materials and flexible duct extensions are certified by their manufacturers as Class 1 materials.

The NFPA Standard 90A allows Class 1 ducts to be used in two applications: (1) in systems where the temperature will not exceed 250°F, and (2) those where they are not used as vertical risers serving more than two stories. Class 0 materials (iron, steel, aluminum, concrete, masonry, or clay tile) must be used for applications where Class 1 ducts are not permitted. Frequently, steel (including stainless) or aluminum is used as Class 0 duct materials.

A factory applied foil or other exterior vapor barrier is usually supplied with glass fiber duct boards. Some now include integral duct liners to limit the possibility of the air picking up and distributing microscopic glass strands into the conditioned space. These liners are also available with an antimicrobial agent. This is to prevent the growth of mold or bacteria if the duct gets wet.

Unlined glass or mineral fiber ducts are rated for air speeds up to approximately 2400 fpm and can handle static pressures up to 2 in. water. They are available in prefabricated (molded) circular sections and flat boards for cutting and assembling in the field. Glass fiber ducts are easy to handle and install. The installer can tape and clip sections together and quickly fabricate an entire duct system. Duct manufacturers and the SMACNA specify details on proper field fabrication of mineral and glass fiber ducts.

Because rigid fibrous duct material has an insulating R/in. of 4, some thermal insulation is provided with these ducts. They are also excellent at controlling noise transmission both through their walls and within the airstream inside.

Do not use mineral or glass fiber ducts where moisture is expected. Return air ducts from humid areas (shower rooms, kitchens, or equipment wash-down areas) often become damp. In these areas metal ducts would be a better choice. Over time moisture causes deterioration of mineral or glass fiber ducts and their fasteners.

Galvanized steel or aluminum is used for most metal duct systems. Stainless steel is used where moisture is present, or as required by codes in food preparation areas for cooking hood exhausts. They are installed in square, rectangular, round, or flat oval cross sections. Some systems use one profile for the main trunks and another for the branches.

Aluminum is commonly used for wet area exhaust and return ducts. It is lightweight and easily formed. Avoid its use where high temperatures are expected. It is acceptable, however, for comfort HVAC applications.

The SMACNA has specific guidelines for selecting proper thicknesses and reinforcing methods used for metal ducts. Correct methods vary depending upon the size of the duct and the operating pressure of the system. Do not use galvanized steel thinner than 28 gauge or thicker than 16 gauge.

Round ducts provide the best efficiency for Airflow. They are also inherently strong and rigid. Lighter gauge metal can be used for round ducts than for rectangular ducts operating at the same pressures. Finally, round ducts provides the greatest internal area compared with the amount of material used. This reduces friction.

Round ducts are available with a spiral or longitudinal (running along the length of the duct) seam. Spiral ducts are stronger because the seam acts as a reinforcement. Prefabricated elbows, reducers, tees, and dampers are available for rapid field assembly and system construction.

Oval ducts provide an advantage over round ducts in some applications. They are not quite as efficient or as strong, but their flat profile lets them fit into low or narrow spaces more easily than round ducts. Oval ducts are supplied with a spiral seam for strength, and a wide assortment of fittings are available.

Rectangular ducts are the least efficient for moving air. Their large perimeter compared to the open area causes more friction against the air moving inside. Rectangular ducts are, however, easily fabricated with automated or manual cutters, brakes, and folders. It is easy to build seams, branches, and take-offs. They are inexpensive to field manufacture, and modify with simple hand tools. They also fit easily into narrow spaces, and their flat faces allow easy installation of standard accessories.

Rectangular ducts are inherently the structurally weakest ducts. The flat sides provide little rigidity to prevent deflection, vibration, and pressure, especially with larger sizes. Strict adherence to the SMACNA reinforcing requirements is essential to prevent duct collapse, ballooning, or section separation. Most rectangular ducts operating above 1 in. water static pressure require reinforcement.

Metal ducts sometimes use sound absorbent liners to reduce noise from the fan and air movement. Materials are available that resist water vapor absorption and mold growth.

Carefully apply liners to prevent delamination and separation from the duct's surface. Fasten the liner with clips or screws and an adhesive (applied with almost total coverage) to positively seal the liner to the duct's surface. If extremely high air velocities are expected (not likely in comfort HVAC applications), use a perforated metal sleeve inside the duct liner.

Accessories

Many components go into a complete air distribution system. These include parts to control the airflow, guide it for least pressure losses, and deliver it to the space quietly and efficiently. Fire and smoke controls must be installed to comply with codes governing mechanical systems.

Dampers. Dampers control the amount of airflow through ducts. They can be manually adjusted for system balancing or operated automatically for dynamic control of the system's operations.

Commonly used for small round or rectangular ducts are simple, flat sheet metal blades. Known as *butterfly dampers*, the internal blade is mounted on a shaft that protrudes through the duct's wall. They restrict the duct by rotating against the airflow. Placing the damper parallel to the duct walls (and parallel to the airflow) provides very little resistance. Positioning the blade perpendicular to the duct virtually cuts off the airflow. A handle outside the duct shows the approximate position of the damper plate inside to help in setting the damper to the desired position. Figure 4.6 provides an illustration.

These dampers are not particularly good for providing precise air control, but they work very well for most heating and cooling system applications. They are most commonly used for round ducts up to 12 in. in diameter.

Butterfly dampers are not easy to set properly. The operating handle usually assumes an approximate position of the damper blade inside the duct. Use care when setting these. Rock the handle back and forth to be certain that the damper is really at the setting the handle indicates. Remember also that the settings between fully closed and 30 degrees to open are critical.

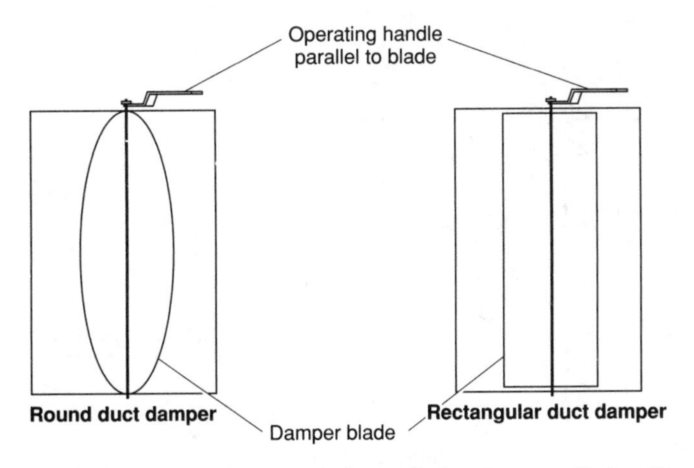

Figure 4.6. Round and rectangular butterfly duct dampers. The handle approximates the position of the internal blade.

More than half the airflow is regulated in the first 30 degrees of opening. A small adjustment makes a large change in the airflow.

Rectangular ducts usually use multiple-blade dampers for adjusting air delivery. Arranged like a set of rectangular butterfly dampers set side by side, these have a set of relatively narrow, flat metal strips (see Figure 4.7). A linkage connected to each blade operates all of them at the same time. airflow is controlled when the blades are rotated to become more parallel (or perpendicular) to the airstream. When perpendicular, the edge of each strip (or blade) contacts the blade edge or the damper frame next to it. This closes the damper and shuts off the air. Some damper blades have soft edging material that provides a positive seal when the damper is closed. These dampers can provide almost total closure to airflow.

Most balancing and control dampers do not require total air shutoff. Simple dampers without edge seals are usually adequate for HVAC applications.

Many larger multiple-blade dampers are designed for automatic controls. An electric motor or pneumatically operated piston drives the damper's linkage to position the blades and provide the desired resistance and airflow. Linkages and blade bearings in these dampers are often subject to binding and wear. They should be inspected if the damper is not operating properly.

Outdoor air inlet ducts often use motorized multiple-blade dampers. Temperature protection controls in the system close them if the temperature in the system falls low enough to risk freezing.

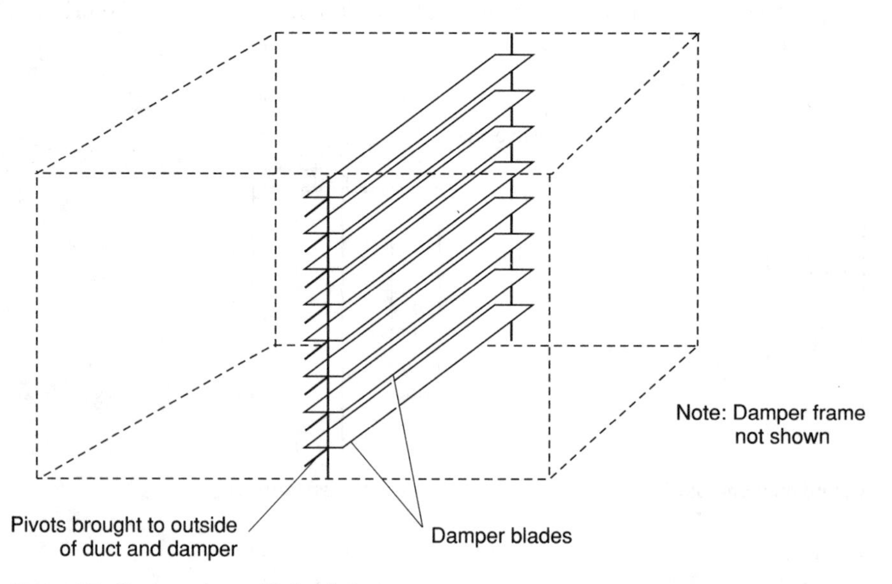

Figure 4.7. Rectangular multiple-blade damper.

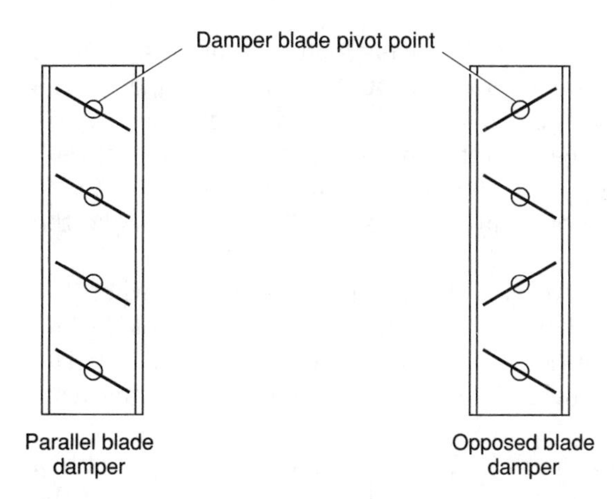

Figure 4.8. Opposed and parallel blade duct damper configurations.

Multiple-blade dampers are available in two types: those that rotate the blades in the same direction simultaneously as they operate, and those where the blades rotate in opposite directions. Figure 4.8 shows how the blades of each are positioned when they are open approximately 50%. The same direction dampers, called *parallel blade* types, are less expensive to fabricate than the *opposed blade* types.

Parallel blade dampers do not provide precise control of Airflow, especially when used to set air delivery below approximately 50% of the wide-open delivery volume. It's better to use opposed blade dampers for applications where accurate control is important. Opposed blade dampers create more resistance with a small rotation of their blades. This effect provides better pressure and volume control throughout their operating range.

Parallel blade dampers cause most of the airflow resistance near the last portion of their operating range (when they are nearly closed). This can make adjustment in this part of their adjustment range very sensitive to small changes.

Automatic control dampers must be sized properly. All components in the air system cause a drop in the air's pressure as it flows through or past them. Automatic control dampers should be selected to develop a relatively high pressure drop compared to the rest of the system. It is common (and acceptable) to have half the total system's air pressure losses occur at the damper when it is wide open.

For example, assume a complete system was supposed to have 1.5 in. water pressure loss at its designed air delivery. A properly sized automatic control damper would develop 0.75 in. water pressure loss in the fully open position. Size the rest of the system (ducts, turns, take-offs, diffusers, etc.) to develop the remaining 0.75 in. water pressure loss.

Selecting control dampers with comparatively high pressure losses helps make them work well as they operate throughout their range. If the dampers were sized to have very small losses when wide open, then significant control could not take place until they were almost completely closed. Then, almost all of the control range (full flow to minimum flow) must take place over a very small range of damper operation. This would make the control unstable and unreliable.

If sized properly so that there is a significant pressure loss when the damper is fully open, they further restrict the air as soon as the blades begin to close. This allows the control to operate smoothly and efficiently. The damper modulates the airflow over its full range of operation and can easily find the best operating point.

Multiple-blade backdraft dampers (Figure 4.9) limit airflow in a duct to a single direction. They are essentially parallel-blade dampers built so the horizontal blades overlap in one direction. Each blade hangs freely from its hinge at the top with its bottom edge in contact with the blade mounted beneath it. This allows air to pass through the damper in a single direction. If air attempts to pass through the wrong way the blades prevent it.

These units are commonly used with louvers to prevent outside air from passing into exhaust ducts installed on the face of a building. Strong winds can force air into relief air vents, allowing the possibility of freezing if the air is below 32°F. A backdraft damper stops the "wrong way" air from entering the building.

Backdraft dampers are subject to extreme wear. Gusts of wind can cause them to open and close many times a minute, causing excessive wear on the

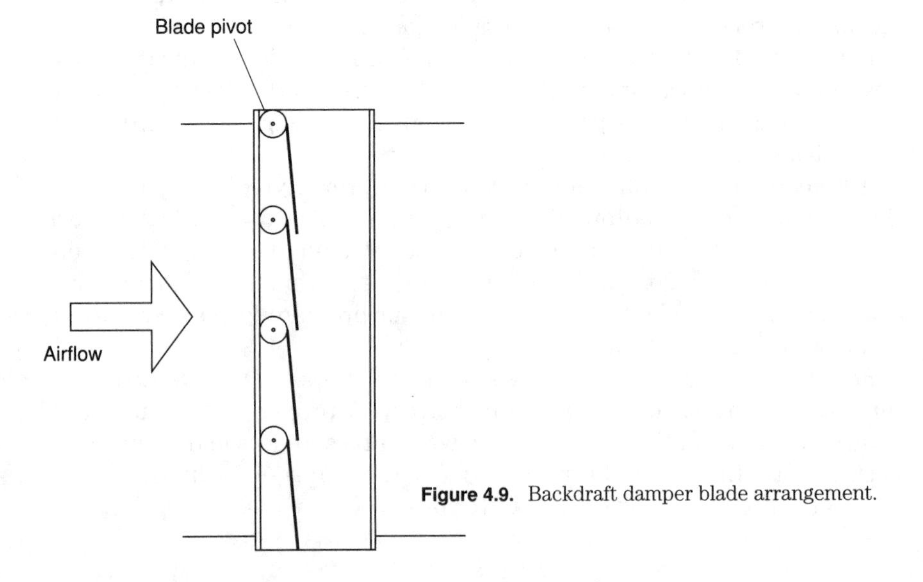

Figure 4.9. Backdraft damper blade arrangement.

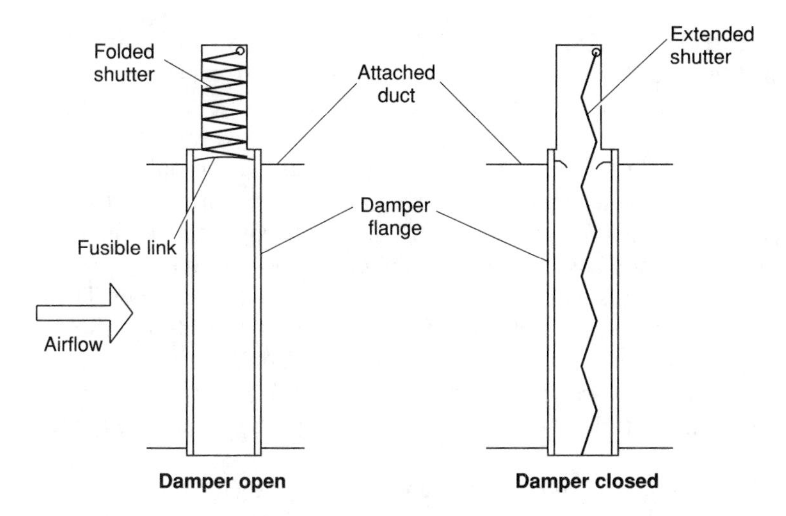

Figure 4.10. Shutter fire damper arrangement.

blade pivots and shafts. It is not uncommon for a backdraft damper to completely wear out within 5 years of its installation if not maintained properly. Regular lubrication and inspection are essential to maintaining proper operation and long service.

In case of fire, automatic dampers close off the airflow. Most codes require fire, smoke, or combination fire and smoke dampers wherever a duct crosses through a building's fire or smoke barriers. Smoke dampers, or combination fire and smoke dampers, are motorized multiple-blade dampers. An electric motor, or a solenoid holding back a spring, closes the damper. The building's fire alarm system, or a duct-mounted smoke detector, starts the damper's motor or solenoid. The fire system causes the smoke or combination damper to close automatically. Combined fire and smoke dampers include both the motorized operator and a fusible (meltable metal) link that closes if excessive temperatures occur within the duct.

Shutter-type dampers are commonly used to completely shut off airflow through a duct. Available for rectangular or square ducts only, these are most often used for fire control, not to modulate or finely control airflow. Duct transitions to rectangular fire and smoke dampers are required if you install them in a project with round ducts.

Shutter-type dampers are kept completely out of the airstream when not activated. When open, they present virtually no resistance to the flow of air. When closed, they provide a fire rated barrier that completely stops air movement and the spread of flames and smoke (see Figure 4.10).

Fire dampers are built with a fusible link. This is a metal strip that melts at a set temperature, usually about 165°F. The damper closes when the link melts from exposure to fire. Duct shutoff is complete, providing a smoketight and firetight seal.

Standard fire dampers are designed for vertical installations. Inserted into a horizontal duct with the shutter upright, gravity drops them into the duct when the link melts, cutting off the flow of air.

Horizontal fire dampers (installed in a vertical duct where the shutter must close horizontally) include one or more springs. These pull the shutter closed across the duct when the fusible link melts. Spring-loaded fire dampers can also be installed in horizontal ducts to drop into the duct like standard units, but they cost more to manufacture and purchase.

Another type of fire resistant damper is the radiant ceiling damper. These shut off airflow and the spread of fire in the event that excess temperatures occur at fire rated ceilings. Installed in horizontal openings at the ceiling to protect vertical ducts, they are closed by radiant heat energy. A fire, melting a fusible link mounted outside or below the damper's housing, closes the damper.

Some radiant dampers are contained in a round case (for circular openings). The damper blades are configured as two half circles mounted like a pair of raised butterfly wings. The fusible link holds the two half circles upward, allowing them to fall when the link melts.

All fire dampers that use a fusible link must be accessible for inspection and replacement of the link if it closes. The dampers must be reset manually, so install access panels to allow reopening of the damper and replacement of the link. Show the location of the fire dampers with a sign at their access panels. Most codes require this, and the sign must be visible to the inspector.

The splitter damper is a very simple device used to divert the airflow in a main duct and feed a branch. They are usually field fabricated from a single blade of sheet metal hinged at the downstream side of the branch opening (see Figure 4.11). The open end is manually positioned with a rod to increase or decrease the quantity of air delivered to the branch. These are ordinarily used to balance a duct system and are locked into place.

Splitter dampers can cause resistance and turbulence in the main duct's flow. Because they protrude into the airstream, pressure losses occur as air is forced to pass around the obstructing damper blade.

Apply splitter dampers carefully, allowing at least 5 ft of duct after the damper for the airflow to become properly established again. Installing two or more splitter dampers close to each other can cause significant problems with airflow. Pressure losses can become excessive and turbulence caused by the upstream splitter will interfere with the proper operation of downstream units. An upstream splitter damper starves another installed close behind it. Downstream splitter dampers have to be opened farther to compensate, and the remaining Airflow in the duct will be limited.

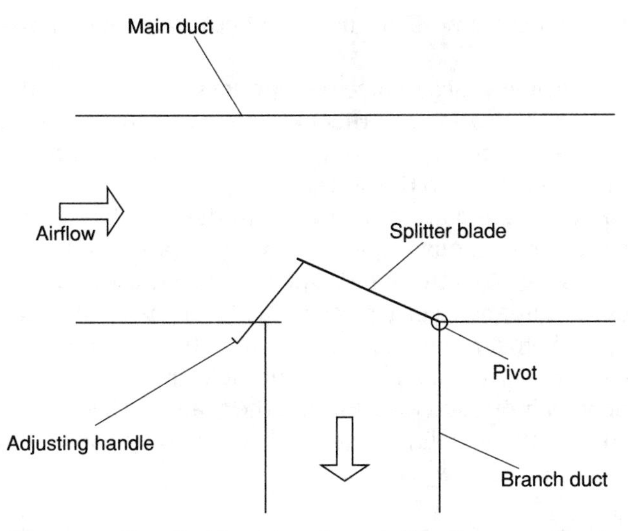

Figure 4.11. Branch splitter duct damper. After the proper flow is established in the branch duct the handle is locked in position.

Turning vanes. Turning vanes reduce turbulence and pressure loss as air turns a corner. Constructed of narrow, curved metal blades, they are installed in a single row along the centerline of a duct corner (see Figure 4.12). The SMACNA handbooks specify their construction and proper application. Turning vanes are important when installing rectangular ducts.

If possible, use rounded turns to reduce pressure losses as the air moves through elbows. The inside radius of the turn should be at least equal to ⅓

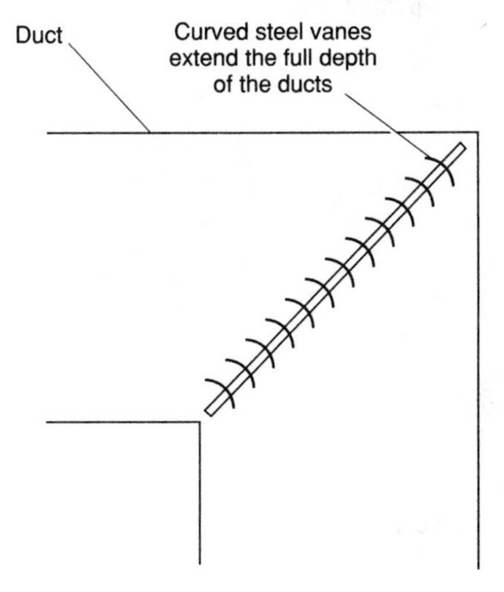

Figure 4.12. Turning vanes (section). These extend the full width of the duct.

the width of the duct. If that is not possible, use a set of turning vanes in the turn.

When used properly, turning vanes reduce the flow loss through a 90-degree duct turn by a factor of 5 to 10. Use them on every rectangular main duct's corners where a sharp turn is necessary. They significantly reduce the pressure required to deliver air to the system.

Pressure losses in a duct system are proportional to the speed of the air squared. Doubling the air's speed quadruples the system's pressure losses. Because most main ducts have relatively high-speed airflows, use rounded turns or turning vanes at right-angle corners, especially in high-speed ducts.

Turning vanes can also help airflow into a branch duct take-off. Installed like splitter dampers, the vanes are mounted on a movable arm. Pushed into the main duct, they scoop air out and direct it into the branch. This type of take-off is superior to the simple splitter damper because they cause less turbulence in the main duct's airflow (see Figure 4.13).

Coils. Coils transfer heat to or from a stream of air flowing through it to fluid flowing inside. Coils are used in many heating and cooling applications. They are essential for air conditioning systems to remove heat from the air in the space.

Coils are most often fabricated from copper or aluminum tubes with thin sheet metal (usually aluminum) fins. The fins are attached firmly to the tube's outside surfaces. All of the tubes are arranged so that air flows around them, passing between them and the fins. This provides a large area of contact between the air and the coil's surface.

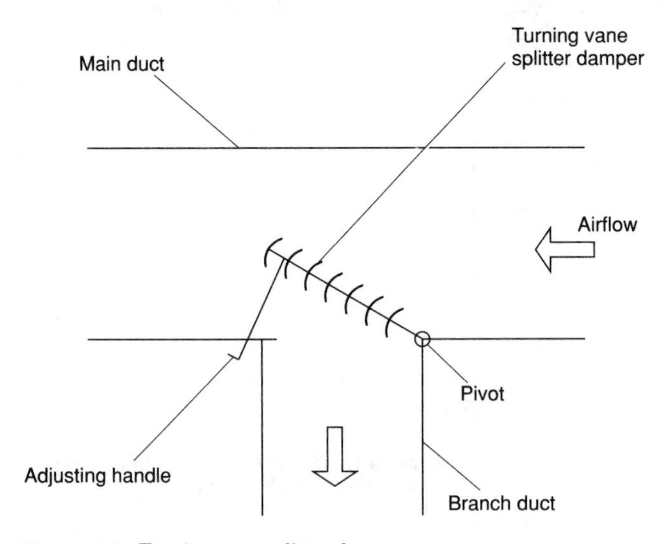

Figure 4.13. Turning vane splitter damper.

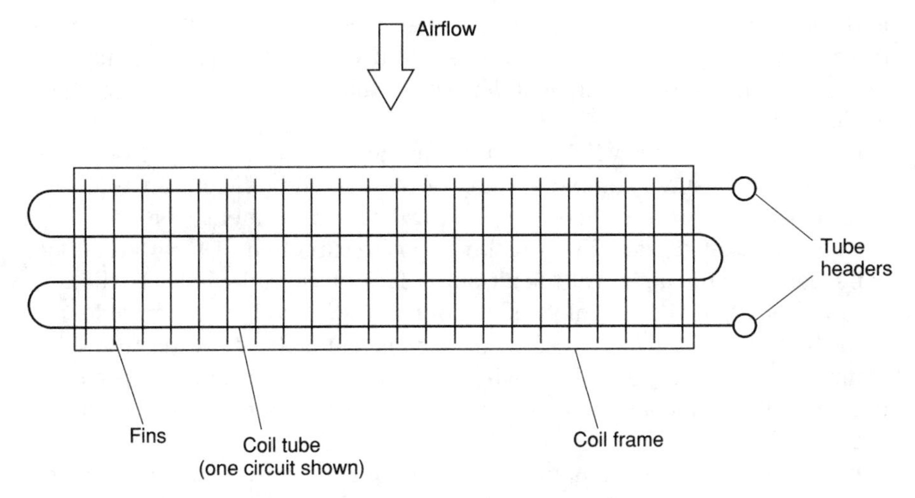

Note: Four-pass coil shown

Figure 4.14. Detail of water coil circuit.

Most HVAC coils have fins spaced 0.25 to 0.05 in. apart. The fin's close spacing helps ensure that the air is likely to remain in contact with them as it passes through the coil. Note that very closely spaced fins are more efficient at transferring heat but cause greater airflow restrictions.

Most coils are designed so the tubes make serpentine, multiple passes back and forth across the face (Figure 4.14). Attached with welded, soldered, or brazed connections, the tubes connect to headers at the inlet and outlet connections. The connections and headers evenly distribute the fluid among all of the tubes. This equalizes flow across all parts of the coil.

Heating and cooling system coils can handle a variety of fluids, including water, steam, or a refrigerant. The specific fluid to be handled affects the design details of the coil.

Water system coils allow moderate pressure drops and fully turbulent internal flow. Most water coils have internal water flow speeds below 10 fps. This is sufficient to ensure good heat transfer and prevent erosion of the tubes. Microscopic grit, common in water systems, can damage the tubes if it travels too fast.

Water coils should be installed level, with the tubes in a horizontal position. This prevents air bubbles from becoming trapped in the upper passes and allows the coil to be drained easily without trapping water. This is most important with coils that are exposed to freezing air. Install them to allow rapid draining.

Some heating water coils are used in duct branches to control humidity or prevent excess cooling in particular zones. These *reheat* coils are of limited use in new system designs. Many energy codes do not allow reheating previously cooled air. Reheat coils are still used in some installations as a primary means of providing heat. These are allowed because the air is not cooled before heating.

Install all water coils with the water inlet at the bottom and the outlet at the top. This ensures complete filling of the coil and efficient air removal. Also, most coils are installed in a counterflow arrangement (Figure 4.15). This means that the air flows in the opposite direction of the water. The coil's water inlet header is at the air outlet face of the coil. Similarly, the water's outlet is on the air inlet face. Counterflow installations provide the greatest amount of heat transfer. Air leaving the coil is exposed to the incoming water. This allows the leaving air to more closely approach the temperature of the incoming water. In this way, heat transfer is maximized between the air and water.

Install hot water coils in noncounterflow configurations where they might be exposed to a flow of freezing air. The air and water flow in the same direction. This arrangement is called *parallel flow*. Although the heat transfer is not as efficient, the coldest air is exposed to the hottest water. This warms the air before it is exposed to water that has been cooled. The risk of a coil freeze is reduced because the water does not become as cold in a parallel flow coil arrangement.

It is possible to learn a bit about the flow of water through a coil while it is operating. This can help you diagnose what might be wrong in a system. Most

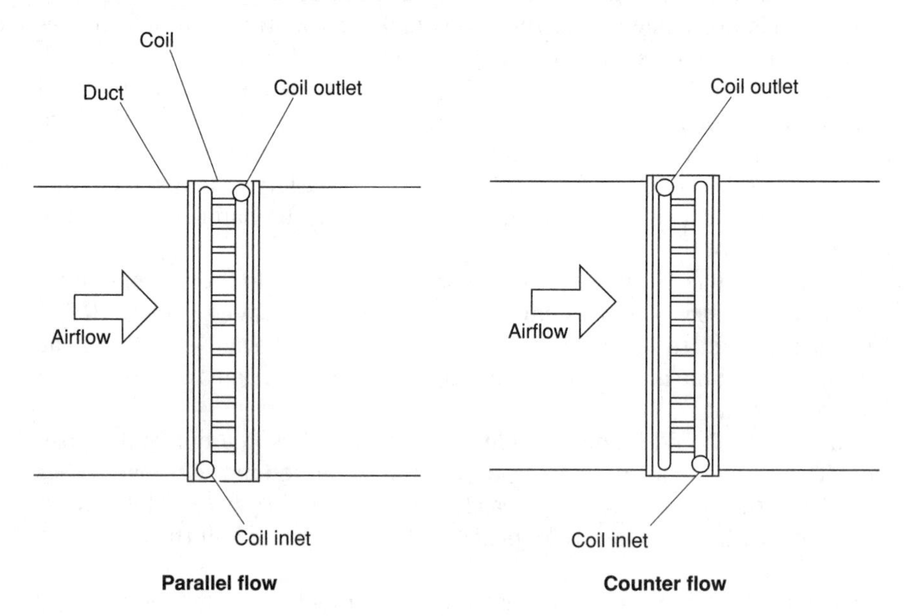

Figure 4.15. Counterflow and parallel flow coil arrangements.

heating coils are designed to operate, while fully loaded, with an approximate 20°F water temperature drop from the coil's inlet to outlet. Most cooling coils are designed to have a 10°F to 15°F water temperature rise from inlet to outlet. If a coil (or coil and control valve) is suspected of malfunctioning, check the difference in the inlet and outlet water temperatures. If there are no built-in gauges, place the probe of an ordinary temperature gauge in contact with each pipe and hold it there with a piece of insulation.

Where the temperature is rising or dropping the proper amount, it is likely that the flow is acceptable. The following can cause too small a temperature change:

- Insufficient airflow through the coil
- Too great a water flow rate
- Dirt on the coil

If the temperature change from inlet to outlet is excessive, the cause might be

- The air flowing through the coil might be at a temperature very different from its design
- Water flow through the coil is insufficient
- Low boiler or chiller output

When heating potentially freezing air with water or steam coils, it is best to modulate the airflow through the coil. A face and bypass damper controls the airflow through the coil. They are set up to allow some air to pass around a coil instead of through it. When more heating is needed, controls reduce the bypass air and increase the amount of air passing through the coil. Links between the dampers in each air path keep the total airflow volume constant. The basic operation is shown in Figure 4.16.

Don't reduce the water or steam flow to change the amount of heat delivered. Restricting the coil's steam or water flow to control the coil's output can cause low flow conditions. This can encourage the coil to freeze. By varying the amount of air moving through the coil and leaving the hot fluid flow fully on, the coil is less likely to freeze.

Steam coils are specially designed to handle the condensate that accumulates inside the tubes as heat transfer occurs. Their tubes have a larger diameter to prevent water hammer from occurring. A restriction of the liquid's flow toward the tube outlet can cause steam and condensate to collide. Larger diameter tubes prevent this.

Install the coils so that the tubes slope slightly toward the outlet header. This assists in condensate drainage from the coil. Slope coils with both headers on the same end toward the common header end.

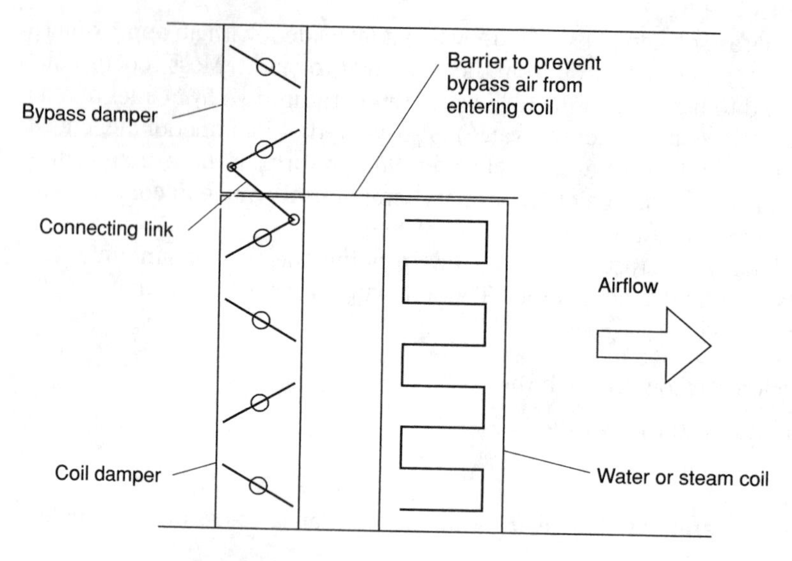

Figure 4.16. Face and bypass damper system. Full water or steam flow is maintained in the coil at all times.

Steam coils are especially prone to freezing when exposed to cold Airflows. Cold air hitting the coil's face can quickly freeze the small amount of water inside. Special freeze-resistant coil assemblies help prevent this. Often the tubes are built with a concentric "tube in a tube" form. The inlet header distributes the live steam to the inner tubes. These are perforated along their length, allowing steam to escape and spray against the inside surface of the outer tube. Here, the steam condenses against the cold inside surface. The condensate formed flows along the inside of the outer tube to the outlet header. The steam (from the inner tube) sprays on the condensate as it drains. This helps to keep it warm and prevent it from freezing.

Arrange the piping and coil to allow the condensate to quickly drain out of a coil that might be exposed to freezing temperatures. Some manufacturers specify that the coils must be installed with the tubes positioned vertically (and the outlet at the bottom) to allow nearly instantaneous draining.

Refrigerant coils are used as air conditioner evaporators in most cooling systems. Called *direct expansion* (DX) coils, they are designed very differently from water or steam coils.

Direct expansion air conditioning coils are fabricated with special header connections. These allow very uniform distribution of the liquid refrigerant to all the tubes (see Figure 4.17). The manufacturer provides a liquid inlet connection for each tubing circuit in the coil. These connections might include integral capillary tubes, thermostatic expansion valves, or attachment points for field supplied valves.

Direct expansion coils have one or more *refrigerant distributors*. These are bundles of individual, small-diameter lines that run from the expansion valve's outlet to each coil tube. The distributor sends the fluid (which might include some bubbles of gaseous refrigerant) uniformly into each tube in the coil. Distributor lines are all the same length and are installed to provide an equal pressure drop for each connection to the coil.

Proper distribution of the refrigerant inside the coil is essential. This ensures that all of the liquid refrigerant evaporates equally in all of the coil's tubes. This prevents the liquid refrigerant from flooding any part of the coil and starving other areas.

Install refrigerant direct expansion coils essentially level, but with a slight pitch down to the outlet (suction) side. This lets the oil carried with the liquid refrigerant flow back to the compressor. If permitted to accumulate in the tubes or the outlet header it could cause erratic operation.

Some direct expansion coils have more than one refrigerant circuit. Multiple-stage air conditioners often use this type of coil. Separate distributors might feed completely different areas of the coil. One circuit feeds the top part of the coil and another circuit feeds the bottom section. Or they might be interlaced (where each circuit feeds alternate tubes) to allow each circuit to partially cool the entire coil. The simplest and most

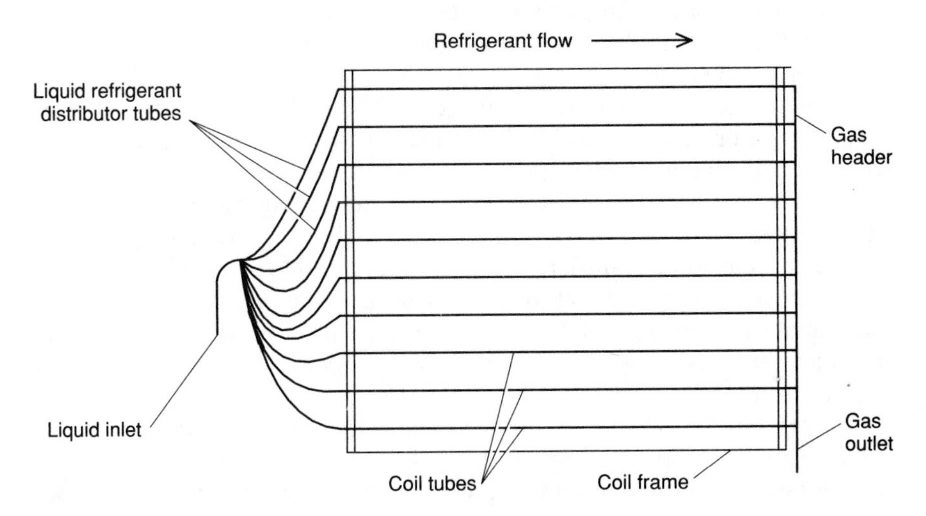

Figure 4.17. Direct expansion (DX) coil with liquid distributor. The thermal expansion valve or orifice feeds the distributor.

common arrangement is to have separate areas of the coil face fed from each circuit.

When operating under design conditions, direct expansion coils allow the refrigerant to completely vaporize inside. With proper operation of the thermostatic expansion valve, the refrigerant has 6°F to 10°F of superheat at the coil's output. Operating this way, the final few inches of each tube do not vaporize refrigerant. They superheat it above its evaporation temperature.

The tubes are connected at their outlets with a simple manifold. Handling only gas and small amounts of oil, no special distributors are required for the outlet.

Most coils used for air conditioning applications have to properly handle the condensed water vapor that forms on their outside surfaces. As latent heat (humidity) is removed from the airstream, water forms on the coil's face. Water accumulates on the fins and tubes and runs in the direction of airflow to the downstream face of the coil. There it runs down the edges of the fins to the coil's bottom frame.

Coils that must handle condensed water should be built entirely of corrosion resistant components. The coil frame is constantly wetted and must be fabricated of galvanized steel (or a material even better able to resist corrosion, like stainless steel). The coil might include a built-in drain pan to collect the condensate. Some air handlers use a separate drain pan.

It is not desirable to allow condensed water to become entrained (picked up and carried) in the airstream. Air velocities through the coil over 600 fpm can cause water droplets to be picked up and carried into the downstream duct system. Size coils and ducts to limit air speed at the coils to a maximum of 600 fpm.

Take care to prevent water from accumulating in the duct system or air handler. Water causes mold, mildew, and bacteria to grow, leading to potential health problems. Follow good practices if adding a cooling coil to a system. Provide a good drainage system and limit air velocities. If necessary, use a larger faced coil and transition fittings to the adjacent ducts. Install no duct liners or other materials that can soak up and retain water near the downstream side of the coil. Also, insulate the duct or air handler downstream of the cooling coil if there is a chance that water might condense on its cold outside surfaces. This condensed water can drip into the duct, leading to a wet interior.

Water draining off the coil must be collected and drained from the air handler. A trap in the drain pan's outlet line is essential. Air handlers with coils upstream of the fan tend to draw ambient air inside through the drain line. This incoming air can move fast enough to prevent the water from draining out. It is blown back inside and accumulates in the air handler. The system might flood, soaking filters and dripping out of the ducts and into the building. A drain trap prevents this (see Figure 4.18). The water's weight in the trap prevents air from being drawn into the drain line. Make

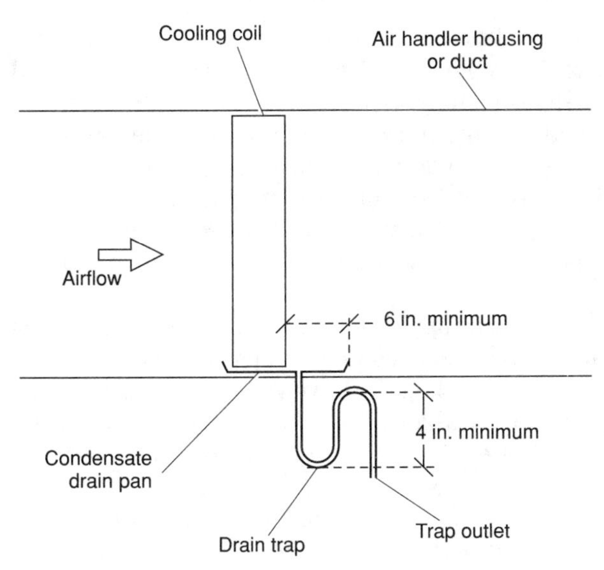

Figure 4.18. Coil water drain trap.

the trap deep enough to prevent the water from being suctioned into the air handler. Usually, 3 to 4 in. deep is sufficient.

Coils must be clean to work properly. Heat transfer is dramatically reduced when coils have even a thin film of dirt on the tube and fin surfaces. There are a variety of methods for cleaning coils. Chemical cleaning is effective, and there are many products on the market to dissolve or loosen the crusted dirt and dust that accumulate on the coil's surfaces. Coils can also be kept clean with a pressure water or air spray.

Always spray clean a coil from the downstream side to the upstream side. This helps dislodge dirt that has become wrapped around fins and tubes. Use high-pressure water carefully to avoid bending the fins. It might take a combination of chemical treatment and high-pressure washing to clean coils that have heavy accumulations of dirt. It is important that the entire depth of the coil be cleaned, so multiple applications of chemicals and washing might be needed to do an effective job.

A method of cleaning coils without water is to use a throwaway filter (or a piece of throwaway filter media) and an air hose. Fasten the filter to the upstream side of the coil's face. It should fit well with no gaps around the edges. Use an air hose to blow into the downstream face of the coil. Dirt is blown off the coil and trapped in the filter. This might not remove some dirt that is tightly crusted onto the fins, but it might be the only way to clean reheat coils in ducts far away from the fan room.

Finally, bent fins should be combed straight. Straight fins maximize heat transfer by allowing air to flow through the entire coil face.

Filters. Air filtration is essential to the proper operation of HVAC systems. Unfiltered air contaminates coils, fan impellers, and smoke detectors, and can cause inaccuracies in duct-mounted temperature sensors. It also introduces and recycles dust and dirt into the space being conditioned, possibly leading to health problems for the building's occupants. Proper filtration is especially important with coils that have narrow fin spacings. These clog easily (especially when wet) with fine dust and lint. Filters help keep airfoil profile fans clean. These fans are easily clogged and lose efficiency with minor dirt buildup.

Most filters rely on mechanical trapping of dust particles in or on a media bed open to the flow of air. As the filter accumulates dust, it creates a higher pressure loss for the air flowing through it. However, a dirty filter does a better job of trapping additional dirt than a clean one. Nevertheless, it is important to change or clean air filters when the pressure drop of the air traveling through it gets too high.

Very few HVAC systems are designed or installed with filter pressure-loss gauges. That is unfortunate because these gauges are able to measure pressure differences of less than 1 in. water, and are the best way to tell when a filter needs changing. It is normal for a filter to triple its "brand new" air pressure loss as it becomes loaded with dust.

Filter pressure-loss gauges have two input lines. One reads static pressure on the filter's inlet, the other reads static pressure downstream of the filter. The gauge shows the difference in pressure between the filter's inlet and outlet. Figure 4.19 shows a typical gauge installation method.

If the air handling system must operate with a very low pressure loss in the filter, the filter should be changed before it is fully loaded. Systems that must have very low filter pressure losses cannot be allowed to load the filter to "normal" dust levels. Full dust loading causes insufficient airflow in these cases.

There are several filter types available for common HVAC systems that provide a wide range of efficiencies. A rating is assigned to filters after testing with procedures developed by the ASHRAE.

The simplest filters, usually made from spun glass, expanded metal, screens, or washable open cell foam units, remove only coarse dust particles from the air. They are rated per ASHRAE's *weight-arrestance* procedure and do a poor job at removing fine dust. Filters of this type might be rated at up to 90% weight removal, but they do a mediocre job of cleaning the air acceptably. Larger HVAC systems use these filters as "roughing" or prefilters upstream of better, higher efficiency units. They trap large particles that would quickly foul the more expensive downstream filters.

Better filters are rated per ASHRAE's *dust spot* test. This test is more critical, as special, graded dust is blown through the filter and then captured on paper. Filters are rated according to the amount of darkening of the paper. The darker the paper becomes from dust that passed through the filter, the lower the filter's rating.

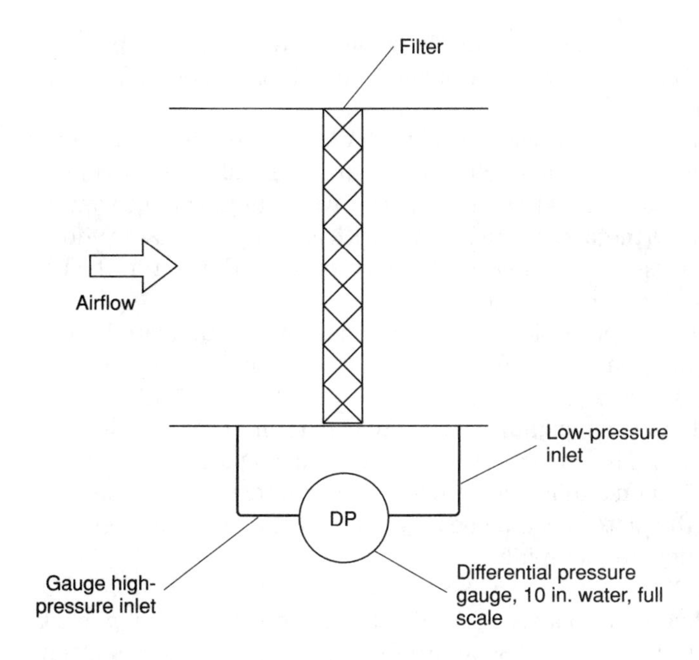

Figure 4.19. Filter pressure drop gauge installation.

These filters are usually made of pleated paper or specially treated cloth bags that present a large surface area to the airstream. They cost significantly more than weight-arrestance filters, so they are often used as final filters downstream of less expensive units.

Very good dust spot rated filters can remove over 30% of the total dust load delivered to them. While 30% efficiency might sound low, in normal HVAC systems they do an excellent job of preventing dirt and dust contamination. They should be used before coils with closely spaced fins or airfoil profile fans.

Note that the dust spot rating of 30% appears much lower than weight-arrestance rating (typically around 90%). The differences between the tests make these rating differences, but dust spot filters are better than weight-arrestance types. Be sure to compare filters rated under the same test when making a selection. Confirm the filter's test type with the supplier to be certain.

The best media filters are *HEPA* (high efficiency, particulate absolute) units. Use these where absolute filtration is essential. They can completely remove asbestos fibers, viruses, and other minute particles from an airstream. Although they are not commonly used in HVAC systems, service technicians should be aware of their use in clean rooms, health care facilities, and in hazardous waste applications. HEPA filters include special edge seals that prevent any material from passing around them. These filters are the most expensive types available, and are used for critical applications. Most comfort HVAC systems do not use them.

Electrostatic filters are sometimes used for HVAC systems. These are extremely efficient and can remove almost as much dust as a HEPA filter. They work by electrically charging dust particles with a high-voltage source and attracting them to metal collecting plates with an opposite charge. Most electrostatic filters use a washable prefilter to prevent them from being quickly loaded with large dust particles.

Electrostatic filters need periodic washing to remove accumulated dust. Some filters have automatic washers (operated by a timer) to prevent excessive dust accumulation.

Electrostatic filters do not significantly increase the air's pressure drop as they accumulate dust. The collection plates are installed parallel to the airflow and the dust film on a fully loaded plate doesn't interfere with the air's movement. When the plates become heavily loaded with dust, however, the filter's total efficiency drops rapidly.

Diffusers. Air delivered into a space for comfort heating and cooling must be slowed properly to prevent uncomfortable drafts. Ducted air is often moving at over 400 fpm, and air at this speed would be extremely disagreeable to the building's occupants.

Diffusers are designed to slow and distribute air delivered from a duct into an open space. They broadcast the air over a large area and allow it to further spread and slow as it enters the room. The delivered air will, through friction, cause room air to become entrained (join) in the flow. This increases the apparent delivered air volume and helps slow the air.

The distance that the air is delivered before it slows to a specific speed is the diffuser's *throw*. The cross-sectional area of the flow is the *spread* of the airflow. Note that the spread includes some induced ambient room air. This occurs when the friction of the delivered air causes some room air to join its stream.

Diffusers installed on a ceiling are usually designed to force the delivered air to flow along the ceiling's surface. Air, once it begins flowing over the surface, tends to "stick" to it, and continues moving over and along it in a thin layer. Air in this layer slowly diffuses into the space with little apparent movement. Even persons directly below the diffuser don't feel a draft.

Take care to prevent the relatively high-speed layer of air running along the ceiling from hitting a nearby wall or other obstruction. This forces it down and into the occupied space and can cause draft problems.

Some ceiling-mounted smoke detectors might be sensitive to air being blown over them. This is especially true when the delivered air is at a different temperature than the air in the surrounding space (or the detector itself). Try to keep ceiling diffusers at least 3 ft away from smoke detectors and 6 ft away from walls.

Some diffusers confine the airflow to a tight cone. These diffusers give the air as long a throw as possible. Use these where air must be blown down from a very high ceiling to the spaces below. This is especially important when delivering warm air. The natural buoyancy of warm air tends to make it rise and it will resist the downward velocity given to it by the diffuser.

Pin-cushion diffusers, types where the outlet is a flat plate perforated with hundreds of small holes, deliver their air very slowly. The entire flow of air from the duct is evenly distributed over the diffuser's face. The large outlet area, often about 4 ft^2, slows the air. They do not entrain much room air because of their low delivery velocity.

Many diffusers include butterfly or flap dampers to allow varying of the delivered air volume. Use these dampers to fine tune a system's operation. Don't use them for system balancing because excessive noise is certain to occur. These dampers create a great deal of turbulence and noise if they are closed almost completely. An almost-closed damper creates a large pressure drop and high velocity for the air moving through it. This high-speed air causes noise.

Diffusers are available that deliver air in select directions. For example, ceiling diffusers are available that deliver air in one, two, three, or four directions. Corner-type diffusers are available that supply air in two streams at right angles to each other. Directional supply diffusers should be chosen that cause the air to be directed away from people and toward open spaces. Use directional supply diffusers on a ceiling to prevent air from washing down a nearby wall and causing a draft.

Another important consideration to remember when selecting air handling diffusers is noise. Units that slow the air as it is delivered tend to be quieter than those that concentrate the air and deliver it at a high speed. If possible, use quiet, low-speed diffusers to prevent complaints.

The pressure drop of the air flowing through the diffuser is important. Diffusers vary widely in their pressure losses. Diffuser manufacturers list the pressure drop for a diffuser delivering various amounts of air. Consider this specification carefully when designing an air delivery system.

There are specific, suitable locations for diffusers if an HVAC system is to work properly. Locate heating air supply diffusers near outside walls, windows, and doors. Because heat is lost in these areas, a barrier of warm air delivered there keeps the rest of the space comfortable. The warm air intercepts cold air infiltrating inside and helps keep the perimeter surfaces from becoming cold.

If possible, warm air outlets should be located low on walls or in the floor. If that is not possible, acceptable performance can be obtained if good ceiling-mounted diffusers (that encourage plenty of air mixing) are used.

Install air conditioning supply diffusers throughout the space. Heat gains occur at both the outside perimeter and in the interior areas, so deliver cool air throughout the space. If there is a particular area that needs spot cooling, locate a diffuser as close to that area as possible.

Install return air grilles (diffusers without internal dampers) to prevent air delivered by a supply diffuser from being "short circuited" into the return system. Unless there is a specific reason for doing otherwise, keep return grilles at least 15 ft away from any supply source.

Provide supply air diffusers and return air grilles to take advantage of the fact that the system's air will naturally rise when heated or sink when cooled. Ideally, deliver warm air near the floor where it can rise into the occupied space. The return grilles should also be installed low to return the air after it has cooled (see Figure 4.20).

Supply diffusers should be near the perimeter of the space and the returns installed in the interior. This forces the air to flow toward the occupied areas, constantly changing the air throughout the space.

Deliver cool air high in the space and allow it to settle over the occupied area. Again, be careful that this does not cause drafts by blowing cool air directly onto the occupants. The return air grilles should be installed high also, but not near a supply diffuser. Because air conditioning is delivered throughout the space, not just at the perimeter, return grilles can be located at both perimeter and interior locations.

Modify these generalizations in smaller buildings to make the installation simpler. Often the same ducts and diffusers are used for both heating and cooling, so consider some compromises. Distribute the supply diffusers around the area, placing some near windows and doors for proper heating system performance and some in the interior area for good cooling. The return grilles can be placed where most convenient.

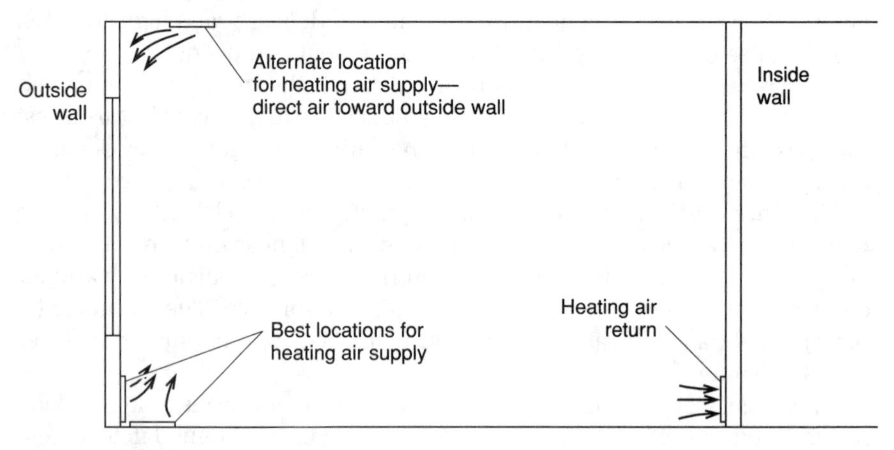

Figure 4.20. Elevation showing locations of heating air supply diffusers.

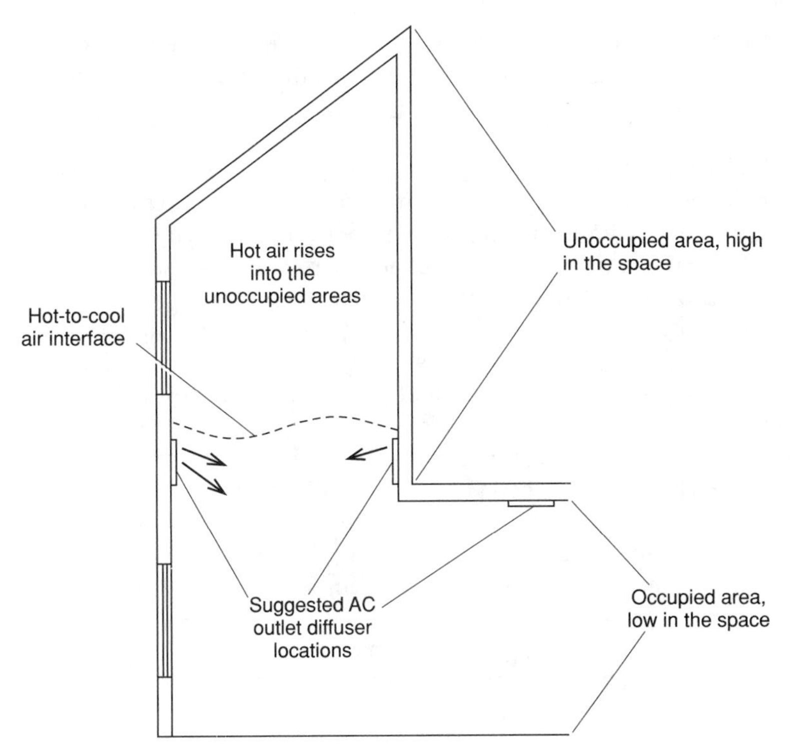

Figure 4.21. Elevation showing cooling air supply locations for properly stratifying an atrium space.

A note about air conditioning areas with very high ceilings, lofts, or atriums: These locations are areas where the occupied space (from the floor to approximately 8 ft above the floor) represents a small portion of the total height. This gives the installer some opportunities for significant energy and equipment cost savings. Instead of attempting to cool the entire volume of space, confine the cooling to the occupied area where it is needed. The natural tendency of air to stratify can be used to the designer's advantage. Install the cooling supply and return diffusers no more than 10 ft above the floor. Take advantage of low soffits and use conventional ceiling diffusers, or use wall-mounted directional diffusers that confine the delivered air to the lower parts of the room. Figure 4.21 shows one suggested installation.

This installation keeps the cooler air on the lowest portions of the space, making the occupied areas comfortable. The upper areas get quite warm, but because they are unoccupied, this won't matter. A smaller air conditioning system can be installed with a stratified air design than would be necessary if the entire space was cooled.

If a high space does need to be uniformly cooled, place return air grilles in the highest locations. Install supply diffusers at various elevations wherever the space is occupied. The high returns receive the air as it is warmed and help ensure a uniform temperature distribution throughout the space.

Louvers. Louvers are fixed-blade devices used to bring air into (or exhaust it from) a building. They prevent rain from entering the duct with a set of horizontal blades angled outward from top to bottom. Rain drips down and away from the blades and runs to the outside. Bird and insect screens are also available to keep vermin out of the building.

Most louvers only prevent rain from entering with the air if the speed of the air being drawn inside is kept below approximately 600 fpm. The designer must be aware that the free area of the louver (that through which air passes) is smaller than the total face area. This area reduction causes the air's speed to increase within the louver. The louver's size should be adjusted to ensure that the air speed is kept below 600 fpm.

The free area versus the face area of a louver varies widely with louver sizes. Check the manufacturer's catalog to find the ratio for a given unit. Free area to total area proportions range from less than 20% to almost 60%.

For example, if a louver is needed to carry 2400 cfm and its free area to total area ratio is 45%, the required total area can be found with the formula

$$\text{Area} = \frac{\text{cfm}}{(\text{Area ratio} \times \text{Speed})}$$

$$\text{Area} = \frac{2400 \text{ cfm}}{(0.45 \times 600 \text{ fpm})} = 8.89 \text{ ft}^2$$

Another example: A 45% open louver fit into a 3-ft by 2-ft opening carries 1200 cfm of air. What is the air velocity?

$$\text{Speed} = \frac{\text{cfm}}{(\text{Area ratio} \times \text{Area})}$$

$$\text{Speed} = \frac{1200 \text{ cfm}}{(0.45 \times 3 \text{ ft} \times 2 \text{ ft})} = 445 \text{ fpm}$$

This would be an acceptable air speed to prevent rain from entering with the air.

Louvers are also used within a building to allow air to transfer from one area to another where a duct is not needed. Common applications include providing a way for air to move from one side of a partition or wall to the other. Sometimes called *transfer grilles* in these applications, their size should be chosen to prevent excessive pressure losses to the system. This also keeps velocities low and helps prevent noise.

Layouts

There are many duct system layouts used for the transport of heating, cooling, and ventilating air. Understanding commonly used systems is required for a good working knowledge of most HVAC systems. These include single- and dual-temperature, constant-volume systems (used in single- and multiple-zone forms) and variable air volume systems. While other types are occasionally used, these represent the vast majority of systems most technicians are likely to encounter in commercial work.

Constant volume, single temperature. This is by far the most common duct arrangement, used for simple heating and cooling systems (see Figure 4.22). Most package rooftop and single-station air handlers are constant-volume, single-temperature devices. A single fan delivers warm or cool air into the space. All supply diffusers deliver the amount of air they were originally balanced for. The volume does not change. All the air in the system is at the same temperature at the same time. The whole space is heated or cooled simultaneously.

Most of these systems provide air for a single zone. One fan draws air (mixed as a combination of return and outside air) through a filter. The air is heated or cooled and blown out into the duct network. The ducts distribute the air into the space via the supply air diffusers.

It is easy to configure these systems to use large quantities of outside air. This can provide two benefits: (1) ventilation air for the building occupants (often required by codes to be a minimum of 20 cfm per person), and (2)

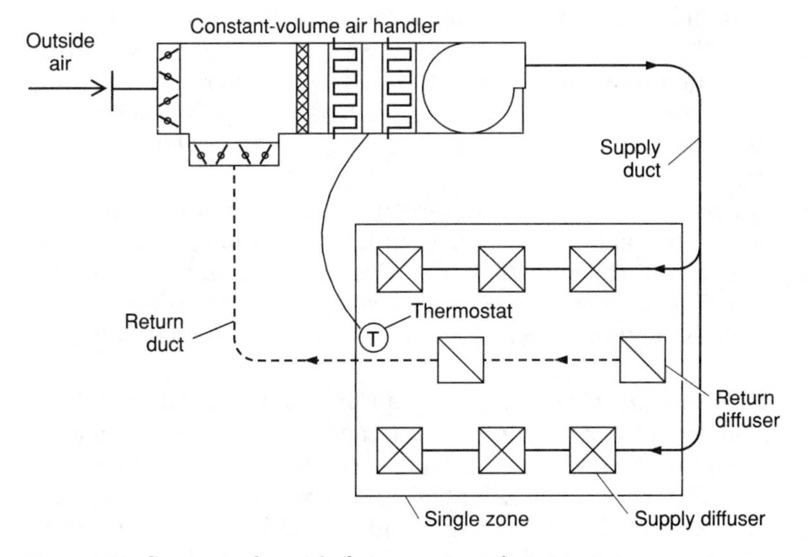

Figure 4.22. Constant-volume, single-temperature air system.

some degree of "free cooling" by using an air side economizer system. Outside air can help to cool a building if it is below about 60°F.

Motorized outdoor and return air dampers are installed in the air handler upstream of the fan. The dampers are connected so that whenever the system uses more outdoor air (by opening its damper), the return air damper is closed. This maintains a constant supply of air. A minimum outdoor air damper setting is chosen to ensure that the building always receives the proper amount of outdoor ventilation air.

The economizer mode is used when the building requires cooling and the outdoor air temperature is low enough to provide cooling. It is common for building interior zones and areas close to closed, sunlit windows to require cooling when the outdoor air temperature is below 60°F. In these circumstances the system opens the outdoor air damper fully and closes the return damper. This provides all of the required cooling without using mechanical cooling. Depending on the climate and the building's cooling load characteristics, significant energy savings are possible.

While operating in the economizer mode, it might be necessary to provide a way for air to leave the building. The 100% outdoor air operation pressurizes the interior space. This might lead to reduced Airflows and operational problems (doors won't close, ceiling tiles flutter, or noise occurs near windows). In these cases, outside relief louvers (equipped with backdraft dampers) or building exhaust fans might be required.

After the air has heated or cooled the space, it is drawn back to the fan via return registers. Most systems use return air ducts, but others use the conditioned space itself or a plenum above a drop ceiling to send the air back to the fan unit. Either way, the air is reused after it mixes with some outdoor air.

Some systems require a dedicated return air fan to overcome pressure drops in the return air ducts. Dedicated return fans are required where low-pressure supply fans are used, or when the zone is a relatively long distance from the fan.

Other than economizer controls and required smoke and fire dampers, there are usually few automatic controls in the single-zone, constant-volume, single-temperature air system. The system thermostat controls the heating and cooling sources. It might start or stop the furnace or air conditioner or regulate the flow of water through a coil. The airflow is either on or off.

Constant-volume air delivery systems can be split into multiple zones. Again, a single fan draws a mixture of return and outdoor air through a filter. After the fan, the airflow is divided into several ducts and sent through heating and cooling coils. Each zone has its own set of coils, so each can have its heating or cooling independently controlled. Every coil can be on or off independent of the others. Figure 4.23 illustrates a two-zone system.

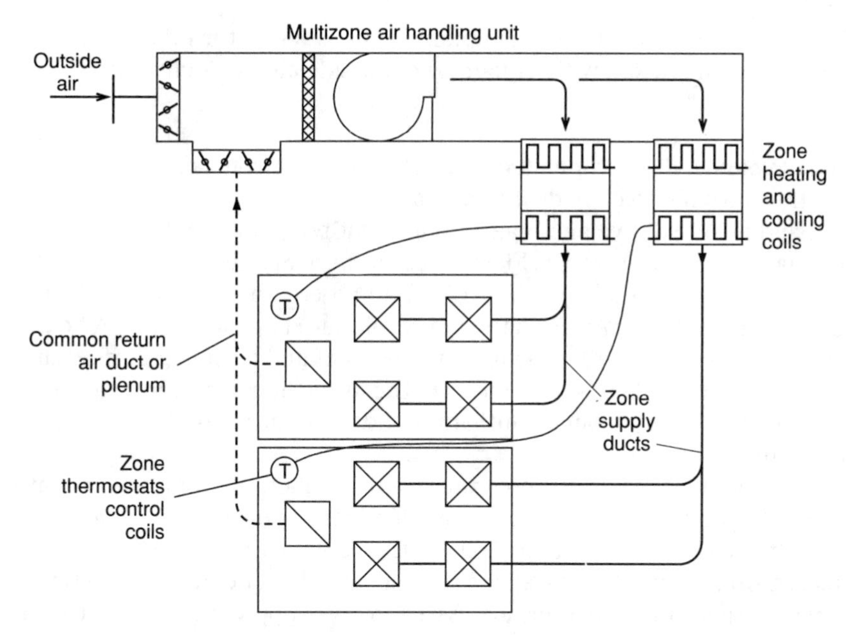

Figure 4.23. Constant-volume, multiple-zone air system.

Separate ducts deliver air to individual zones within the space. Zones are chosen to represent different heating or cooling loads (or building areas) that require separate control. Although each zone can have different air delivery temperatures, this system is still considered a single-temperature design. All of the air in a given duct is at the same temperature and all of it is delivered to its zone. There is no air volume control (other than manual balancing dampers).

For example, one zone might cover perimeter parts of a space that requires heating on a cold day. Another zone might be an interior area that would generally require cooling. Each zone has its own thermostat to allow individual temperature settings and regulation. The perimeter zone might have its heating coil open to keep the temperature high, while the interior zone might have cool air delivered. This could be done either by using a cooling coil or by delivering unheated air cooled with return air from a quantity of outside air.

Multiple-zone, constant-volume systems are relatively versatile. Units can be made with any number of zones, but six or eight is usually the practical limit. Because a single fan feeds all of the zones, they all receive air at the same time. Often these systems run the fan continuously to ensure that all zones are satisfied. Like the single-zone system, air controls are usually limited to outside air dampers and code-required fire and smoke dampers.

The return air from all zones is mixed in the single return duct. It is returned to the fan, mixed with a quantity of outside air, and delivered to the space again.

Constant volume, dual temperature. Another multiple-zone air delivery system is the constant-volume, dual-temperature system. These rely on the selecting and mixing of two airstreams of different temperatures to satisfy the requirements of multiple zones. Figure 4.24 shows a dual-temperature layout.

These systems generally use a single fan and create air supplies of two different temperatures. Air from the fan is split into two duct paths. A heating device (coil or burner) is installed in one duct and a cooling device (direct expansion or chilled water coil) is in the other. These systems can also use outdoor air for the same ventilation and economizer purposes as the single-temperature designs.

Controls on the heater and cooler maintain the ducts' air temperatures at their design point. This is usually about 120°F for the heating air (when operating the heating source) and 50°F for the cooling airstream (during the cooling season). Note that the space thermostats do not control these temperatures. Duct-mounted thermostats keep the supply air temperatures constant by operating the heating and cooling systems.

It is common to run both heating and cooling during the spring and fall. During the winter and summer, however, the "not needed" system can be

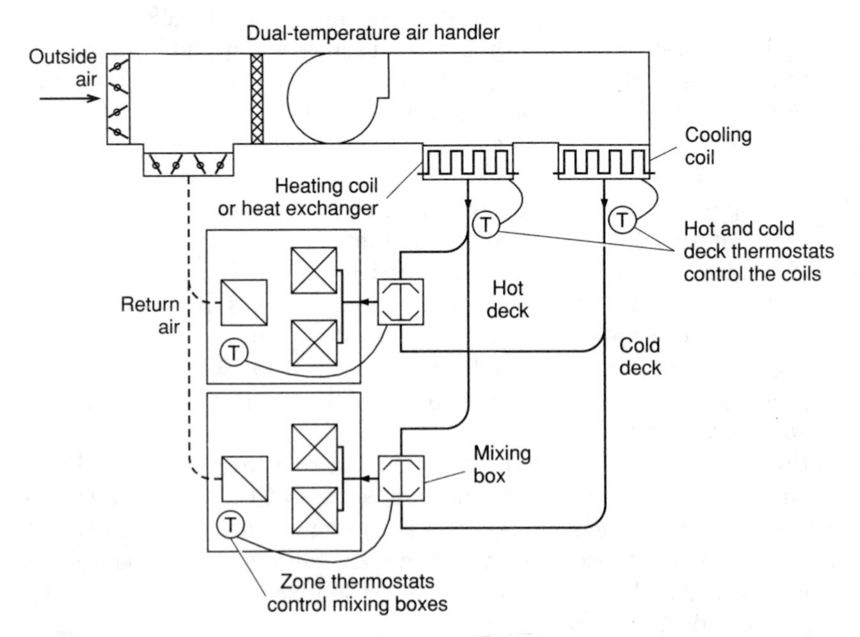

Figure 4.24. Constant-volume, dual-temperature air system.

shut down and its function replaced by using return air. A mix of outdoor and return air provides the cold air supply in the winter. The cold duct thermostat modulates the economizer dampers to maintain a 50°F internal temperature. This can provide significant energy savings compared to running the heating or cooling systems when they aren't needed.

The hot air plenum (or main duct) is the *hot deck* and the cold air plenum is, naturally, the *cold deck*. A branch duct runs from both decks to each zone's mixing box. The mixing boxes are at the beginning of the zone's duct system.

All zones have their own mixing boxes for temperature regulation. These boxes have dampers installed at the hot and cold air inlets. As air of one temperature is selected by the box, its damper opens and the other damper closes. A common linkage operates the two dampers together. Air from the hot and cold inlets mix inside the box and the tempered air is discharged out of the single outlet into the zone's duct system.

Mixed air can assume any temperature from the coldest to the hottest available. The zone's space thermostat controls the amount of air admitted from each deck to the mixing box. By changing the mix of hot and cold air delivered to the mixing box, the thermostat adjusts the supply duct's temperature.

When the thermostat requires heat, the amount of hot deck air increases and cold deck air decreases. Conversely, when the thermostat calls for cooling, it mixes in more cold deck air and reduces the hot air delivered. A dead band in the thermostat's action keeps the dampers from moving when the temperature is at the proper point. The total air volume delivered to the space is always approximately constant throughout the full range of hot and cold air deliveries.

These systems are very flexible and economical. While they use only a single set of heating and cooling coils, each zone has access to air from both the hot and cold supplies. The fan's static pressure requirements for these systems are higher than those supplying single-temperature, constant-volume systems.

Mixing boxes, dampers, and the complex duct systems each add resistance to the airflow. The controls are more complex, as well. These systems need a set of automatic controls to maintain each deck's temperature and controls for each mixing box to maintain each zone's temperature.

It is comparatively simple to add zones to a dual-temperature system. The fan speed might need to be increased to provide additional air volume, but little else must be added beyond new ducts, a mixing box, and a thermostat control.

Variable air volume. A very different approach to regulating heating and cooling systems is the variable air volume (VAV) system. Usually used for systems where cooling loads predominate, they are extremely versatile and can save significant amounts of energy compared to constant-volume systems.

Heating or cooling energy available from an airstream depends upon two things: the temperature of the air compared to the space it is conditioning, and the volume of the air delivered. Changing either parameter changes the energy delivered. Constant-volume systems vary the air's temperature (often by turning the heating or cooling system off or on), while VAV systems maintain a constant air temperature. Instead, VAV systems change the quantity of air delivered.

Variable air volume systems work well in the cooling mode. As discussed in Chapter 3, many spaces have "natural" heat gains from people, lights, and equipment and require constant cooling. Unless there are perimeter heat losses, most spaces require cooling whenever they are occupied.

Variable air volume systems use a fan to deliver air into the system. It is cooled with conventional equipment and supplied to the duct system for distribution into the space.

Each zone has a damper (called a VAV "box") that can close the supply duct leading to the diffusers. In the cooling mode, as the zone heats up (from its intrinsic heat sources), the space thermostat calls for cooling. The thermostat opens the zone's box, causing more cool air to be delivered. This increased flow of cool air conditions the space, satisfying the thermostat. When the space thermostat is satisfied, it closes the VAV box. Each zone's thermostat decides how much air is required. Very little airflow is required or delivered if all zones in the system are satisfied. Plan for this situation. It can lead to problems with the fan and cooling system.

Most VAV systems do not fully close the boxes. A minimum amount of air is always supplied to provide some ventilation of the space.

When most of the VAV boxes in the system demand very little air, the supply system's pressure must be controlled. If all of the VAV dampers close, the system's static pressure becomes very high. This leads to increased noise as air is forced around the almost closed dampers. Direct expansion air conditioning coils can freeze if insufficient air is flowing through them. The fan tries to drive air into a closed volume, leading to instability and poor efficiency. Excessively high static pressure can damage some fans. A pressure control system must be used in VAV systems to prevent these problems.

The simplest type of VAV damper is the *dump box*. These allow the air not required by the zone to be discharged (dumped) into the space surrounding the box. They are installed in return air plenums where the dumped air is drawn back into the system's return air inlet. This air bypasses the space to be conditioned. Dump boxes are simple and reliable, but their installation limitations can prevent their use in many applications. They work best when installed in a return air plenum, where the dumped air won't cause drafts or noise as it discharges (see Figure 4.25).

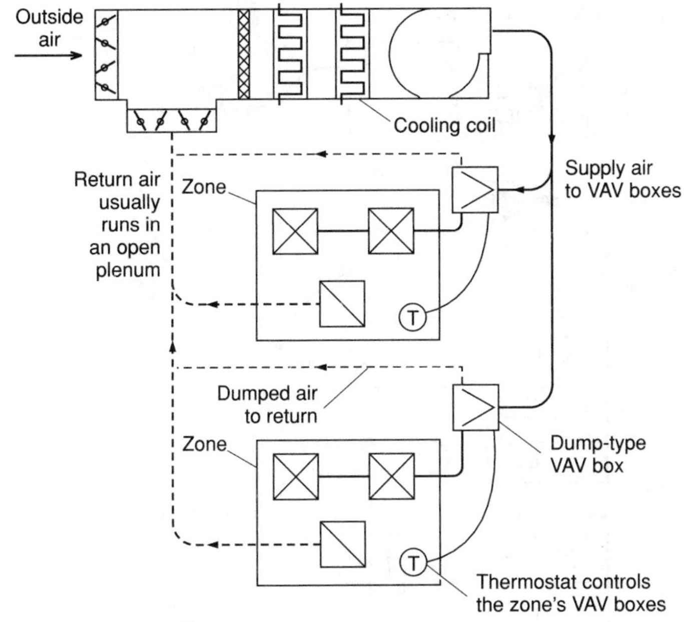

Figure 4.25. Variable air volume dump box system.

Another method of preventing insufficient flow conditions in the system is with a bypass duct (see Figure 4.26). Connected between the system's supply and return plenums, the bypass includes a motorized damper and pressure sensor. The damper opens when the static pressure difference between the system's supply and return plenums becomes higher than wanted, allowing some air to avoid passing through the system. The pressure sensor is usually installed in the supply duct about two-thirds of the way down the main duct from the fan. A position is chosen to represent an "average" duct pressure for the entire system.

Dump box and bypass duct VAV systems save energy when compared with constant-volume systems. Conditioned air returns directly to the fan without warming or cooling the space. No heat is gained, so the bypassed air imposes no burden on the heating or cooling system. The fan is, however, consuming energy in moving the bypassed air unnecessarily. Better VAV systems reduce the air volume delivered by the fan. Additional energy can be saved in this way only if a direct expansion cooling coil is not used in the system. These coils need a full Airflow to avoid freezing.

These systems can use boxes that do not dump the unneeded air, but simply close the supply ducts. Automatic controls monitor the system's static pressure to avoid moving any more air than required, saving the most energy (see Figure 4.27).

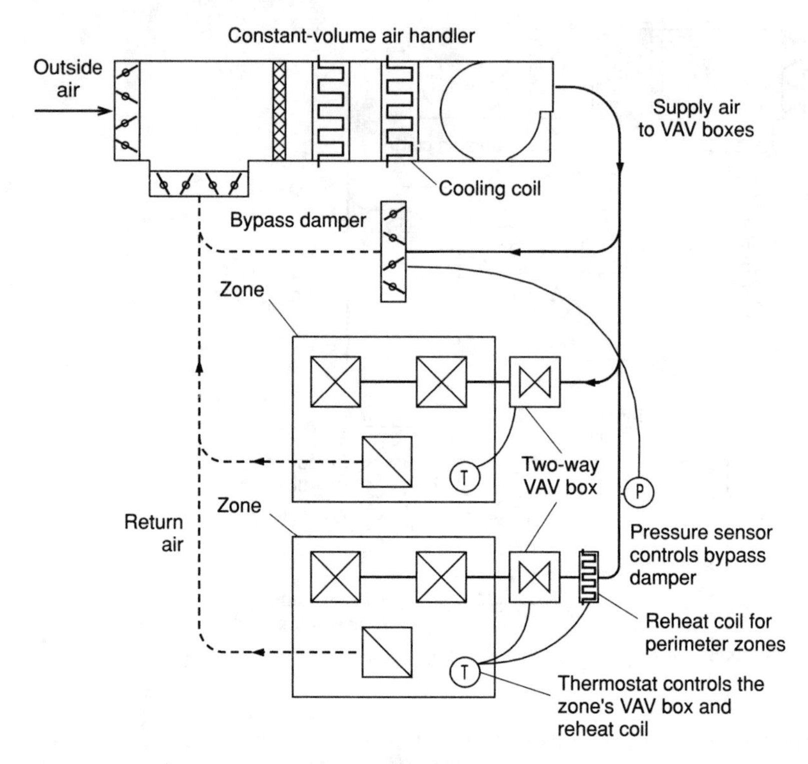

Figure 4.26. Variable air volume bypass duct system.

One method of varying the total air volume delivered is to equip the fan with a variable, motorized scroll. The internal geometry of the fan housing changes in response to the pressure difference between the system's supply and return air plenums. If the pressure becomes too great, the scroll's position changes, placing it closer to the impeller. This reduces the amount of air the fan can deliver, reducing the system's pressure.

Rotating vanes installed in the fan's inlet throat can make similar reductions. This reduces the area available for airflow, decreasing the load on the fan.

Another method of varying the capacity of the system is to change the speed of the fan. As noted earlier, the pressure a fan can deliver varies with the square of the impeller's revolutions per minute. Slowing the fan to 70% of its full speed cuts the delivered pressure in half. The system's pressure control signals an electronic variable-speed drive (connected to the fan's motor) to adjust its speed. As the pressure increases (due to the VAV boxes closing) the fan is slowed.

Slowing the fan is an excellent energy saver. Because the power required by a fan varies with the cube of a change in its speed, the power savings can be significant. Consider the 30% speed reduction noted previously (running at 70% of its original speed is the same as a 30% reduction). The fan supplies 50% of the original static pressure, but the power consumed by the fan drops to 34% of its original value. This is a 66% savings.

Variable-speed motor drives use electronic variable-frequency generators. Instead of supplying the steady 60 Hz of power obtained from the power company, they generate any frequency from zero to 66 Hz. The rotational speed of motors used for HVAC fans depends on the frequency of the power supplied. Variable-frequency alternating current supplied to the motor changes its speed.

Speeds can be selected automatically from a full stop to 10% over the motor's rated revolutions per minute. These drives can also control the motor's starting and stopping rates, limiting wear on drive belts and sheaves. Electrical interfaces are available to connect these drives to the system's pressure sensors, allowing them to accurately maintain proper internal duct static pressure.

Make special provisions for heating when using a VAV system. Their normal "cooling only" operation often requires a separate heating system. If the building uses a timed temperature setback during unoccupied hours, the whole VAV system should provide heat. Include standard hot water coils or electric

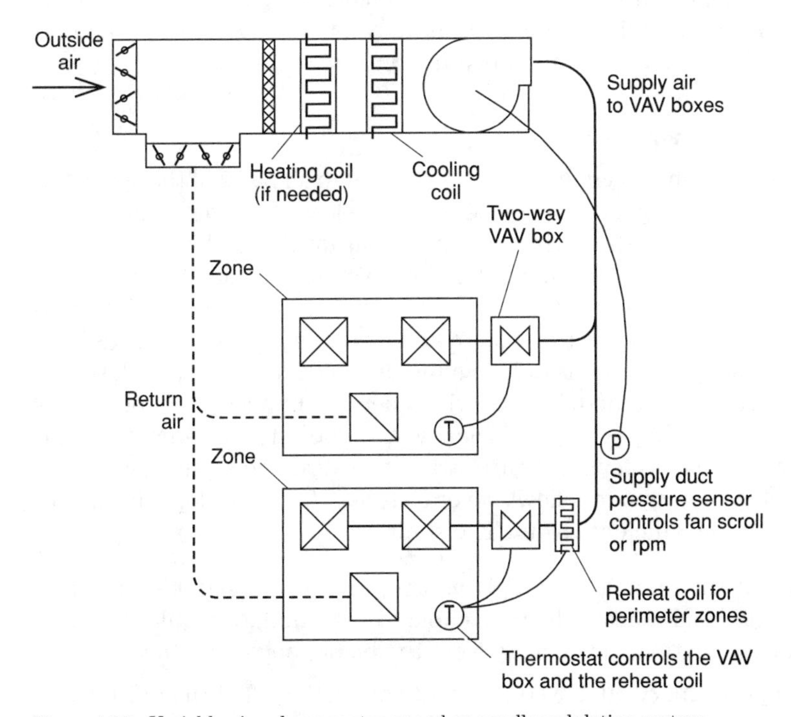

Figure 4.27. Variable air volume motor speed or scroll modulation system.

heat strips in the air handler. The building can be warmed rapidly by running the air and keeping the VAV boxes open during warm-up. However, when the building is occupied, the air handler coils are usually not used for heating.

Perimeter zones usually use hot water or electric reheat coils installed at the VAV box or an entirely separate fin tube (or other) heating system. The VAV box and perimeter heating are both controlled by the zone's thermostat. The thermostat smoothly operates both systems to provide heating or cooling. If a duct-mounted reheat coil at the VAV box provides the heat, the thermostat automatically opens the box and the coil when heat is needed. If separate fin tube heat is provided, the thermostat sets the box to its minimum air delivery setting and opens the fin tube's valves. The thermostat's controls include a dead band. This prevents the heating and cooling systems from operating simultaneously.

Ventilation air must be carefully handled with VAV systems. Outdoor air requirements (often up to 20 cfm per person) might be difficult to supply with a VAV system operating in its lowest air delivery modes. Don't set the minimum VAV air delivery to provide required ventilation air for a space's maximum occupancy. It provides too much air for times of minimum occupancy and the room will be overcooled. The system wastes energy, thus you lose one of the major benefits of using a VAV system.

Many buildings with VAV systems use completely separate ventilation air supplies. As discussed in Chapter 10, these can be energy efficient and integrated with the VAV system to prevent control problems.

Estimating losses and sizing

Proper duct system design provides air delivery to all parts of the space being conditioned with good balance between the loads and the air delivered. Each diffuser delivers the correct amount of air for the load it is heating or cooling. Pressure losses throughout the system are balanced and noise is minimized. Proper noise control is achieved by preventing excessive pressure loss and turbulence in the system and by limiting duct air speeds.

There are several methods of designing duct systems. The simplest way is the equal friction method. It is described here. It allows reasonable air velocities, low noise, high energy efficiency, and good flow balance throughout the system. The calculations are straightforward and easily managed.

The duct system design follows several steps. These develop a balanced, quiet, and energy efficient layout. The steps are

1. Figure out the number of supply and return air diffusers required and their locations. Estimate the total cubic feet per minute required at each diffuser for heating and cooling. Note the higher air requirement.
2. Make a preliminary duct system layout connecting all of the diffusers to the air handler. Note limitations on duct sizes, required fire and smoke

dampers, and possible obstructions. Make the layout as direct as possible to limit pressure losses.

3. Using the air handler unit's manufacturer's data, establish the proper external pressure loss associated with the airflow required for the system. Split the loss between the supply and return systems, and estimate a system target loss for each.
4. Add the cubic feet per minute flowing through all duct branches and mains to get a total required throughout the system.
5. Start sizing the ducts with the supply system. Establish the critical path as the longest route air must travel from the fan to the farthest supply diffuser. Calculate the duct sizes all along the critical path from the fan to the farthest diffuser. Pressure loss in the critical branch, including losses from friction and at fittings, equals the total system's target pressure loss. Cumulative losses from all of the turns, reducers, fire dampers, take-offs, and ducts in the critical path should be the highest in the system.
6. Size the rest of the supply branches. Provide dampers in noncritical branches to balance the airflows and develop the proper pressure losses in all noncritical paths.
7. Size the return air system. Again, figure out its critical path and select the proper pressure loss along it. This is the total return air system target. Use dampers in noncritical return air paths to prevent system imbalance and excessive noise.

The first step in the design process is to locate the supply air diffusers and figure out the amount of air required from each. The total air volume the air handler will supply to the system must equal the sum of air supplied to each diffuser.

It is important to estimate how the loads vary from location to location within the building. The amount of air delivered to a particular diffuser depends on the local heating or cooling load at the diffuser's location. Allow 400 cfm/ton of cooling capacity required and about 20 cfm/1000 Btu of heating required. Choose the larger of the airflows calculated, heating or cooling. The total building load determines the size of the system (and the total cfm required from it).

The total volume delivered from all of the supply diffusers must equal the airflow the fan is to provide. If introducing ventilation air (additional outdoor air) into the space, the total cubic feet per minute including those requirements must be estimated. Size the system to provide the greater of the required forced air heating or air conditioning flow. In cold climates it is possible that the heating volume will exceed the cooling volume.

Deliver more heat to those perimeter locations that admit infiltration air and to areas prone to excessive conductive losses. These areas include locations near doors and windows. Provide additional cooling air for south and west facing windows, near heat producing machinery, and at doors used during occupancy.

Design a preliminary duct layout after the diffuser locations and supply capacities are established. Don't attempt to estimate the size of the ducts at this time. The layout should provide for as direct a route for the air as possible from the air handler to all of the supply diffusers. Most layouts include one or more main ducts that carry air out to the space. Branch ducts and diffuser take-off ducts run away from the mains to feed diffusers.

Proper duct design and installation should follow a few simple rules. These help optimize the air delivery with good efficiency, balance, and little noise. The guidelines are

- Install ducts in as straight a line as possible. Avoid unnecessary turns. Transitions from one size to another, or from round to square or rectangular cross sections, should be made smoothly and with gradual angles. Sharp-edged expansions or reductions increase turbulence, losses, and noise.
- Use flexible duct connection collars at fans and other equipment that produce noise. This reduces the sound transmitted into the ducts.
- Duct turns should be as gradual as possible. Rectangular ducts should have elbows with radii no tighter than ⅛ the width of the duct. If it is necessary to make tighter turns than this, use turning vanes.
- Locate branch connections to the duct mains at least 4 to 5 ft past any turn, transition, or butterfly damper. This allows space for the airflow to return to a less turbulent condition and reduce noise.

It is important to properly maintain air speed throughout the system. The air's momentum contributes to its proper delivery and distribution. While very low air duct velocities do lead to quieter systems and lower pressure losses, insufficient speeds can prevent the air from being delivered to distant branches and outlets. Proper velocity also helps keep the air moving around corners. Keep velocities within the ranges given in Table 4.2 to ensure proper distribution and balance.

Make the duct layout as simple as possible. Minimize the number of turns and attempt to any keep rectangular duct's *aspect ratios* (the width divided by the height) as close to one as possible. Round ducts are more efficient than rectangular, so consider using them where possible. Plan for required smoke detectors and smoke and fire dampers (including necessary damper inspection access doors).

Most codes require the installation of smoke detectors in main supply ducts providing over 2000 cfm. Connect them so they shut down the air handler's fan if smoke is present in the duct. Check with the local building

TABLE 4.2 Air Velocities at Various Duct Sizes and Flows (cfm)

Round duct diameter (in.)	Rectangular duct area (in.^{3})	cfm	Air speed (fpm)
4	13	50	573
4	13	75	859
4	13	100	1146
5	20	50	367
5	20	100	733
5	20	150	1100
6	28	100	509
6	28	150	764
6	28	200	1019
8	50	150	430
8	50	200	573
8	50	250	716
8	50	300	859
8	50	350	1003
10	79	300	550
10	79	350	642
10	79	400	733
10	79	500	917
12	113	400	509
12	113	500	637
12	113	600	764
12	113	700	891
12	113	800	1019
14	154	600	561
14	154	800	748
14	154	1000	935
14	154	1200	1123
14	154	1400	1310
14	154	1600	1497
16	201	800	573
16	201	1000	716
16	201	1200	859
16	201	1400	1003
16	201	1600	1146
16	201	1800	1289
16	201	2000	1432
16	201	2200	1576
18	254	1500	849
18	254	2000	1132
18	254	2500	1415
18	254	3000	1698
18	254	3500	1981

inspector for specific requirements, but most follow the National Fire Protection Agency (NFPA) codes. A connection from the building's fire alarm panel to the fan control might also be required to stop Airflow if a fire occurs in the building. However, this cannot be substituted for a duct-mounted smoke detector when codes require one.

Include dampers in the design. Allow for fire and smoke dampers as required in the supply and return duct systems. Codes require a fire or smoke damper at each of the building's fire and smoke barriers.

Next, select the type of return air system. Plenum return systems might be the simplest. These use the space between a drop ceiling and the floor or roof deck above to carry air from the space return grilles back to the fan. These systems are regulated by the materials used in the space. Proper electrical wiring (rated for plenum service) and other materials used there might make it costly to use this type of return. Also, most codes require that large plenums be divided into multiple smoke or fire zones with nonflammable barriers. These can prevent proper airflow, so it will be necessary to install smoke or fire dampers.

Ducted returns are usually easy to install. A large duct runs from the air handler's return air inlet to one or more return grilles installed in the conditioned space's wall or ceiling.

Most codes prohibit that corridors in buildings be used as return air plenums. This prevents fire exit paths from filling with smoke. Individual returns (a grille to the return duct or plenum) must be used for each space or room. Place return air grilles in all rooms connecting to the corridors. This ensures that air can return to the fan from all areas without having to enter an emergency exit path.

Multiple individual rooms connected to each other, but not to an exit corridor, can share a single return grille. Small amounts of air will pass under doors and flow from one room to another to reach the common return grille. Don't rely on under-door flow to handle more than 100 cfm of return air. Install a separate return grille in areas where large amounts of air must be returned.

Once the supply and return duct plans are roughed out, including any duct size limitations, estimate the duct system's sizes. The air handler fan will have a specific external static pressure capability at the design volume (supplied to the space). Check with the manufacturer for this figure.

The fan's allowable external static pressure is the design point for the complete duct system. Total pressure loss from the entire duct system should equal this requirement. Too high a static pressure loss won't allow the required air to flow in the system. Too low a pressure loss increases the airflow excessively. This can cause increased energy use, excessive noise, drafts, and insufficient temperature changes as the air passes through the heating or cooling coils.

If there is little or no return duct system, allow almost all of the static pressure to be used by the supply system. Provide only enough static pressure on the return to overcome losses from the diffuser. Allocate about 75% of the total pressure loss to the supply when the return system is simple (one or two return diffusers and a small amount of duct). If there is a complex return air duct system, plan on 50% of the total pressure loss for the supply. Use the remaining pressure drop allowance for the return duct system.

Most HVAC systems use low-velocity ducts, selected to keep pressure losses at or below 0.1 in. water per 100 ft of straight, galvanized steel duct. Noise is minimized, as this low pressure loss keeps the speed of the air slow enough to prevent excessive turbulence.

Low-velocity heating and cooling system ducts typically carry the airflows shown in Table 4.2. This table gives the cubic feet per minute handled by various round duct diameters and equivalent rectangular duct areas. Equivalent rectangular duct sizes listed are 10% larger than the equivalent square inch area of the round duct. This is because rectangular ducts are not as efficient and have more friction losses due to their larger surface area.

Table 4.2 does not include all of the possible combinations for duct sizes and airflows. If a velocity or volume needed is not in the table, calculating the air velocity given the duct's size and the air volume is simple. The formula for finding the velocity in a duct is

$$\text{fpm} = 144 \times \text{cfm}/\text{Area (square inches)}$$

Find the area of rectangular ducts by multiplying the width by the height. Add approximately 10% to the area calculated for rectangular ducts. This compensates for their higher friction levels.

The area of a round duct is found by using the formula

$$\text{Area (square inches)} = 0.785 \times \text{Diameter}^2 \text{ (inches)}$$

Velocity corresponds to duct diameter, therefore, with the formula

$$\text{fpm} = \frac{144 \times \text{cfm}}{(0.785 \times \text{Diameter}^2)}$$

The above formula can be rearranged to provide the duct size (square inches or round duct diameter) that corresponds to a given air speed and volume. This formula is

$$\text{Area (square inches)} = 144 \times \text{cfm}/\text{fpm, or}$$

$$\text{Diameter (inches)} = \sqrt{(183 \times \text{cfm}/\text{fpm})}$$

Table 4.2 shows that, at constant friction levels, larger air volumes flowing in bigger ducts results in higher air velocities. This speed increase should be considered when sizing ducts. High air speeds can create more noise and turbulence and cause increased pressure loss at duct fittings (turns, splits, etc.).

Size main ducts (that lead from the fan to the branch ducts) to provide 0.1 in. water friction loss per 100 ft of duct. Noise might be a concern with very short branch ducts feeding supply diffusers near the main, or with diffusers installed directly onto the main duct. The velocity in these ducts should be limited to 1500 fpm. Higher air speeds are permissible if there are no diffusers or short branch connections tied directly to the main duct.

Size branch ducts to limit friction losses and air velocities. A duct size should be selected to keep the air's speed below 650 fpm and pressure losses at or below 0.1 in. water per 100 ft. Lower velocities help provide quieter air delivery to diffusers and hold pressure losses to a minimum.

High air velocities in duct mains cause greater pressure losses from duct obstructions or poor design. As noted earlier, pressure losses are proportional to the square of the air speed: doubling the speed quadruples the loss. It is very important to use good duct layout practices in high-speed mains.

Duct fittings (turns, cross-sectional area changes, branches, splits, etc.) contribute the majority of a system's air pressure losses. It is common for over 75% of the total pressure losses in a system to occur as the air negotiates turns, splitters, wyes, and converging ducts. This is why fittings should be carefully applied—small differences in their details can have a significant impact on total system losses. See Table 4.3 for a listing of some common duct fittings and the equivalent duct length associated with them. For example, note that a 90-degree rectangular duct turn (with turning vanes) has the same pressure loss as 7 ft of duct. The same turn without vanes is equivalent to 67 ft of duct.

TABLE 4.3 Equivalent Fitting Pressure Losses in Feet of Duct

Fitting description	Equivalent length (ft)
Abrupt duct exit perpendicular to wall or ceiling	49
Abrupt duct exit angled 30–150 degrees along wall	56
Smooth radius 90-degree round elbow, tight turn	
with no inside turn radius	39
with inside turn radius = ½ duct diameter	12
Four-piece, 90-degree elbow with inside turn radius = ½ duct diameter	21
Two-piece, 90-degree mitered elbow (round duct cut to "square" turn)	67

Fitting description	Equivalent length (ft)
Rectangular 90-degree elbow, smooth radiused outside and inside walls. Duct height 1½ times its width, no inside radius	61
inside radius = ½ duct width	11
inside radius = duct width	7.8
Two-piece, 90-degree mitered elbow (rectangular duct cut to "square" turn), no turning vanes	
height = width	67
height = 1½ width	61
Two-piece, 90-degree mitered elbow (rectangular duct cut to "square" turn) turning vanes	6.7
Abrupt rectangular duct size increase,	
doubling area	21
quadrupling area	39
Abrupt round or rectangular duct size decrease,	
area decreases to 50% of original	6.7
area decreases to 25% of original	9.4
Converging round wye. Branch joins main flow at 45 degrees, has 25% of the main duct's area and carries air at the same velocity as the incoming main flow.	
effect on branch flow	0.83
effect on main flow	2.2
Converging 90-degree round tee. Air flows in one straight-through leg and the branch fitting and leaves the remaining straight branch. Branch flows 25% of the incoming main flow	
effect on branch flow	9.4
effect on main flow	15
Converging 90-degree rectangular wye. Branch is radiused to smoothly blend incoming flow into the same direction as the main flow. Branch carries 25% of the incoming main flow at the same velocity.	
effect on branch flow	0
effect on main flow	17
Converging 90-degree rectangular tee. Air flows in one straight-through leg and the branch fitting and leaves the remaining straight branch. Branch flows 25% of the incoming main airflow.	
effect on branch flow	–21
effect on main flow	15
Diverging 90-degree rectangular wye. Branch is radiused to smoothly blend departing flow out of the main flow. Branch carries 25% of the departing main flow at the same velocity.	
effect on branch flow	28
effect on main flow	–1.7

TABLE 4.3 (Continued)

Fitting description	Equivalent length (ft)
Diverging 90-degree tee. Air flows into one main branch and out of the side tap and the remaining main. Branch carries 25% of the departing main flow at the same velocity. A rectangular main duct is assumed.	
Round tap: effect on branch flow	73
effect on main flow	1.1
Rectangular tap: effect on branch flow	67
effect on main flow	1.1
Diverging rectangular tee with splitter damper. Air flows into the main and past the open splitter. Area before and after the splitter is equal. Branch flows 25% of the leaving main flow at the same velocity.	
effect on branch flow	54
effect on main flow	6.7
Diverging rectangular tee with turning vane extractor. Air flows into the main and past the open extractor. Area before and after the extractor is equal. Branch flows 25% of the leaving main flow at the same velocity.	
effect on branch flow	61
effect on main flow	6.7
Fully open round butterfly damper	11
Fully open rectangular butterfly damper	2.2
Fully open rectangular parallel blade damper	29

Excerpt from *ASHRAE Handbook; 1985 Fundamentals,* reprinted with permission

Table 4.3 lists only some of the duct fittings and combinations available. The ASHRAE *Fundamentals Handbook* has listings for many more that can be used in ventilation systems. A copy can be ordered directly from ASHRAE (see Appendix A for their address and telephone number).

Add all of the system's supply diffuser airflow requirements together along each branch of the layout. Figure out the total amount of air flowing through each duct, ending with the main duct at the fan. Double check that the fan can provide the total cubic feet per minute for all of the diffusers.

Begin sizing the layout with the supply ducts. Figure 4.28 shows a typical supply duct system. Decide which path is farthest from the fan to a diffuser. Size this route (the *critical path*) first. Its pressure losses are the highest because the air travels the longest distance, or through the greatest number of fittings, from the supply fan to the diffuser.

A target pressure loss should be selected for the critical path. The critical path's pressure loss equals the total supply system loss the fan "sees." While

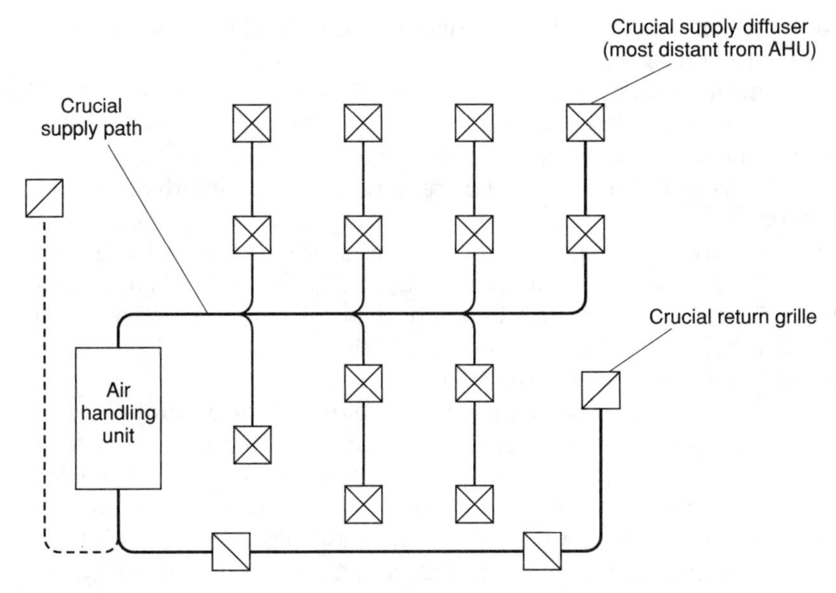

Figure 4.28. Duct system critical path. This is usually the only path that must be carefully estimated for pressure losses.

the fan delivers the total system airflow at this pressure, the critical path only determines the fan's pressure requirements. Proper flow through the critical path is almost automatic if its pressure drop equals the total supply system pressure drop.

Remember that, if installing an automatic control damper in the system, the pressure drop across the damper should equal half the total pressure drop. Select the damper and the rest of the system sizes to make their pressure losses equal.

Size the main duct and any major branches along the critical path to allow a 0.1 in. water per 100 ft pressure drop. Size the final branch to the diffuser to keep the air velocity below 650 fpm. Use Table 4.2 to estimate the proper sizes.

Add all of the pressure drops along the critical path. Add the losses caused by straight ducts. Multiply each duct's length by its pressure drop per 100 ft. Divide the number obtained by 100 to get the actual pressure drop.

For example, a 25-ft-long, 8-in.-diameter duct is carrying 175 cfm. What is the pressure drop?

Pressure drop for the duct is approximately 0.07 in./100 ft.

$$\frac{25 \text{ ft} \times 0.07 \text{ in.}}{100 \text{ ft}} = 0.0175 \text{ in. water pressure drop}$$

Total all of the pressure drops caused by fittings in the critical path. Each turn, take-off, tee, damper, and diffuser in the critical path contributes to the total pressure drop. Add the losses from the straight ducts' friction to those from the fitting losses.

If the critical path's pressure drop is very close to the target amount, all of those ducts are properly sized. If the pressure drop is more than the target, adjust the duct design. Enlarge some ducts or straighten the route to eliminate turns. Recalculate the total pressure loss to confirm that it is close to the target.

If the pressure loss is below the target, resize the system to increase it. Smaller ducts are more economical to install, so their use should be considered first. Where duct resizing is impractical, install manual dampers in the main ducts or the critical branch to increase the pressure loss. Don't install a damper immediately behind the fan outlet, however.

If the fan delivers too much air it is better to reduce its output by varying its speed, adjusting its pulley sheaves or the motor's winding taps. Contact the manufacturer for advice if the air handler's external pressure requirement is far above what is available from the system. Remember, the goal is to have the pressure loss for the critical branch within 10% of the target.

Once the critical path is dimensioned, and all of the ducts and fittings along it have been sized, calculate the rest of the branches and submain sizes. Each of these need to have dampers installed to prevent an excess of airflow and system imbalance. Install the dampers at the take-offs where they separate from the mains or submains. Locating them very close to their supply diffusers might cause excess noise if the pressure losses at the damper are high.

Most supply diffusers are available with integral butterfly dampers. Do not rely on these dampers to provide system balance. Use them to fine tune the system after it is running. If the diffuser's dampers are used to excessively limit airflow, they develop unacceptable noise levels as the high-pressure air attempts to rush into the space around the damper blade. Install balancing dampers where the branch ducts split away from the main duct. These should be far enough upstream from any supply diffusers to limit noise transmitted into the diffuser.

Size noncritical branches and mains the same as for the critical path, limiting the air's speed to 650 fpm. It is permissible to allow noncritical main ducts to have pressure drops higher than 0.1 in. water per 100 ft of duct. As long as their total pressure losses do not exceed the already sized critical branch, there will be no problems. This allows smaller ducts to be installed. The air speed in these noncritical path mains should be limited to less than 1800 fpm to prevent excessive noise and turbulence.

After sizing all of the supply ducts and fittings, do the same for the return air system. The methods are identical. Add all of the airflows along each path to establish the cubic feet per minute flowing in each duct. Use the target loss for the return air system (established earlier) as the goal for the critical return air path. Size the critical ducts and fittings to develop the proper pressure drop. Equip noncritical paths with dampers to prevent excess airflow and noise.

Noise control

Duct system noise comes from a variety of sources. The fan causes noise from its motor, drive, and bearings. Movement in the duct is the source of much noise. Air makes rumbling, rushing, or hissing noises as it travels confined within a duct. Additional noise can come from the fan's drive motor and pulley system.

Air handler noise is controlled by proper installation. The motor must be securely mounted with the drive sheaves properly aligned. The fan's drive belt should track straight from the motor's pulley to the fan's. The motor and impeller shafts should be parallel to reduce noise and friction and improve efficiency.

Connections from the air handler to the connected ducts must be properly isolated. When using sheet metal ducts, install flexible, fire- and smoke-rated couplings between the ducts and the air handler's housing. These prevent noise and vibration in the air handler from being directly coupled into the ducts. If they aren't used, excessive noise is transmitted into the building. All air handlers should be isolated. Many residential furnaces are installed without these couplings. The owners of these systems tolerate noise without needing to.

Large air handlers might need to be isolated from the fan room's structure (floor or other supports) when sound attenuation is important. Multiple-floor office buildings and retail establishments require fan isolation if air handlers are installed above occupied spaces. Typical air handler vibration and noise isolators are synthetic rubber shear mounts and compression spring mounts. Rubber isolators are built as hollow cones. The weight of the air handler distorts the cone, causing it to support the unit in shear. This greatly reduces the sound and vibration that goes through them. See Figure 4.29 for an illustration of a rubber shear mount under pressure.

Compression spring mounts use steel springs to hold the fan unit's frame. The dead weight of the air handler partially compresses the coils. Vibration from the fan and drive causes the springs to extend and compress, absorbing the movement and transmitting very little of it to the base.

All vibration isolators must be properly matched to the load they are to support. Select isolators to match the dead weight of the fan. Properly applied, both rubber shear and compression spring isolators work equally well.

Some fan noise is transmitted within the duct. Sound is broadcast inside the duct and is carried along with the air. As the sound moves along, it passes out of the ducts through their walls, escaping into the surrounding space.

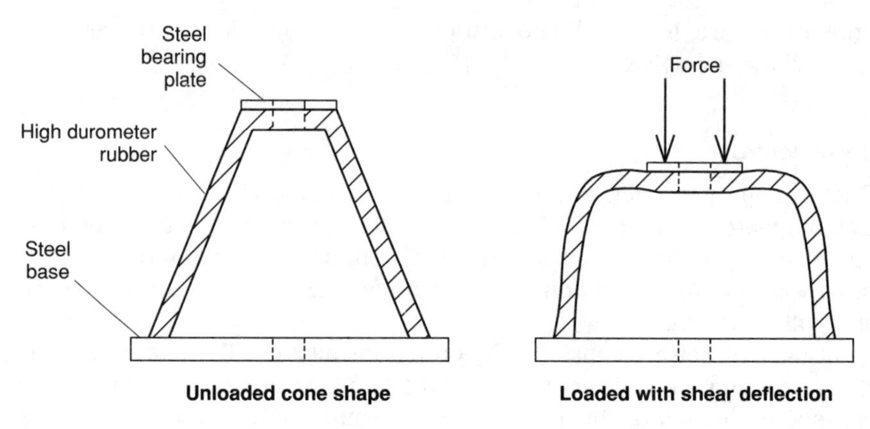

Figure 4.29. Shear-type vibration isolator.

Proper handling of the main duct and branches away from the fan helps prevent much of the fan's noise from propagating down the ducts. Attention to detail is important, as small differences in layout can significantly reduce fan noise transmission.

Noise carried within metal ducts can be reduced with liners. As noted earlier, they help quiet sound as it moves within the duct and as it passes through the duct's walls or out of a diffuser. Usually, liners are not needed more than 30 ft from the fan. After that distance, fan noise is reduced sufficiently to allow the use of bare metal ducts. Glass fiberboard ducts do not need liners. The rough internal surface acts as a liner and quickly reduces noise.

Fan noise can be reduced by ensuring that there is no line of sight path for sound to take directly from the fan to a diffuser. The duct should include a right-angle turn at least once (preferably twice for ducts less than 50 ft long) after the fan. Liners should be used on the turns and duct immediately after the fan to absorb sound. If the entire duct is going to be unlined, it is good practice to at least line the first turn or two after the fan. This significantly reduces noise propagated down the duct system.

Additional noise is caused by the movement of the air inside the ducts. These sounds are caused by turbulence as air rushes around turns and past obstructions. It also occurs wherever air passes through small openings under pressure. Partly open dampers are especially prone to noise.

Noise caused by moving air is prevented by attention to detail during system planning and installation. As explained earlier, proper duct system design limits air pressure losses and duct velocities to minimize noise. Smooth turns, size transitions, and branch connections are essential to limiting noise produced in the duct system.

Keep air speeds in all ducts adjacent to diffusers below 700 fpm. Try not to install diffusers within 10 ft of duct branch connections or dampers. Turbulence from these restrictions creates areas with high noise levels.

Don't use the supply diffuser's internal dampers for system balancing. Excess noise, especially near the fan, is virtually guaranteed. The large pressure drop needed at the diffuser's dampers to reduce the airflow increases the air's speed as it tries to get out of the system. The high-speed jet of air, while possibly at the proper total cubic feet per minute, causes an unacceptably loud rushing sound. The building's occupants will be dissatisfied, and the problem cannot be easily fixed after the ducts are installed.

Balance the system with properly sized ducts and dampers installed at the branch duct's connections to the mains. This allows the diffuser's dampers to be used to fine tune the system.

Duct sealing

Preventing duct leaks is an important part of installation. Leaks in duct systems cause many problems. They contribute to system capacity losses, imbalance, noise, drafts, and even water problems. Systems with few leaks work best, delivering air where it is needed.

All duct systems leak. Typical systems lose 10% to 20% of the air delivered into them from the fan. Higher static pressure systems leak more than lower pressure ones if they are sealed equally well. Every seam and connection is a potential leak source. The metal-to-metal contact of typical duct systems does not seal perfectly.

Given that all duct systems leak some air, many installers feel that sealing to minimize leaks is not important. This is not so. Preventing as many leaks as possible contributes positively to total system performance.

Excessive air leaks can cause many problems. The system's performance is reduced because the air that leaks out takes heating (or heat removing) capacity with it. Leaking air causes the system pressure to drop, disturbing the system's overall balance. Pressure needed to drive air to the distant parts of the system is lost and the critical path receives insufficient air.

Possibly worse than leak-caused performance problems are moisture problems that can occur from air leaks. Cooled air leaking into a warm, moist plenum can cause condensation. Water forms wherever the cooled air is below the ambient air's dew point. This water, whether it is confined to the duct's insulation or forms on building surfaces, causes damage. It might leak into the occupied area, damaging ceilings and building contents.

Because higher pressure ducts leak more, and main ducts have more pressure closest to the fan, concentrate on sealing here first. The main duct connection to the air handler (which should have a resilient vibration isolating collar) must be well sealed. All turns, branches, transitions, dampers, and other fittings on the main duct should be carefully sealed.

Next, seal the larger branches that carry air to the diffusers close to the fan. Keeping these tight ensures that air needed by the critical path and other more distant diffusers is not lost. Finally, seal the more distant branch ducts.

Duct seams should be taped or caulked. Caulk is more permanent; tape adhesive often becomes brittle and can fail after repeated heating and cooling cycles. Close all gaps and seams. Good workmanship is essential to prevent leaks. Use materials properly rated for fire and smoke protection. Check with the local building inspector for specific requirements.

Chapter

5

Water Flow

This chapter describes some basic considerations for hydronic systems, including the components and materials used and guidelines for sizing them. Proper water treatment is also covered. Descriptions of specific piping layouts are provided in later chapters dealing with heating systems, chillers, and heat pumps.

The flow of water is used in many heating and cooling systems. Many buildings use hot water heating and chilled water cooling systems. Using water has several advantages.

As discussed earlier, water can carry more heat than air. Its higher specific heat and density allow a moderate flow of water to distribute a large amount of heat far from the source. Water systems are quiet and require physically smaller sources than primary air systems.

Water heating systems usually operate at higher temperatures than air systems because it is easier to insulate pipes than ducts. The higher temperature usually does not help the hot water system carry more heat though. The entire system operates at higher temperatures (both the supplies and returns are hotter), so the temperature differences are smaller than with air heating systems. Nevertheless, water heating systems deliver more heat than air heating systems. Consider the following examples.

How much 125°F air, in cfm, must be delivered to a 70°F space to supply 25,000 Btu/hr? Use standard conditions for the air calculations.

$$\text{cfm} = \frac{\text{Btu/hr}}{(1.08 \times T_d)}$$

$$\text{cfm} = \frac{25{,}000 \text{ Btu/hr}}{(1.08 \times 55°\text{F})} = 421 \text{ cfm}$$

How much 180°F water, in gpm, must be delivered to a 165°F outlet heating coil to supply 25,000 Btu/hr?

$$\text{gpm} = \frac{\text{Btu/hr}}{(498 \times T_d)}$$

$$\text{gpm} = \frac{25{,}000 \text{ Btu/hr}}{(498 \times 15°\text{F})} = 3.35 \text{ gpm}$$

A much higher horsepower fan motor is needed compared with the pump's motor to deliver the same amount of heat. The air duct should be 10 to 12 in. in diameter, while a ¾ in. pipe would be sufficient to handle the same heat flow.

Chilled water systems also deliver more cooling capacity when compared to air systems. Smaller pipes and pumps can cool more space than larger fans and ducts. Water is an efficient carrier of heat energy.

Components, Pipes, and Fittings

Various pipes and fittings are used in hydronic HVAC systems. They carry water and regulate its flow, providing the proper amount of heat energy to each piece of equipment. Major components and their applications are outlined in the following paragraphs.

Pipes

Hydraulic laws similar to those governing airflow in ducts control water flow through pipes. Water is a noncompressible fluid, which makes its flow and pressure relationships in a pipe follow the equation

$$\text{psi} = \frac{0.0807 \times f \times \text{Length} \times \text{Velocity}^2}{\text{Diameter}}$$

The pipe's length is given in feet, its diameter is in inches, and the water velocity is in feet per second. The term f is the friction coefficient, related to the internal roughness of the pipe. The rougher the pipe, the more friction the water encounters as it flows through the pipe. This increases its pressure drop.

The previous equation can also be expressed as

$$\text{psi} = \frac{0.0135 \times f \times \text{Length} \times \text{gpm}}{\text{Diameter}^5}$$

Note that the pressure drop is related to the diameter of the pipe raised to the fifth power. This means that nothing influences the amount of pressure required to drive water through a pipe as much as the pipe's diameter. A very small change in pipe diameter makes a large change in pressure drop. For example, doubling a pipe's diameter cuts the pressure drop by a factor of 32.

These equations assume that there is fully turbulent flow in the pipe. As discussed in Chapter 2, fluid flow is laminar at very low flow rates. It becomes turbulent as the flow increases past a critical value. This not only improves heat transfer, but it changes the flow characteristics, making them more consistent.

Pressure losses in pipes increase with greater length and water speed. Increasing the pipe's diameter reduces the pressure loss.

A variety of pipe materials are used for most HVAC systems. They include black and galvanized steel, copper, and plastic. These materials each have specific advantages and disadvantages that make them appropriate for certain installations.

Steel. By far the most common type of pipe used in hydronic heating and cooling systems is black steel. It is the material of choice for many reasons. Some advantages of black steel pipe include

- Low cost. Black steel pipe is the least costly material available.
- Durability and strength. This pipe can withstand moderate physical handling before and after installation. Larger sizes are particularly resistant to abuse.
- Noise. Steel pipe is quieter than copper with the same amount of water flow. The thicker walls absorb more of the sound.
- Familiarity. Most installers are comfortable with this material. Smaller sizes (up to 2½ in.) are joined by threaded connections. Most technicians can reliably make these joints without special tools or skills. Note that larger sizes (over 3 in.) are usually joined by welding, but these sizes are not usually used in most commercial and residential systems.
- Available accessories. Many components are available to simplify steel pipe installations. Almost all valves, fittings, and components are available with threaded fittings compatible with steel pipe.
- Regulatory acceptance. Most codes accept black steel pipe for almost all applications. Some require it for natural gas piping.

Black steel pipe is moderately corrosion resistant. Very little internal corrosion takes place if proper water treatment and oxygen control is used in the system. The pipe is, however, susceptible to external corrosion. If constantly wetted from leaks, external sources, condensation, or burial, the pipe will fail. Black steel has little inherent corrosion resistance to prevent it from failing from outside moisture. Paint and protect them when they are installed in areas where they are likely to become wet.

Other disadvantages of using black steel pipe, at least compared with copper or PVC plastic, are

- Weight. Steel pipe requires more supports than other materials.
- Smaller flow area. Steel pipe has thicker walls than copper, giving it less internal area for water flow. This leads to higher pressure drops over the same distance.
- Rougher internal surface. The smoothness of the internal surface of the pipe affects the pressure drop of the water flowing inside. Steel pipe has a rough surface, causing a comparatively high pressure loss.

Galvanized pipe has many of the same properties as black steel pipe. The primary difference is the zinc coating applied to the inside and outside surfaces. This enhances corrosion resistance but, for HVAC systems, this provides few advantages. While the external corrosion resistance added by galvanizing might be helpful when installing pipes in wet areas, the internal coating gives no benefit. Proper water treatment effectively controls rust and corrosion.

Galvanized pipes cost more than black steel and are often used for domestic water systems. There, the internal coating helps prevent damage from the constant supply of fresh water and oxygen.

Some common properties of steel pipes are listed in Table 5.1. The table gives the properties for both schedule 40 and schedule 80 pipes. These designations refer to the weight of the pipe and the thickness of the walls.

Schedule 40 pipe is the "standard" for heating and cooling systems, with schedule 80 pipe used in very high-pressure or corrosive systems. Most schedule 80 pipes are installed in steam and condensate systems, not hydronic heating and cooling systems.

Copper. Copper pipes are often used for domestic heating systems. The material is corrosion resistant, smooth, and easily handled. It is usually joined by soldering slip joints and couplings together. Many fittings and accessories are available for copper pipe.

Technically what is commonly thought of as copper pipe is actually rigid tubing. It is available in two common weights, designated as K and L. Type K pipe has the greater wall thickness, with type L pipe being thinner.

Note that a type M copper pipe is also available. Most building codes do not allow the use of type M pipe. Its extremely thin walls are easily damaged. These thin walls also make type M pipe noisier than types K or L. Using type M copper pipe might invite customer complaints because of a "too noisy" system.

TABLE 5.1 Select Properties of Steel Pipe

Nominal pipe size (in.)	Schedule	Outside diameter (in.)	Inside diameter (in.)	Wall thickness (in.)	Pipe weight (lb/ft)	Weight of water (lb/ft)	Weight of full pipe (lb/ft)	Volume of water (gal/ft)	Support spacing (ft)
½	40	0.840	0.622	0.109	0.851	0.132	0.983	0.0159	5
	80	0.840	0.546	0.147	1.088	0.101	1.189	0.0122	5
¾	40	1.050	0.824	0.113	1.131	0.230	1.361	0.0277	6
	80	1.050	0.742	0.154	1.474	0.188	1.662	0.0226	6
1	40	1.315	1.049	0.133	1.679	0.374	2.053	0.0451	7
	80	1.315	0.957	0.179	2.172	0.311	2.483	0.0375	7
1¼	40	1.660	1.380	0.140	2.273	0.648	2.921	0.0781	7
	80	1.660	1.278	0.191	2.997	0.555	3.552	0.0669	7
1½	40	1.900	1.610	0.145	2.718	0.882	3.600	0.1063	9
	80	1.900	1.500	0.200	3.631	0.765	4.396	0.0922	9
2	40	2.375	2.067	0.154	3.653	1.455	5.108	0.1753	10
	80	2.375	1.939	0.218	5.022	1.280	6.302	0.1542	10
2½	40	2.875	2.469	0.203	5.793	2.076	7.869	0.2501	11
	80	2.875	2.323	0.276	7.661	1.837	9.498	0.2213	11
3	40	3.500	3.068	0.216	7.580	3.200	10.780	0.3855	12
	80	3.500	2.900	0.300	10.250	2.864	13.114	0.3451	12

Table 5.2 shows some common properties of copper pipe. Note that the inside diameters are larger than those for steel pipes, allowing them to hold and carry more water. This, and its smooth internal surface, give it lower flow resistance than steel pipe.

Copper pipe is fairly expensive, but many customers (particularly residential clients) prefer its use to steel piping because of its corrosion resistance. While internal corrosion is rarely a problem with steel piping in small, closed hydronic heating systems, many customers do not want steel pipe. Their concerns about it eventually leaking, or internally closing from corrosion, are unfounded with most hydronic systems. However, many customers are willing to pay for copper piping.

When using copper pipe, some precautions to prevent galvanic corrosion should be considered. Where copper pipe attaches to steel or iron it establishes an electrically conductive circuit. The iron and copper generate an electric current when filled with water. The assembly functions like a battery, with electrons flowing from the iron, to the copper joint, and into the copper pipe. Once in the copper pipe, the electrons circulate through the water and return to the iron.

This circulating current causes severe corrosion of the iron. Metal is removed from the iron by the electric current, causing rapid thinning and deterioration. Prevent this by using *dielectric* (insulating) couplers and unions when joining copper to steel or iron. These couplers block the current flow through the joint and prevent the corrosion from starting.

This problem can also occur with copper pipe installed against steel supports or on improper (noncopper) hangers. Use either copper or brass materials to support the pipe, or electrically insulate the surface of the pipe from any ferrous supports.

Plastic. Plastic is sometimes the material of choice for heating and cooling systems. Polyvinyl chloride (PVC) and its high temperature rated derivative copolyvinyl chloride (CPVC) are easy to handle and install. Table 5.3 provides some properties of CPVC pipe, which is most often used in HVAC applications. Its rating for up to 180°F allows it to be used with water source heat pumps and low-temperature radiant hydronic systems.

Installers often prefer plastic piping systems, especially for smaller systems. The pipes are lightweight and the glued joints are simple to fabricate and install correctly. The pipes can be cut with a hacksaw to required lengths, and one or two people can quickly install a system. Their internal surfaces are very smooth, yielding low pressure losses. The material is noncorrosive, both inside and out. Finally, water flow in plastic pipe is usually very quiet compared to the same flow in metal pipe. PVC pipe's flexibility allows it to act as an effective sound damper.

TABLE 5.2 Select Properties of Copper Pipe

Nominal pipe size (in.)	Thickness	Outside diameter (in.)	Inside diameter (in.)	Wall thickness (in.)	Pipe (lb/ft)	Water (lb/ft)	Weight of full pipe (lb/ft)	Volume of water (gal/ft)	Support spacing (ft)
½	K	0.625	0.527	0.049	0.344	0.094	0.438	0.011	5
	L	0.625	0.545	0.040	0.285	0.101	0.385	0.012	5
¾	K	0.875	0.745	0.065	0.641	0.188	0.829	0.023	6
	L	0.875	0.785	0.045	0.455	0.209	0.664	0.025	6
1	K	1.125	0.995	0.065	0.839	0.335	1.174	0.040	6
	L	1.125	1.025	0.050	0.655	0.356	1.011	0.043	6
1¼	K	1.375	1.245	0.065	1.040	0.525	1.565	0.063	7
	L	1.375	1.265	0.045	0.884	0.542	1.426	0.065	7
1½	K	1.625	1.481	0.072	1.360	0.743	2.103	0.089	8
	L	1.625	1.505	0.060	1.140	0.767	1.907	0.092	8
2	K	2.125	1.959	0.083	2.060	1.300	3.360	0.157	9
	L	2.125	1.985	0.070	1.750	1.334	3.084	0.161	9
2½	K	2.625	2.435	0.095	2.930	2.008	4.938	0.242	10
	L	2.625	2.465	0.080	2.480	2.058	4.538	0.248	10
3	K	3.125	2.907	0.109	4..00	2.862	6.862	0.345	10
	L	3.125	2.945	0.090	3.330	2.937	6.267	0.354	10

TABLE 5.3 Select Properties of CPVC Pipe

Nominal pipe size (in.)	Outside diameter (in.)	Inside diameter (in.)	Wall thickness (in.)	Pipe weight (lb/ft)	Weight of water (lb/ft)	Weight of full pipe (lb/ft)	Volume of water (gal/ft)	Support spacing (ft)
½	0.625	0.485	0.070	0.085	0.079	0.164	0.0096	3
¾	0.875	0.713	0.081	0.140	0.172	0.312	0.0207	3
1	1.125	0.921	0.102	0.218	0.288	0.506	0.0346	3
1¼	1.375	1.125	0.125	1.330	0.429	0.759	0.0516	4
1½	1.625	1.329	0.148	1.460	0.599	1.059	0.0721	4
2	2.125	1.739	0.193	0.790	0.025	1.815	0.1230	4

But plastic pipe also has limitations. Keep the following considerations in mind when planning a project or repair with PVC or CPVC pipe:

- Temperature rating. Use the proper material. CPVC pipe is rated for higher temperatures. Never use ordinary PVC pipe for any heated water service.
- Durability. Plastic pipe can be fragile. Rough handling during installation and later impact can cause it to fail. Pipe supports can cause damage if they cut into the plastic. Use enough hangers and cushion them if they might scuff the pipe.
- Thermal expansion. Plastic pipe expands much more than metal pipe as it heats. The lower strength of the plastic makes it tend to bend, sag, and distort as it heats and expands. It is critical to use proper expansion compensation to prevent damage to the piping system.
- Cost. Plastic pipe is more expensive to purchase than black steel. Its higher purchase price can be offset, however, by the ease of installation in smaller piping systems.

Pipe installation considerations

Most piping system installations are straightforward. Lay out the system routing and calculate the pipe sizes. Sizing is done to maintain reasonable pressure drops and internal water velocities. (See the section "Flow rates and economical sizing" later in this chapter.)

Properly support pipes so they will deliver dependable service. Many installations do not allow for proper control of thermal expansion and vibration. These conditions can stress piping long after it is installed, leading to system leaks.

The simplest pipe supports are saddle clamps that hold the pipe against a flat surface. They do a good job and are easy to use. Saddle clamps are usually attached with nails or screws. Use clevis hangers, hoop clamps, or ring supports when supporting pipes below an overhead anchor. There are even plastic clip-on pipe supports for pipes up to 1 in. in size. These use a small clip to hold the pipe in place. It is secured by pressing it into a tang on the support.

Don't rigidly support hydronic piping in place. Allow it to slip so that it can expand and contract as needed. Let pipe hangers swing to give a degree of movement to the system. At the same time, pipes should be supported well enough to prevent excess motion. Water flow transmits force to the pipe as it moves around turns and past restrictions. If an automatic valve opens and closes, the water flow starts and stops quickly. In extreme cases, water stopped very quickly can cause *water hammer*, a severe pressure spike driven through the system at the speed of sound. These forces cause pipe to shift and sway. The supports should be rigid enough to prevent gross movement, but should also prevent excessive stress.

When supporting insulated pipe, the hangers or clamps should not be a heat leakage path. Don't attach the hanger directly to the pipe. Leave room for insulation between the hanger and the pipe. The goal is to have the insulation continuous at all pipe supports. Use oversize saddles, hangers, or clamps for insulated pipe. If the insulation might be crushed by the weight of the pipe, use a sheet metal shield between the support and the insulation to protect it. This also provides enough flexibility in the installation to allow for thermal expansion.

Hydronic piping, particularly heating piping, changes dimensions significantly as it heats and cools. The thermal expansion of common piping materials is given by its expansion coefficient. These coefficients are

Steel	6.5 microinches/°F/in.
Copper	9.5 microinches/°F/in.
CPVC	34.0 microinches/°F/in.

These coefficients allow us to calculate the amount a pipe expands as it heats. For example, find the expansion of 50 ft of copper pipe as it heats from 70°F to 180°F.

First, figure out the length in inches:

$$50 \text{ ft} \times 12 \text{ in./ft} = 600 \text{ in.}$$

The temperature changes by

$$180°\text{F} - 70°\text{F} = 110°\text{F}$$

Finally, the expansion is

$$\frac{9.5 \text{ microinches} \times 600 \text{ in.} \times 110°\text{F}}{1{,}000{,}000 \text{ microinches/in.}} = 0.627 \text{ in.}$$

This pipe will grow approximately ⅝ in. from the heating. Allow for this expansion to prevent the pipe from being excessively stressed. If the pipe is rigidly connected or supported at one end, all of the length expansion accumulates and takes place at the other end. If both ends are rigidly supported, the pipe is forced to adapt to its expansion internally. It either buckles to absorb the length increase or compresses to compensate for its unchanged length.

Most thermal expansion can be accommodated by a combination of slipping in the supports and directional changes in the layout. The pipe bends to the side as it expands and contracts, changing its shape for the least total stress.

Support pipes at regular intervals to prevent them from sagging and stressing the pipe's joints. Tables 5.1, 5.2, and 5.3 show suggested support distances for pipes of various sizes. These distances are the minimums needed to properly support the pipe.

Heat exchangers

Heat exchangers are used to transfer heat energy into or out of a flow of water. They include the heat transfer surface in a boiler, air coils, plate and frame heat exchangers, and shell and tube units. The boiler's heat exchanger is an integral part of the total unit, and the manufacturer designs it for that exact application. The other types of units require some description.

Air coils were described in Chapter 4. They heat or cool a flow of air delivered to a space. As noted, specific coil types are designed for various applications.

Plate and frame, and shell and tube, heat exchangers transfer heat from one fluid flow to another. They are used to physically isolate the flows from each other while allowing most of the heat to move from the hotter fluid to the cooler one. Isolation might be required for various reasons: to prevent contamination of one fluid with the other; to keep dirt out of sensitive systems; or to prevent pressure differences from driving one fluid into the other. Heat exchangers allow two separate systems to share energy without any material transfer.

Plate and frame exchangers (Figure 5.1) are very efficient. As their name implies, they consist of sets of parallel metal plates (usually steel, but copper is the most efficient) held within a frame. The total assembly resembles a stack of rectangular plates spaced apart with bolts installed along the edges. Water flows through the spaces between the plates and the flows are isolated from each other by the plates. Flows alternate, with one system's water flowing between the first and second, third and fourth, and fifth and

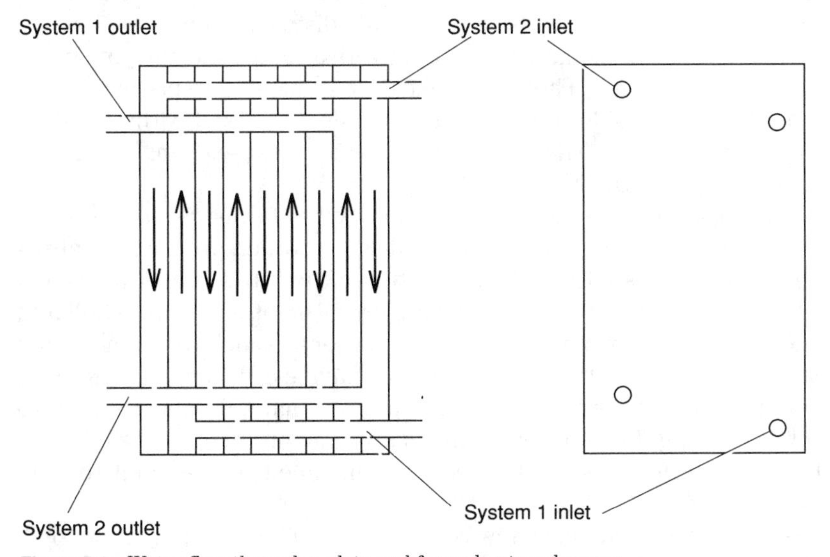

Figure 5.1. Water flow through a plate and frame heat exchanger.

sixth plates. The other system's water flows between the second and third, fourth and fifth, and sixth and seventh plates.

Water from each system flows in opposite directions. If one system's flow is from top to bottom, the other is from bottom to top. This counterflow arrangement maximizes the unit's efficiency by exposing the water leaving one side of the plates to the incoming water entering the other side.

Plate and frame exchangers are also very efficient because of their large surface to volume ratio. The large, flat plates provide a lot of heat exchange surface area in a compact package. Also, if the system needs more heat exchange surface area, it is usually easy to add more plates to the stack and extend the manifolds.

The large flow areas can easily handle fouled water with little chance of clogging. Most foreign material flows right through the exchanger with little chance of becoming stuck.

All of the gaskets and joints in plate and frame heat exchangers increase the chance of leaks. If all of the peripheral bolts are not evenly tightened, or if the manifold tubes are not properly installed, a leak is almost certain. It is common to install these exchangers over a drain to dispose of the water that leaks from these units.

Disassembling, cleaning, and reassembling a plate and frame heat exchanger is not difficult. Properly bolting the plates and manifold tubes back together is the main concern. Follow the manufacturer's tightening sequence and use a torque wrench to prevent warping the plate stack. This minimizes the chances of causing a leak. If the sequence and tightening specifications are not known, contact the manufacturer for directions. It is easy to overtighten the bolts and permanently distort the unit.

Shell and tube heat exchangers are built with a set of fluid carrying tubes suspended within a shell. The shell is like a large-diameter pipe that carries water around the outside of the tubes. Its ends are closed with bolted-on covers or cast closures. The tubes' walls are the heat exchange surfaces. Figure 5.2 illustrates their construction.

Shell and tube exchangers are built in several forms. Some have the tubes running from one end of the shell to the other, with external fittings on both of the shell's ends connected to the tubes. Water flows in a single pass, from one end of the exchanger to the other. Other shell and tube exchangers use two pass tubes. They enter the shell from a manifold at one end, run to the other end where they make a U-turn and run back to the entry end. There they pass into another manifold and connect to the outside piping. If the internal tubes are long or narrow, one or more *tube sheets* (perforated steel plates) are installed in the shell to help support them.

These heat exchangers have few gaskets or bolted connections. The shell ends bolt to the body, providing little likelihood of leaks. The gaskets are

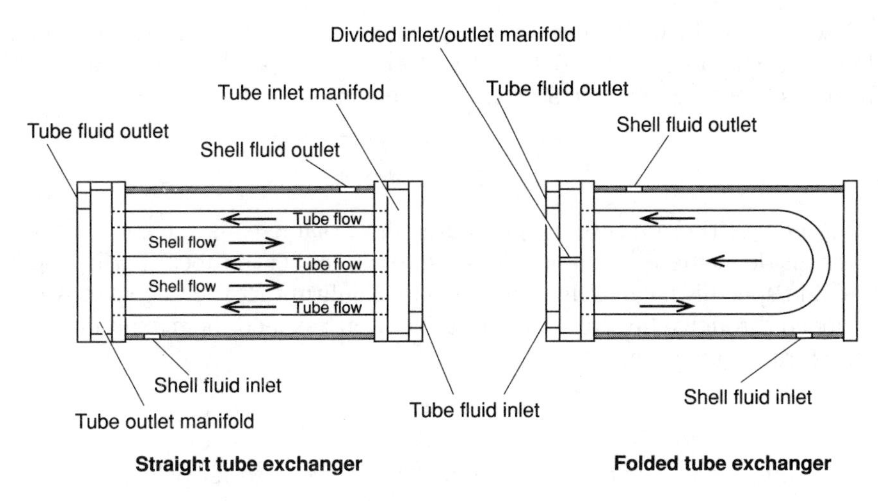

Figure 5.2. Shell and tube heat exchanger flows.

similar to those used with large diameter, flanged pipe. They can easily withstand high pressures and temperatures.

Maintenance of these heat exchangers is simple. They are easily disassembled at the shell heads and the tubes can be withdrawn. The tubes can be blown out with a high-pressure washer, and straight (single pass) tubes can be cleaned out with a rod or brush. Holes in the tubes are easily soldered or brazed.

When installing shell and tube heat exchangers, consider future maintenance needs. Provide enough clear space in the room beyond the end of the exchanger to allow the tubes to be withdrawn. Leave a clear space equal to the length of the shell for future service.

Shell and tube heat exchangers do not have surface to volume ratios as high as plate and frame units. Therefore, a physically larger heat exchanger is required to provide an equivalent heat transfer. Efficiency is enhanced if the exchanger uses many long, thin-walled tubes. Many smaller diameter tubes provide more heat transfer than a few larger ones.

Using small tubes increases the potential for clogging. If using a shell and tube heat exchanger with water that might be contaminated, keep the clean water inside the tubes and the fouled water in the shell. The shell has greater clearances that can pass larger contaminants.

Water velocity should be considered when selecting a heat exchanger. Water flowing too fast through an exchanger can cause erosion, leading to leaks. Most manufacturers list maximum water speeds for their exchangers, but some generalizations can be made.

Steel can handle water flow up to approximately 10 fps, while copper and aluminum should be limited to 5 fps. The water speed through an exchanger's tubes can be estimated with the equation

$$\text{Speed} = \frac{0.408 \times \text{gpm}}{(\text{Diameter} \times N)}$$

where gpm is the total exchanger water flow, diameter is the exchanger tube's inside diameter in inches, N is the total number of tubes sharing the flow, and speed is given in feet per second. Confirm that the expected water speed through a heat exchanger's tubes won't exceed the maximum velocity for its material.

Valves

Valves control the flow of water. Manual valves are used for balancing and shutting off flow and automatic valves divert water to where it is needed with electric or pneumatic operators. Manual valves are classified by their internal construction and how they operate.

The most common manual water valves are gate valves (Figure 5.3). As their name implies, an internal flat plate slides into the valve as it is closed. This shuts off the water flow when closed and has a low pressure drop when wide open. Gate valves are excellent for turning flow fully on and off, but are not suitable for fine flow control. The sliding gate does not give smooth flow control when it is partially opened. Flow varies rapidly with the first portion of the gate's movement and less when the gate is over half open. This can lead to difficulty in precisely setting the flow.

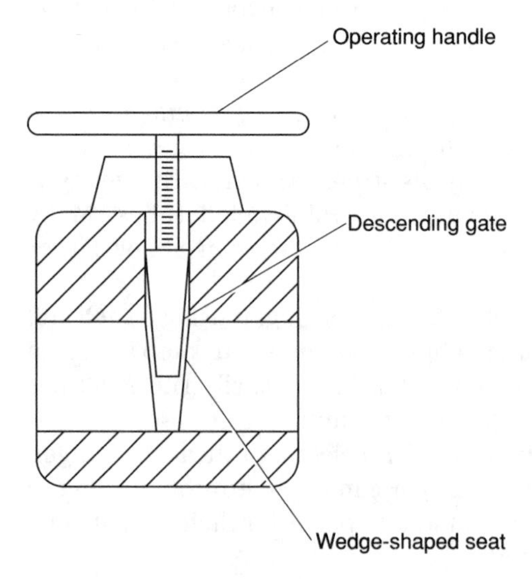

Figure 5.3. Gate valve.

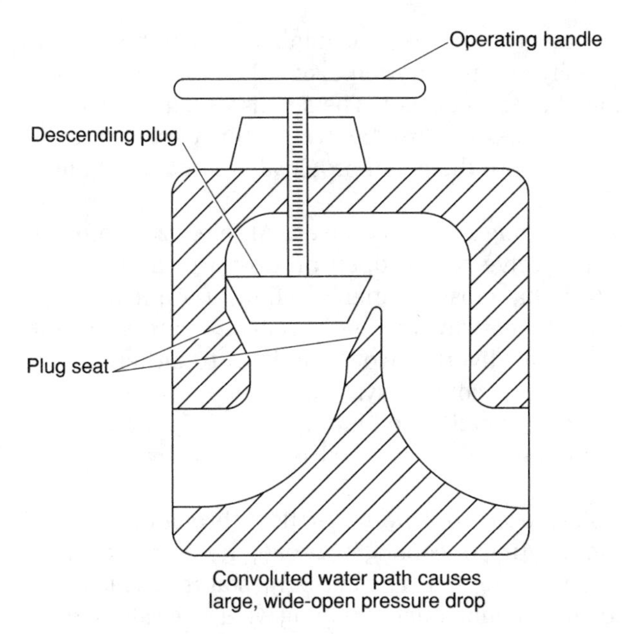

Figure 5.4. Globe valve.

Valves suitable for precise flow adjustment are globe valves. As the handle is turned a stopper lifts out of a seat (as shown in Figure 5.4), allowing more water to flow. Common lavatory valves are globe valves. Globe valves are more restrictive to flow than gate valves when wide open.

Other manual valves sometimes used in HVAC systems are plug, ball, and butterfly valves. These provide the least flow restriction when wide open; almost as low as straight pipe. They cost more than the other valves noted previously, but they have special characteristics that make them suitable for specific uses.

Plug and ball valves have a machined (usually tapered) cylinder or sphere with a hole drilled through its diameter. The cylinder or ball is installed in the housing so that, as it rotates, its hole aligns with holes in the housing (see Figure 5.5). These valves open and close with only a quarter turn and provide an extremely tight closure. These valves need no resilient seats, and will close virtually perfectly over years of service.

Plug valves are sometimes used for system balancing. Their housing has a scale and the plug has a pointer. When turned, the pointer gives a repeatable setting on the scale, showing how far the valve is opened. Finally, a locknut is tightened to prevent the plug from accidentally turning after it is set. The settings are repeatable and are stable over long periods.

Butterfly valves use a thin, flat disk installed in a cylindrical housing. The disk turns to face either at a right angle to the flow (shutting it off) or parallel to it. The disk has a tapered profile that minimizes flow restriction when open. Water flows easily around the thin disk. When closed, the disk seats against a ring installed in the housing. The seat is usually a resilient material or brass, allowing the disk to close tightly. Butterfly valves are often used on larger pipe sizes where their thin profile allows them to fit into narrow spaces.

Most automatic valves are plug or globe valves. An electric motor or air-driven piston moves the valve to the open or closed position. Most systems use two-way valves that close or open the flow of water through a pipe. Other systems use three-way valves. Three-way valves do not close the flow of water off, but divert it from one inlet to either of two outlets. These are often used in control systems where a temperature control diverts water flowing through a coil into a bypass pipe. Total water flow is unchanged, but the flow in the coil is reduced by diverting water into the bypass.

Automatic valves are often used to zone water flow through a system. When a thermostat calls for heating or cooling in a zone, the valve is opened and water flows to the coil or fin tube. After the thermostat is satisfied, the valve closes. Because the pump might still be running when the valve closes (and water still flowing), the valve must be designed with special components to prevent *water hammer*.

Water hammer is caused by water flowing in a pipe. The moving water has momentum and resists any attempt to slow or stop it. If the water is suddenly forced to stop flowing, its motion causes it to momentarily "pile

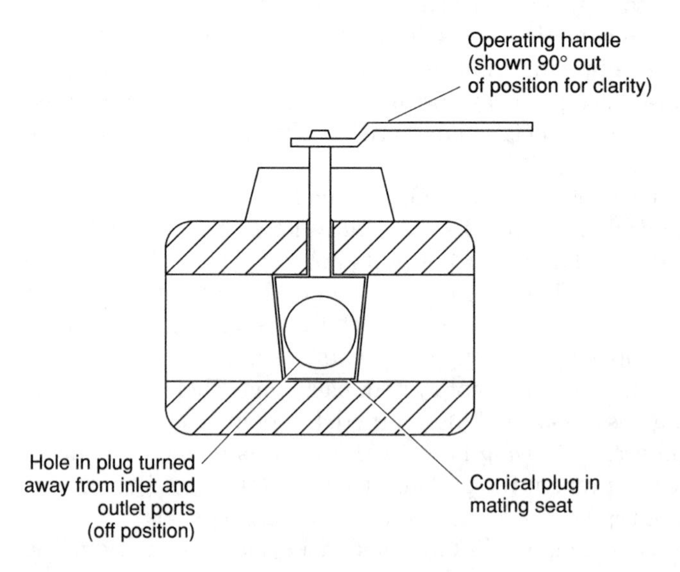

Figure 5.5. Plug valve. Ball valves are similar, with spherical plugs.

up" at the point where it is stopped. It wants to continue flowing but cannot. A large pressure pulse develops that courses through the entire pipe system from the stopping point back to the source. At the very least, water hammer causes annoying noise. At worst, it can split pipes or damage equipment, hangers, or supports.

Automatic valves used in HVAC systems must be designed to prevent water hammer. The valve's plug closes slowly, allowing the water flow to diminish gradually. This dissipates the flow's momentum and prevents a significant pressure pulse. The water hammer handling capability of some valves depends on the direction of water flow through the valve. Install the valve with the flow in the direction the manufacturer intended to help prevent excessively fast closure. Many valves have an arrow stamped into the housing showing the intended flow direction.

Special valves used in HVAC systems include check and automatic flow control valves. Check valves allow flow in one direction only. Most check valves use either a hinged metal gate or a floating plug. Install hinged gate valves with the hinge horizontal and at the top of the valve. Gravity helps close the gate when there is little back pressure. If gravity holds the gate open (which can happen if the valve is installed upside down), significant reverse flow can occur.

Install floating plug check valves vertically, usually with the intended flow from bottom to top. Gravity helps close these valves, too. Both these valves must be installed correctly or they won't function properly.

Check valves are not immune to reverse flow leaks. Most allow some water to backflow under even moderate pressure. Don't depend on them to hold back water in critical situations where safety is a necessity.

Use automatic flow control valves for "install and forget" system balancing. They have an internal piston and spring to maintain a preset flow over a wide pressure range. They hold their flow constant at the design point at any pressure within their rated range. Obtainable in a variety of sizes to handle many flow rates, they change their pressure loss as needed to maintain their constant design flow. If the pressure from the pump is less than expected, they open farther to keep a constant flow.

These valves are available with built-in strainers and pressure taps to keep them clean and confirm their operation. Consider using these valves when installing hydronic systems. They make system balancing automatic, preventing customer complaints because of imbalance. The water in systems with automatic flow control valves must be kept clean. If dirt or sediment builds up on the valve's seat the internal piston can jam. Use internal or external strainers in the system, or confine the use of these valves to closed recirculating systems.

Note that automatic flow control valves can greatly ease system design. They restrict the pressure delivered to equipment installed downstream to maintain the design flow. The only location that is not appropriate for these valves is in series with two-way automatic control valves. If a control valve attempts to reduce flow by increasing its restriction, the automatic flow control valve will attempt to compensate. It will open farther as the control valve closes and very little flow reduction will occur. The control valve will eventually close far enough to reduce the flow, but by then it will have used much of its control range.

The effect is the same as having a control valve mismatched (oversized) for the system: poor control and wide swings in flow because of its excessive sensitivity. Always use a manual flow control valve in circuits with two-way automatic control valves.

Control valve selection. Control valves must be selected to match the flow expected in the piping system. To properly control water flow over the valve's entire range, it must impose significant flow restriction when wide open. An incorrectly sized valve, particularly one too large for the application, operates unsatisfactorily.

Oversized valves, with low wide-open pressure drops, don't modulate flow over their entire operating range. When the valve moves from wide open toward the closed position, very little pressure drop occurs over most of its operating range. It must close to approximately 25% of the wide-open point before the pressure drop becomes significant. The pressure drop at that point finally begins to reduce the water flow. This means that the valve has used up three-quarters of its operating range and movement with no effect on the water flow. The remaining operating range must control the flow from full to zero. Instability and oversensitive operation occurs in this condition. Very precise adjustments must be made to get proper flow, and a small setting error can cause large flow problems.

At least 25% to 50% of the total circuit's pressure drop should take place at the wide-open valve. As the valve closes, its pressure drop becomes even higher. This allows the entire range of the valve's operation to control the flow. As the valve moves from wide open to 50% closed, the flow rate drops significantly (perfectly matched valves allow 50% flow when 50% closed). The control works smoothly, reliably, and repeatably.

Select a valve's size based on a coefficient called its C_v, which can be found with the following equation:

$$C_v = \frac{\text{gpm}}{\sqrt{\text{Pressure}}}$$

where gpm is the full flow in gallons per minute through the system and pressure is the system's full flow pressure drop measured in psi. If given the pressure drop in feet of water (head), the equation is

$$C_v = 1.52 \times \left(\frac{\text{gpm}}{\sqrt{\text{Pressure}}} \right)$$

Choosing a valve with a C_v equal to this calculation provides a pressure drop through the valve equal to the system's pressure drop. This is a good target. Valves are available with various C_v values, so select a valve that is as close to the target value as possible.

For example, find the C_v for a system that has a flow of 45 gpm and a pressure loss of 25 ft water.

$$C_v = 1.52 \times \left(\frac{45 \text{ gpm}}{\sqrt{25}} \right) = 13.7$$

Choose a valve with a C_v between 13 and 20 to achieve the proper control. Note that the pump must handle a larger pressure drop with the control valve in the system.

Avoid oversizing a control valve when the flow or pressure drop isn't known by following a simple rule of thumb. Use a valve with connections at least one or two pipe sizes smaller than the system's piping. For example, if a system uses 1 in. pipe, the control valves should have fittings no larger than ¾ in. A ½ in. valve might be an even better choice. Use reducer bushings or couplers to match the valve to the pipes.

Note that valves intended for system shutoff or isolation do not need to be sized according to system flow. Because they are used to turn the flow on or off, select a valve that matches the adjacent pipe's size.

Pumps

Pumps are similar to fans. Both convert motor energy to the increased energy in a moving stream of fluid. The internal construction of some pumps is similar to some fans. The most common pump used for HVAC systems—the centrifugal pump—is analogous to a centrifugal fan (see Figure 5.6).

Centrifugal pumps use a rotating impeller to increase the speed and pressure of a flow of water. The impeller spins inside the housing, shaped to confine the water flow so it can only move from the impeller's inlet fitting to the outlet. Water enters the pump at its center, along the axis of the impeller's rotation. Centrifugal pump impellers change the water's direction, spinning it out to their outside diameter. Water collects in the pump housing where it goes to a common outlet connection.

Note the key similarity to a centrifugal fan: both use impellers that change the fluid's flow direction. The impeller designs are different, however. Most pump impellers used for HVAC systems are not built like a squirrel cage, with slots for the water to flow through. Most pump impellers are built in the form of a round, flat plate with a set of raised ridges cast onto one or both faces. If one face is flat, it is installed facing the motor and does not handle water. Figure 5.7 illustrates a single-face pump impeller's profile.

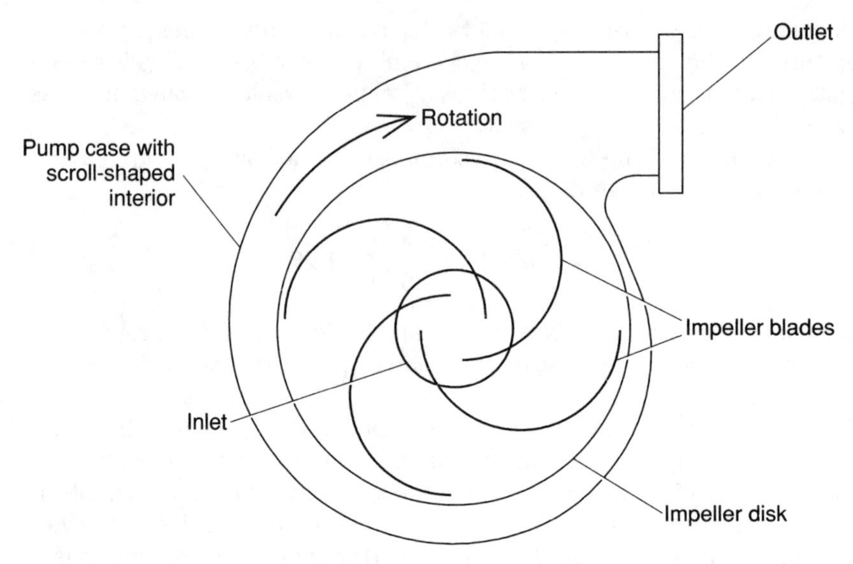

Figure 5.6. Centrifugal pump impeller and case arrangement. Impeller diameter can be varied within the case to limit flow and pressure capabilities. This also reduces horsepower requirements.

The ridges run from the approximate center of the face out to its outside diameter. They are tallest close to the center of the face, tapering to a shorter height at the impeller's outside diameter. The ridges usually do not run straight from the disk's center to the outside edge. Most have a "swirl" shape running opposite to the impeller's rotation. Like a backward inclined

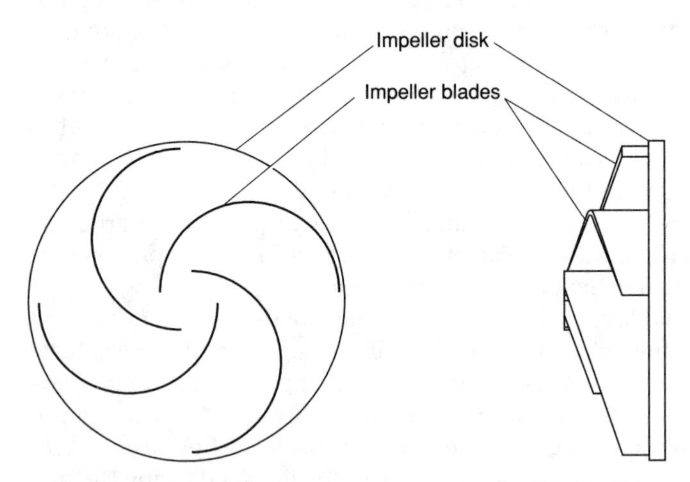

Figure 5.7. Pump impeller shape. Note the long, tapered blades. These allow the pump to maintain a constant internal water volume from the inlet to the outlet.

centrifugal fan, the pump impeller increases the water's pressure. The backward tilt to the ridges reduces the water's outlet velocity and increases its static pressure. This puts more of the motor's energy into the water with higher efficiency. Energy is added to the water as additional flow and pressure.

Typical pump efficiencies range from 40% to 70% for most applications. The extra energy that does not contribute to increased water flow or pressure is added to the water as heat. Like fans, some water heating occurs in the fluid as it runs through the pump. It is common for a pump to add some heat to water, but usually it is not as significant as fan heating with air. The amount of heating caused by a pump can be found from the equation

$$\text{Temperature rise} = \text{Head} \times \frac{(1 - \text{Efficiency})}{(778 \times \text{Efficiency})}$$

where temperature rise is in degrees F, head is pump head in feet of water, and efficiency is the pump's efficiency.

The equation shows that pump heating is very sensitive to efficiency, and not at all dependent (directly) on flow rate or horsepower. The more efficient the pump, the less water heating will occur.

This equation only gives the amount of heating that occurs in the pump while it is running. In fact, however, all of the energy supplied by the pump motor eventually adds heat to the water. Energy added to the water as pressure changes to heat when its flow encounters friction in the system. This can have implications when pumping chilled water. Chilled water systems should use high-efficiency pumps. This allows you to use a smaller horsepower motor and minimize pump-induced heating.

Centrifugal pumps are available in a wide range of forms and flow capacities. Units range from small circulators of less than 15 gpm to large engine-driven units of several thousand gallons per minute. Most HVAC pumps require flow rates in the lower ranges, usually less than 1000 gpm.

Pumps in these ranges are available as pipe supported in-line units, base mounted, and vertical types. In-line pumps (used for lower flows) have a special inlet connection to avoid changing the flow direction. Water flows through the inlet connection, goes past the housing to the pump's inlet, and is discharged. The pump fits into a straight length of pipe with no flow direction change. The piping carries the weight of the pump and might require some additional support.

Motors are close coupled to in-line pumps. They can be disassembled and replaced, but there is no special alignment of the motor and pump shafts when they are put back together. The housing guides and aligns the parts adequately. Internal flexible couplers between the motor and pump shaft reduce vibration and noise transmitted into the piping.

Base-mounted pumps usually handle higher flow rates. The pump and motor are installed separately on a single steel base. The coupling is open and, sometimes, requires alignment when replacing the motor or pump. These pumps do not have special, direction-changing inlets. Water enters along the pump's axis and discharges out of the side.

Vertical pumps work with the motor standing upright away from the pump, with the shaft pointing straight down. The pump itself is at the bottom with the impeller rotating in the horizontal plane. They usually draw water up from a sump and discharge it under moderate pressure. Some are available with extended shafts, allowing the pump to be immersed in the sump while the motor is held out of the water (see Figure 5.8). Vertical pumps handle larger HVAC flow requirements. They often carry flows of over 1000 gpm.

The flow characteristics of a pump are determined by its operating speed, impeller size, and the shape and other attributes of its housing. The effect of pump speed should be considered. Pumps used in HVAC systems typically operate with either 1750- or 3450-rpm motors. Higher speed motors can deliver water at higher pressure than the slower speed units, but their flow curve has a more pronounced "peak." Restrictions in flow through a high-speed pump cause greater disturbances than they would in lower speed units. This should be considered when a pump must run under variable pressure or flow loads. Situations where the flow can vary widely (because of possible zone or control valve closures) are best handled with

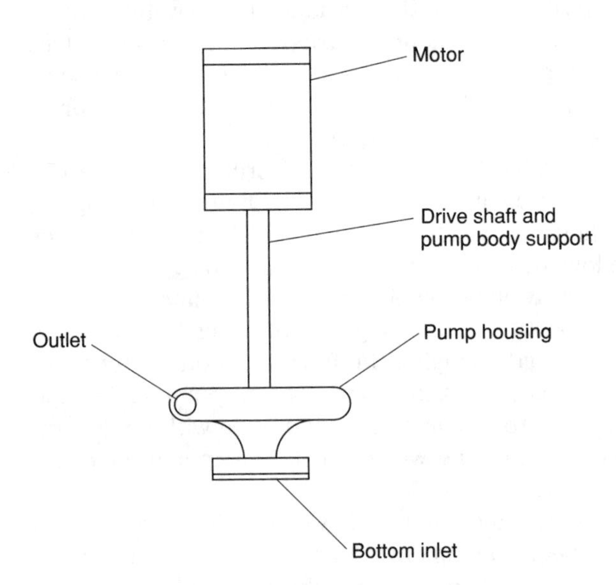

Figure 5.8. Vertical pump arrangement. The pump is often supported by the shaft housing extending down from the motor.

low-speed pumps. Their flat pressure and volume curves allow them to compensate for restrictions better than high-speed pumps.

Install piping upstream of the pump's inlet with care. Turbulence in the inlet piping limits the pump's capacity and can lead to damage. Turbulence can cause flow into the inlet to become unstable. On a microscopic scale, it might cause the pressure at the impeller's inlet to fall below the water's vapor pressure. This can cause serious problems. Low-pressure water boils into vapor and forms bubbles of steam. This is *pump cavitation*. These bubbles cause the impeller to wear and erode prematurely. Avoid excessive inlet turbulence with attention to installation details.

An elbow or other fitting immediately before a pump inlet imparts swirl to the incoming water flow. This leads to the kind of turbulence that causes pump damage. If possible, use a minimum of 3 ft of straight pipe before a pump's inlet connection. Avoid tees, elbows, and valves within a few feet of the pump's inlet.

If it is impossible to use straight pipe immediately before the pump inlet, use a *diffuser* elbow. These have internal vanes that prevent the flowing water from spinning inside. The water leaves these elbows with very little swirl; they can be connected directly to the pump's inlet. Some pump diffusers include integral strainers to reduce the complexity of the system.

Pump housings must not be stressed by the piping. Pipes and the pump must be supported to prevent the pump's case from warping because of the pipe weight or misalignment. A pump's cast iron or bronze housing cannot withstand excessive stress caused by poor installation, and might crack or warp enough to cause leaks. Sometimes the impeller bearings bind, leading to excessive internal friction.

Water must be able to flow through the pump. If the flow is excessively restricted, the pump only heats the water. Much of the energy consumed by the pump motor goes into the water. The water can heat to over 212°F when a pump is running *dead headed*. The water might not boil immediately, however, if its pressure is high enough. The heated water explosively flashes into steam when the pressure is released. This can completely cut off flow and cause severe water hammer that can damage the system.

Don't install a pump that is significantly oversized for the system's requirements or use controls that can shut off water flow through a running pump. Instead, use a pipe system that allows the water to bypass the pump.

Centrifugal pumps are manufactured using a variety of materials. Most are made of iron, steel, or bronze. Some pumps use a combination of materials for various internal parts to limit wear or reduce costs. Use all-bronze pumps when the water is potable (drinkable). Do not use steel or iron pumps where the water might be consumed by people. They are entirely suitable, however, for HVAC applications. With proper corrosion control there are no special problems with iron pumps.

Pumps are chosen for the proper flow rate and pressure increase, just as fans are. Associated with centrifugal pumps are manufacturers' *flow curves*. These illustrate how the pump works over its operating range. Most pumps are available with a variety of impeller diameters and with at least two motor speeds. This allows a single pump to provide a variety of capacities. Increasing the diameter of the impeller (while keeping the housing the same) provides higher flow capacity. Similarly, increasing the pump's speed improves its pressure capability. Figure 5.9 shows a typical pump curve.

Pump flow curves also show the unit's varying efficiencies when operating under different conditions. When selecting a pump for a specific use, it is likely that the manufacturer has more than one workable unit. Select the pump that gives the highest efficiency for the application.

The operating horsepower of a pump can be estimated if its flow and working head are known. The formula for centrifugal pump horsepower is

$$\text{hp} = \frac{(\text{gpm} \times \text{Head} \times \text{Specific gravity})}{(3960 \times \text{Efficiency})}$$

where the pump's head is expressed in feet of water, and the efficiency is a fraction between zero and one. Water's specific gravity is 1.0.

Various concentrations of ethylene glycol, used for freeze protection, give the following specific gravity values:

30%	41.03
40%	1.05
50%	1.06

Pump efficiency must be assumed during the design process. Typical installed values range from 0.5 to 0.75. A value of 0.6 is generally safe for estimating horsepower.

An important consideration when selecting a pump for a given application is that it must have adequate *net positive suction head* (NPSH). This is the absolute pressure (measured as feet of water) available at the pump's inlet. Centrifugal pumps are not designed to operate with the inlet at any more than a slight vacuum. Sufficient pressure is required to ensure that the water won't vaporize as it passes through the pump. Net positive suction head is a measure of the available pressure. If it is adequate, the pump starts and runs with no problems. The NPSH at a pump inlet can be found with the equation

$$\text{NPSH} = 33.8 + \text{Static head} - \text{Friction head} - \text{Vapor pressure}$$

Static head is the elevation difference (in feet) between the pump inlet and an open point in the system. If the entire system is closed, this term becomes

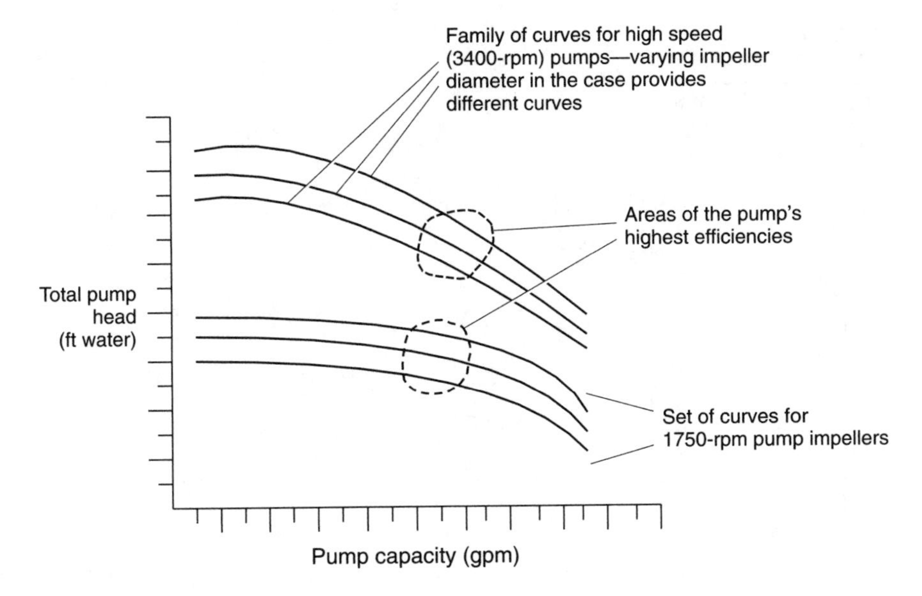

Figure 5.9. Typical pump flow curves. Efficiencies drop on either side of the curves away from the zones indicated.

zero. Systems with the pump inlet below the open water level have a positive static head. Use the difference in their heights in feet. Systems with the pump located above the water level have a negative static head. Again, use the distance the pump is above the water's level, but subtract it from 33.8.

Always subtract friction head. It is the pipe friction in feet of water loss from the open water area (or a closed system's expansion tank) to the pump inlet. Don't include friction on the pump's outlet to the open level or from the outlet to the expansion tank.

Finally, the water's vapor pressure varies with temperature. Table 5.4 gives the vapor pressure of water at various temperatures.

Confirm that the pump being chosen for the application works with the available NPSH. Pump manufacturers list the required NPSH for most pump lines. Meeting or exceeding the requirements helps prevent cavitation and starting problems.

Strainers and filters

It is important to keep the water in hydronic systems clean. Water that recirculates without significant exposure to air usually remains clean. Water that is exposed to the atmosphere, however, picks up a significant amount of dirt. This must be removed. Dirt can foul the heat transfer surfaces of boilers, chillers, and coils. It can clog tubing and reduce the amount of flow through systems. Grit can erode system parts as it flows inside and abrades metal. Finally, dirt can jam automatic valves and controls, preventing them from opening and closing completely.

TABLE 5.4
Vapor Pressure of Water

Temperature (°F)	Pressure (ft water)
40	0.2811
50	0.4112
60	0.5919
70	0.8388
80	1.171
90	1.614
100	2.194
110	2.947
120	3.913
130	5.139
140	6.679
150	8.596
160	10.96
170	13.85
180	17.36
190	21.59
200	26.64

Most water strainers use a fine cylindrical metal screen to filter the water as it flows through (see Figure 5.10). Designed like a Y, the screen filter is installed inside one arm. Water flows into another branch and is guided into the arm containing the screen. There it flows through the screen from its outside to inside. The water then flows to the remaining branch and out the strainer's outlet. Hopefully, the dirt remains on the outside surface of the screen.

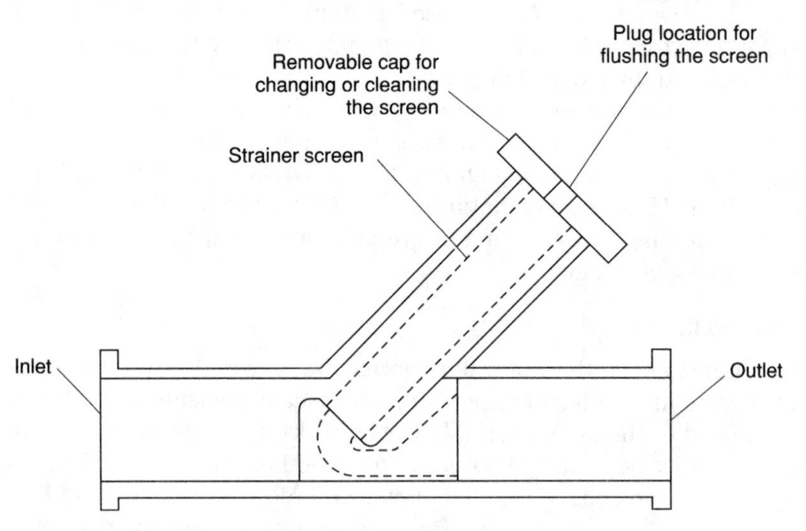

Figure 5.10. Water flow through a Y-type strainer.

A small outlet is installed on the tip of the branch containing the screen. Used for flushing the screen, most installations have a gate valve installed on the flush outlet to allow regular screen cleaning. Opening the valve allows water to wash in reverse through the screen, removing the dirt and washing it out the outlet.

These strainers do a good job, and a variety of screens are available to filter the water as finely as needed. As they accumulate dirt the pressure drop increases. Clean them regularly to prevent the system's flow from deteriorating.

Centrifugal strainers work differently. Water swirls inside a large housing. Dirt suspended in the water drifts to the center of the swirl where it concentrates. The strainer allows the water in its center to move too slowly for the dirt to remain suspended. It settles to the bottom and accumulates in a collection chamber. Dirt can be removed by opening a flush valve on the outlet.

Centrifugal strainers must be matched to the system. Each size properly operates only over a specific flow range. Flows either above or below the manufacturer's specification cause the strainer to pass excessive amounts of dirt. These strainers are physically larger and heavier than screen strainers, usually requiring special supports for the larger sizes.

On the other hand, they operate well with very low pressure losses. Accumulated dirt doesn't increase the pressure loss, and they can hold much more than screen-type strainers. They are usually used with water systems exposed to large amounts of dirt. Open cooling tower systems are typical applications.

Air removal

Air is a potential problem for all hydronic systems, and a real problem for many. It has a natural affinity for water and accumulates in hydronic systems. Water readily absorbs air at lower temperatures and releases it at higher ones. The released air forms bubbles that circulate through the system with the water. When these bubbles reach a high point in the system, they can coalesce and form an air lock. The pump might not be able to dislodge the air, and in extreme cases it can entirely prevent water flow. No heating or cooling water will flow to that part of the system.

Air also contains oxygen that can damage a hydronic system. Corrosion is greatly accelerated when heated oxygen is present with ferrous metals (common in hydronic systems). Oxygen causes premature wear to the system.

There are several components available that remove air from systems. These include momentum removers, centrifugal units, special boiler fittings, and screen-type microbubble removers.

Momentum removers depend on slowing the flow of water and allowing bubbles to float to the remover's top. Water enters the remover at a moderate speed, turns inside, and moves to the outlet. Air bubbles stay in the remover near its top. A fitting guides the air to a separate outlet pipe. The air rises into the outlet pipe and collects in a separate tank. Opening a valve on the tank's top, or on a special fitting with a riser tube, removes the air.

Momentum removers do a fair job, removing larger bubbles that easily float to the surface. Very small bubbles, however, can sweep along with the flow of water and continue into the system.

Centrifugal air removers function like centrifugal strainers. Water enters the air remover around its outside diameter and is caused to swirl. Bubbles separate out of the water and collect at the top of the remover. Like with momentum removers, a fitting at the remover's top guides the air out to a tank.

Centrifugal air removers do a better job than momentum removers. The active swirl of the water allows even small bubbles to be collected in the tank. Like all strainers, however, they should be used within specific flow ranges.

Special boiler fittings are very simple (see Figure 5.11). A large-diameter fitting on the boiler's top has a smaller diameter dip tube extending down into the boiler. All of the boiler's outlet water flows up and out of the fitting through the dip tube. The fitting has two top connections: a water outlet connects to the top of the dip tube and an air outlet is connected to the fitting's outer diameter.

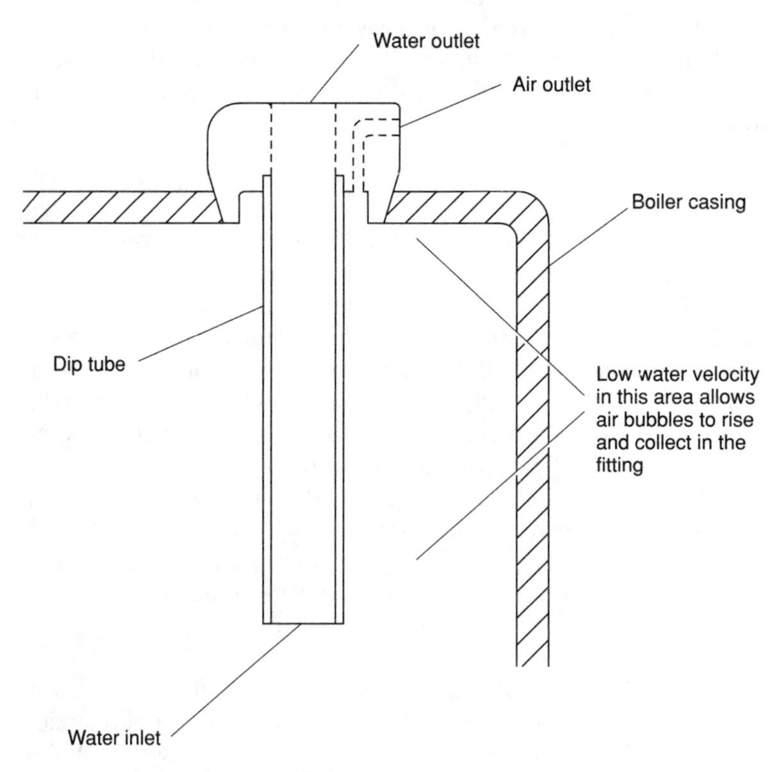

Figure 5.11. Boiler air removal fitting.

Water flow speed at the boiler's outlet is usually the lowest in the system. At this point the boiler's tubes connect to the common header and the water moves slowly toward the outlet. Air bubbles (which might have formed in the heated tubes) tend to bypass the water entrance at the bottom of the dip tube. Instead, they rise to the fitting where they collect in a pocket between the threads and the tube. The air is guided to the fitting's air outlet where it rises into a tank.

Screen-type microbubble removers function exactly like a strainer. Air bubbles stick to and coalesce against a screen. After they get large enough, they break away and rise to the top. A small float chamber there collects the air. Normally the chamber is flooded with water and the float stays at the top. However, when air accumulates in the chamber, the float sinks. The float is linked to an air outlet valve at the remover's top. When the float sinks it pulls the valve open. This releases the air trapped in the chamber to the atmosphere. There is no need for a pressurized tank to catch and hold the air.

Screen-type air removers work very well. They remove almost all of the fine air bubbles suspended in the water, and don't need to use a tank to collect the air. The water flow rate should be matched to the unit to permit the bubbles to accumulate on the screen without being dislodged.

There is no "best" place to install air removers in a system. While some types (such as special boiler fittings) are designed for a specific location, others can be installed almost anywhere. Typically, a manual relief air vent (usually a plug valve) is installed at the system's highest point to aid in air removal during initial water filling. It is opened while the system is first filled with water and then permanently closed.

The best place to eliminate air in a system is where it naturally forms and accumulates. Locating these places requires an understanding of how bubbles form in hydronic systems.

Air bubbles form when air molecules dissolved in water are driven out of solution. This happens when water is heated or when the pressure on it is reduced. Cooler water can hold more gas as a solution than warmer water. Also, water at higher pressure holds more air than lower pressure water. Therefore, the best place for air to come out of solution with water and form bubbles is where the water is hottest and at the lowest pressure.

Unfortunately there might be no single location in the system where both conditions exist. Most systems have the lowest absolute water pressure at the physically highest point on the suction (return) side of the flow. The hottest water is at the boiler's discharge. Which is the best location for air elimination? It depends.

Many systems use the circulating pump to draw water away from the boiler and discharge it into the system. Figure 5.12 illustrates this arrangement. The pump's inlet is connected to both the boiler's outlet and the expansion tank. Water is pulled by the pump through the boiler and past the expansion tank's

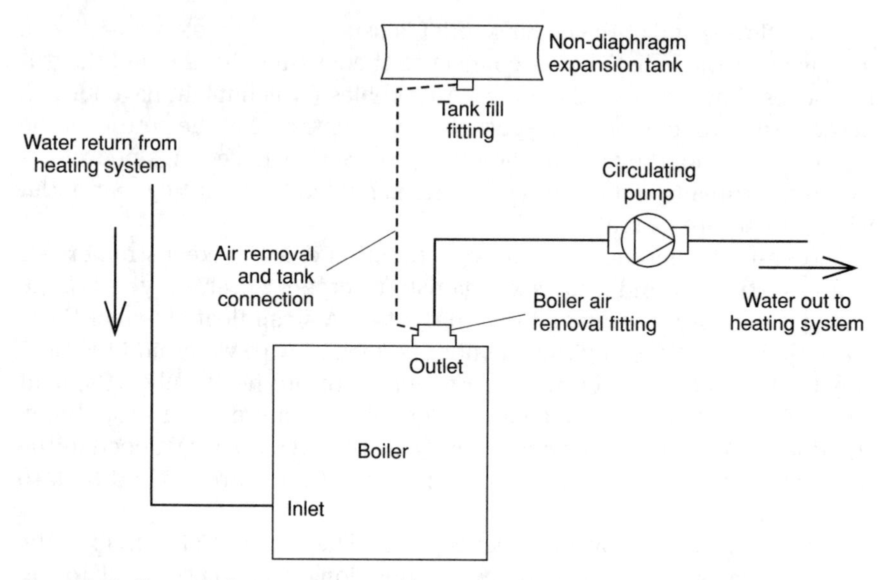

Figure 5.12. Proper boiler arrangement with expansion tank and pump. The expansion tank should be connected to the pump's inlet.

connection to the system. This arrangement is best for a boiler outlet air eliminator. The pump inlet (which is near the high-temperature boiler outlet) is at a lower pressure and higher temperature than most other parts of the system. These conditions encourage the air to form bubbles.

Systems that have the pump discharging into the boiler (and expansion tank) are good candidates for air elimination high in the system. These systems have the pump upstream of the boiler and reduce the pressure in the system. There is usually no pressurized tank available away from the boiler room to accept the air. Use an eliminator with an internal float and valve to allow the air to discharge to the atmosphere.

Expansion control

Hydronic systems depend on water temperature changes to transfer energy. As water changes temperature its density changes, causing its volume to change. Within the range of most HVAC applications, as water heats, it expands. As it is cooled, it contracts. The only exceptions to this are systems that use very cold water or an antifreeze mixture.

When water is cooled below 39°F, its volume increases. The colder water requires more space at the same pressure. This expansion continues as the water freezes, causing ice to float. However, most systems do not use water far below 39°F. Typical air conditioning applications keep the water temperature at 40°F or above. When designing or working on a very cold hydronic system, however, remember to consider the expansion of the chilled water.

The uncontrolled expansion of water with decreasing temperatures can cause a damaging pressure increase. Most systems use expansion tanks to give the water a place to increase its volume. There it can expand and contract in a controlled environment with little chance of damage, leaks, or contamination.

Expansion tanks should be chosen to keep the system's pressure at a safe level under all operating conditions. At the highest temperature the system can achieve, the pressure should remain below the operating point of safety relief valves. At the lowest temperature, there should be enough residual pressure to ensure that the highest parts of the system are always under positive pressure (usually at least 5 psi). Also, some pressure should always be maintained on the water to reduce its ability to liberate dissolved air.

Many small, older hydronic systems used open expansion tanks. A tank installed at the highest location in the system was partially filled with water. The weight of the water pressurized the entire system. The expansion tank was left open to the atmosphere so that the water level inside could rise and fall with temperature. No significant pressure change took place in the system as it heated or cooled.

There are several limitations when using an open tank. The physical height of the open tank decides the system's operating pressure. Equipment located lower in the building is at a significantly higher pressure than equipment located near the tank. Equipment at nearly the same elevation as the tank operates at approximately atmospheric pressure. This can cause problems with valves and flow restrictions.

Flow restrictions cause local pressure drops. With the system's water at nearly atmospheric pressure without flow, the pressure drop caused by flow through a valve can cause a vacuum to develop. This can cause water there to boil and encourage air dissolved in the water to come out of solution. The air, being near the top of the system, might accumulate and could block the flow.

Another problem with open tanks is the large open surface of water exposed to "fresh" air. The tank has an unlimited amount of air and oxygen available, and the water can absorb it. As water moves into and out of the tank (from its heating and cooling) it absorbs a fresh load. This causes the water to stay saturated with air, thus oxygen is available to cause corrosion.

Most expansion tanks used in modern systems are closed and pressurized. Many are located near the boiler and assist with air control. As noted earlier, some air eliminators must discharge their trapped air into a closed, pressurized tank. The system's expansion tank is often used for this purpose. Pressurized tanks are available with and without internal bladders. Bladder tanks can handle water expansion only, while nonbladder tanks can also be used to absorb removed air. Figure 5.13 illustrates both types of tanks.

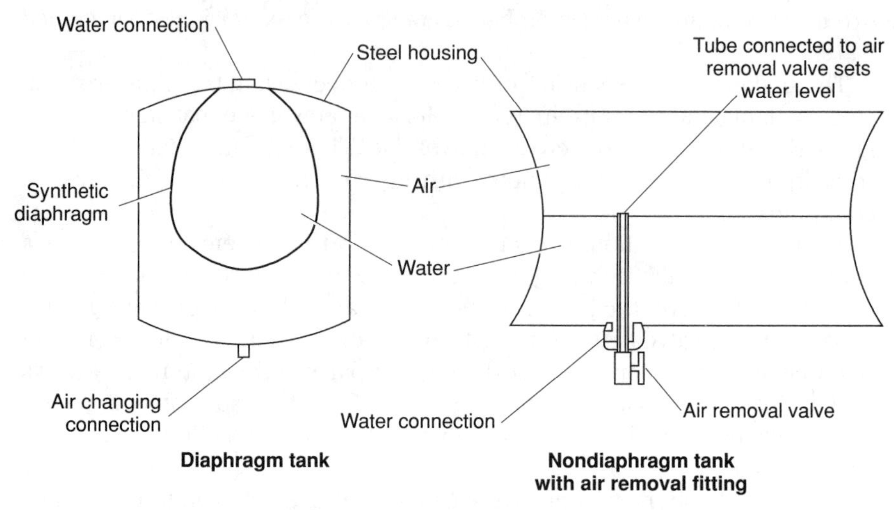

Figure 5.13. Bladder- and nonbladder-type expansion tanks.

Bladder-type expansion tanks separate the air and water with a synthetic rubber sheet. The air side of the tank can be pressurized with a fitting similar to a tire inflation valve. Pressurizing the tank before the system fills with water allows the tank to absorb a larger amount of expanding water. The tank can be physically smaller than nonbladder types. As the system fills with water, no water goes into the tank until the entire system equals the tank's pressure. More space remains for the water as it heats and expands.

Air in the tank cannot contact the system water in a bladder-type tank. This isolation prevents the system from absorbing air and oxygen, helping to keep the water clear and noncorrosive.

Bladder-type tanks can cause problems if the bladder leaks. Many of these tanks are installed with the air side of the bladder at the bottom and the water side on top. The tanks are often placed on the boiler room floor. This allows the wet side of the tank (the top) to remain free of air. However, if the bladder fails, water runs into the lower part of the tank and air enters the system. While the air might be removed by the eliminator, the tank floods and there is no expansion space in the system. The next time the system is heated its pressure is likely to become very high. It can easily become high enough to cause weak piping or equipment to fail. A leak is almost certain.

Most systems that use bladder-type tanks use float-operated microbubble air removers. These do not need to discharge their air into a pressurized tank.

Nonbladder expansion tanks are simple vessels designed to hold air and water. They are usually installed near the boiler room ceiling. Most are horizontally mounted, round, steel tanks. A connection on the bottom allows air to enter the tank and water to flow in and out as it expands. Special fittings allow the tank to partly fill with water and still keep a sufficient amount of air to work effectively. These fittings have an air relief tube extending up inside from the bottom of the tank toward its top. The length of the tube is matched to the tank's capacity. Larger tanks use fittings with longer tubes. This keeps an air cushion above the top of the tube when the system and tank are first filled. An integral drain allows periodic removal of either water or air from the tank, allowing a safe cushion of air to remain inside.

Some of these fittings also partly restrict the exchange of the tank's water with the rest of the system's water. This helps to limit the amount of dissolved air and oxygen introduced into the system. Water in the tank is usually saturated with air. By preventing that water from entering the rest of the system, the total amount of air circulating in it is kept low.

Some nonbladder expansion tanks have sight glasses installed at one end. This allows monitoring of the water level for necessary adjustment.

Most nonbladder expansion tanks are not pressurized before installation. The system pressurizes the tank as it fills with a quantity of water. Water fills the tank from the bottom and forces the air into the remaining space. This causes the tank to use part of its volume for pressurization, with less available volume to absorb water as it expands from heating. The result is that nonbladder tanks must be physically larger than bladder-type tanks to fit the same system capacity.

Expansion tanks are sized based on the total system volume and the expected temperature swing. The larger the system volume or change in temperature, the more expansion occurs and the more air space is needed. See Table 5.5 for expansion tank capacities for heating and cooling systems. This table covers normal temperature and pressure ranges. Select the maximum and minimum pressure needed at the tank. Find these pressures on the table's rows and columns. Then, find the proper size tank by multiplying the factor where these numbers intersect by the system's total water volume.

TABLE 5.5 Expansion Tank Capacities

To find the number of gallons required for the expansion tank, multiply the factor given in the table by the total system capacity in gallons. Copper piping is assumed. If steel pipe is used, multiply the tank volume found by 0.68.

Capacities for closed, pressurized nondiaphragm tanks
Heating systems

Minimum temperature = 40°F
Maximum temperature = 220°F

Maximum pressure (psig)	Minimum pressure (psig) 10	15	20	25	30	35	40	45	50
15	0.416								
20	0.243	0.585							

TABLE 5.5 (Continued)

Capacities for closed, pressurized nondiaphragm tanks
Heating systems

Minimum temperature = 40°F
Maximum temperature = 220°F

Maximum pressure (psig)	Minimum pressure (psig)								
	10	15	20	25	30	35	40	45	50
25	0.185	0.334	0.781						
30	0.157	0.251	0.440	1.006					
35	0.139	0.209	0.326	0.560	1.260				
40	0.128	0.184	0.269	0.411	0.693	1.542			
45	0.119	0.168	0.235	0.336	0.505	0.841	1.852		
50	0.113	0.156	0.212	0.291	0.410	0.608	1.004	2.191	
60	0.105	0.140	0.184	0.240	0.316	0.421	0.579	0.843	1.371
70	0.099	0.130	0.167	0.212	0.268	0.341	0.438	0.574	0.777
80	0.095	0.123	0.155	0.194	0.240	0.297	0.367	0.458	0.579
90	0.092	0.118	0.147	0.181	0.221	0.268	0.325	0.394	0.480
100	0.089	0.114	0.141	0.172	0.208	0.249	0.297	0.353	0.421

Cooling systems

Minimum temperature = 35°F
Maximum temperature = 100°F

Maximum pressure (psig)	Minimum pressure (psig)								
	10	15	20	25	30	35	40	45	50
15	0.050								
20	0.029	0.070							
25	0.022	0.040	0.094						
30	0.019	0.030	0.053	0.121					
35	0.017	0.025	0.039	0.067	0.152				
40	0.015	0.022	0.032	0.049	0.083	0.185			
45	0.014	0.020	0.028	0.040	0.061	0.101	0.223		
50	0.014	0.019	0.026	0.035	0.049	0.073	0.121	0.263	
60	0.013	0.017	0.022	0.029	0.038	0.051	0.070	0.101	0.165
70	0.012	0.016	0.020	0.025	0.032	0.041	0.053	0.069	0.093
80	0.011	0.015	0.019	0.023	0.029	0.036	0.044	0.055	0.070
90	0.011	0.014	0.018	0.022	0.027	0.032	0.039	0.047	0.058
100	0.011	0.014	0.017	0.021	0.025	0.030	0.036	0.042	0.051

Capacities for closed, pressurized diaphragm tanks
Heating systems

Minimum temperature = 40°F
Maximum temperature = 220°F

Maximum pressure (psig)	Minimum pressure (psig)								
	10	15	20	25	30	35	40	45	50
15	0.125								
20	0.083	0.167							

Capacities for closed, pressurized diaphragm tanks
Heating systems

Minimum temperature = 40°F
Maximum temperature = 220°F

Maximum pressure (psig)	Minimum pressure (psig)								
	10	15	20	25	30	35	40	45	50
25	0.069	0.104	0.208						
30	0.063	0.083	0.125	0.250					
35	0.058	0.073	0.097	0.146	0.292				
40	0.056	0.067	0.083	0.111	0.167	0.333			
45	0.054	0.063	0.075	0.094	0.125	0.188	0.375		
50	0.052	0.060	0.069	0.083	0.104	0.139	0.208	0.417	
60	0.050	0.056	0.063	0.071	0.083	0.100	0.125	0.167	0.250
70	0.049	0.053	0.058	0.065	0.073	0.083	0.097	0.117	0.146
80	0.048	0.051	0.056	0.061	0.067	0.074	0.083	0.095	0.111
90	0.047	0.050	0.054	0.058	0.063	0.068	0.075	0.083	0.094
100	0.046	0.049	0.052	0.056	0.060	0.064	0.069	0.076	0.083

Cooling systems

Minimum temperature = 35°F
Maximum temperature = 100°F

Maximum pressure (psig)	Minimum pressure (psig)								
	10	15	20	25	30	35	40	45	50
15	0.015								
20	0.010	0.020							
25	0.008	0.013	0.025						
30	0.008	0.010	0.015	0.030					
35	0.007	0.009	0.012	0.018	0.035				
40	0.007	0.008	0.010	0.013	0.020	0.040			
45	0.006	0.008	0.009	0.011	0.015	0.023	0.045		
50	0.006	0.007	0.008	0.010	0.013	0.017	0.025	0.050	
60	0.006	0.007	0.008	0.009	0.010	0.012	0.015	0.020	0.030
70	0.006	0.006	0.007	0.008	0.009	0.010	0.012	0.014	0.018
80	0.006	0.006	0.007	0.007	0.008	0.009	0.010	0.011	0.013
90	0.006	0.006	0.006	0.007	0.008	0.008	0.009	0.010	0.011
100	0.006	0.006	0.006	0.007	0.007	0.008	0.008	0.009	0.010

Use the "Volume of water" columns in Tables 5.1, 5.2, and 5.3 to estimate how much water the system's piping and standard fin tube radiation holds. Multiply the numbers given in those tables by the length of pipe to arrive at the number of gallons held by the pipe. Add all of the volumes of the individual pipes together to get the total.

Allow 5 gal of water for boilers up to 150,000 Btu/hr. Consult the manufacturer of larger boilers to get their specific water capacity, but if this information is not available make as accurate an estimate as possible. Allow 1 gal of water for each 30,000 Btu/hr capacity.

If the estimate is slightly oversized, no harm is done. The tank might be larger than needed. This is better than installing a tank that is too small. Pressure can become excessive without sufficient expansion space in the system.

An example of sizing an expansion tank: A hydronic heating system is designed to operate at a minimum pressure of 25 psi and a maximum pressure of 40 psi. The design calls for a bladder-type expansion tank. The system contains a 250,000 Btu/hr boiler with an internal water capacity of 12 gal. The system contains the following type L copper pipe lengths and sizes:

Pipe	Feet
1¼ in.	235
1 in.	268
¾ in.	92

From Table 5.2, the pipes' volumes are

Pipe	Volume
1¼ in.	235 ft × 0.065 gal/ft = 15.28 gal
1 in.	268 ft × 0.04 gal/ft = 10.72 gal
¾ in.	92 ft × 0.025 gal/ft = 2.3 gal

The total amount of water in the piping is 28.3 gal. Add the boiler's 12 gal, and the total system volume is 40.3 gal.

From Table 5.5, the tank factor is 0.111. Multiply the system capacity by the tank factor:

$$40.3 \text{ gal} \times 0.111 = 4.47 \text{ gal of tank capacity}$$

In this application, a 5-gal tank would be adequate to control the pressure.

Flow diversion tees

A few pipe fittings specially designed for the hydronic industry are available. A very useful fitting is the flow diversion tee. These help the flow of water through branch hydronic systems to balance the flow in certain types of pipe circuits. They look like regular pipe tees, but there is a restricting orifice installed to limit the amount of water that can flow "straight through" the fitting. Water entering one side of the tee encounters a restriction as it attempts to flow straight through the other leg. The restriction diverts some water into the branch leg.

The internal orifice causes a higher pressure drop for water traveling straight through the fitting. This pressure drop makes more water than usual flow into the branch. These fittings act something like balancing valves, preventing too much water from bypassing parts of the system.

Unlike balancing valves, diversion tees are not adjustable. While that is true, they do have a place in some systems. They are less expensive than balancing valves, and when used properly can provide well-balanced, low-cost flow diversion. The appropriate uses of diversion tees are covered in Chapter 7.

Flow Rates and Economical Sizing

Properly estimating flow rates for systems is not difficult. Once the heating or cooling load for the building is determined, and the boiler chosen, follow the simple procedure outlined below. There are six steps to hydronic system design and component selection.

1. Find out the total heat loss or gain for the entire building and for each zone.
2. Figure out the amount of water flow required in all parts of the system. Determine the required pipe sizes based on flow and heat capacity requirements. Plan on using balancing or automatic flow control valves on all branches. Place isolation valves at all equipment to ease future service work.
3. Estimate the water pressure drop from the pipe network.
4. Add pressure losses from fittings in the system.
5. Calculate pressure losses from control valves and add them to the total loss.
6. Size the pump and expansion tank.

First, flow through the boiler or chiller should be selected to give a reasonable temperature change. Most heater manufacturers prefer that their equipment have a maximum 20°F internal temperature increase. Chillers should have a 10°F temperature drop. Use the flow equation to select the proper flow to get this change:

$$\text{gpm} = \frac{\text{Btu/hr}}{(498 \times T_d)}$$

Because we are planning on 10°F and 20°F temperature changes, the proper flows are

$$\text{Heating: gpm} = \frac{\text{Btu/hr}}{(498 \times 20°\text{F})} = \frac{\text{Btu/hr}}{9960}$$

$$\text{Cooling: gpm} = \frac{\text{Btu/hr}}{(498 \times 10°\text{F})} = \frac{\text{Btu/hr}}{4980}, \text{ or}$$

$$\text{gpm} = \text{tons capacity} \times 2.4$$

Adjust the flow rate up or down to change the temperature throughout the system, if necessary.

An example: Find the proper flow rate for a 5-ton chiller.

$$\text{gpm} = 5 \times 2.4 = 12 \text{ tons}$$

Next, find the flow rate required for a 750,000 Btu/hr boiler.

$$\text{gpm} = \frac{750{,}000 \text{ Btu/hr}}{9960} = 75 \text{ gpm}$$

The flow through the boiler or chiller equals the required pump flow. Next, figure out the necessary pump head (pressure increase). This process is similar to figuring out the air pressure loss in a duct system. First, make the system's piping layout. A variety of piping systems are covered in Chapter 7. Select the type of system that best meets the job's requirements and route the piping for the shortest, most direct runs possible.

Estimate the flow rates to various parts of the system based on the Btu per hour requirements of each. From the flow rates, figure out the pipe sizes. Table 5.6 gives appropriate pipe sizes for specific maximum flow rates. These flow rates have been chosen to allow no more than 0.050 ft of head pressure loss per foot of pipe. This pressure drop is appropriate for heating and cooling systems. It allows reasonable losses for the pump to handle and keeps noise to acceptable levels. These values "fit" most HVAC applications.

Starting with the main pipes connecting the boiler or chiller to the pump, use Table 5.6 to select all of the system's pipes. This gives a pressure loss of 0.050 ft of water per foot of pipe in each branch. Find the pressure drop (in feet of water) in each pipe by multiplying its length by 0.050.

Where there is more than one branch for water to flow through, the total pressure loss is less than the loss in any one branch. Use a value of two-thirds of the lowest pressure branch for all of the branches. This compensates for the fact that the water has more than one path it can flow through to get through the system.

When using flow diversion tees figure the pressure drop equal to that of all branches off of the main pipe. Flow diversion tees increase the pressure loss in hydronic systems and this allows for it.

Add the pressure losses for pipes that are in series with each other. Water that flows through series connected pipes experiences a loss from each of them. Also, be sure to include the pressure losses from the supply and return pipes. If both the supply and return pipes follow the same path, double the distance of one of them before multiplying the loss per foot to find the total pressure drop.

Estimate pressure losses for fittings. Estimate this by allowing a fitting loss that is proportional to the total pipe loss. For systems with few fittings compared to the length of the piping, figure that the fitting losses are 75% of the pipe loss. Multiply the pipe's losses by 1.75 to get the losses including the fittings.

TABLE 5.6 Economical Maximum Pipe Flow Rates and Heat Transfer Capacities

Friction loss = 0.050 ft of water per foot of pipe.

Nominal pipe size (in.)	Flow (gpm)	10°F (temperature difference) heat capacity (Btu/hr)	20°F (temperature difference) heat capacity (Btu/hr)
½	1.6	8000	16,000
¾	4.1	20,400	40,800
1	8.5	42,300	84,600
1¼	16.0	79,700	59,400
1½	25.0	125,000	50,000
2	52.0	259,000	18,000
2½	85.0	423,000	46,000
3	150.0	747,000	94,000

Systems with an "average" number of fittings have about half the total loss from the fittings. Double the pipe-only pressure drop to get the total. Finally, complex systems with many fittings have most of their losses caused by the water's twisting and turning in the system. Multiply the piping loss by 2.5 (or more if there are many fittings) to allow for the higher losses.

If using a modulating control valve, this is the time to estimate its C_v. Use the equation to find the coefficient after the flow and the pressure losses are known. This applies whether calculating the C_v for a branch or the entire system. Remember that the control valve has a pressure loss approximately equal to the system when it is fully open. Because of that, the total pressure loss (system and valve) is double that of the system alone.

If the entire system is using the control valve, find the total pressure drop by doubling the total system-only pressure loss. If a branch is using a control valve, the valve's pressure loss only affects that branch. The control valve should have a wide-open pressure drop equal to the drop in the rest of the branch. Total system losses might have to be recalculated once the branch's control valves are considered.

Select the pump when all of the pressure losses and the flow are known. There is a wide range of available pump styles, speeds, and capacities that cover any conceivable need. Select the type of pump that is the easiest to install (in line or base mounted) for the application. Select a unit, based on the calculated head and flow requirements, that gives the highest efficiency on its pump curve.

Finally, estimate the required volume of the system expansion tank. Use Table 5.5 and select an appropriate sizing factor based on the system volume, pipe material, and pressure design.

Use balancing valves or automatic flow control valves on all branches to allow balancing of the system. When designing the system, it is not necessary to plan on the pressure drop produced by these valves. They are expected to be "wide open" when the system is running, reducing the flow to the required level.

The Bell & Gossett System Syzer calculator

One of the handiest devices for properly designing a hydronic system is the System Syzer calculator from the Bell & Gossett division of ITT Fluid Technology Corporation. This circular slide rule makes performing the following common hydronic calculations very easy:

- Figure out the friction of water flow in a pipe given the gallons per minute or velocity and the pipe size. It gives the friction in terms of feet of head and can be used to find the total pressure loss for a given length of pipe.
- Find the velocity of water in a given size pipe if the gallons per minute are known.
- Figure out the required valve C_v knowing the system's flow and the valve pressure loss required (in psi or feet of head).
- Calculate the temperature change in water from its flow and Btu exchange rate. Or, find the heat or water flow needed to attain a specific temperature change.

These are the most common system calculations required, and this calculator makes finding solutions easy. The calculator operates with a set of scales printed on a rotatable disk. By entering the known information onto one set of scales, the unknown quantity is found on the other.

Rugged and easily cleaned, the System Syzer should be on every technician's reference shelf. It's rugged enough for years in the field, and it makes hydronic system math almost automatic.

The System Syzer is available from your local Bell & Gossett dealer to professionals in the heating and cooling industry. Refer to your telephone directory for the Bell & Gossett representative near you. If one isn't listed, call Bell & Gossett at the telephone number listed in Appendix A. They will give you the name and number of the representative authorized in your area.

Water Treatment

Proper water treatment is necessary in many hydronic systems. While closed hydronic systems do not require routine maintenance, open systems do.

Water in most closed systems becomes chemically stable shortly after the system begins operating. Oxygen dissolved in the water reacts with metals in the system, causes a small amount of oxidation, and goes out of circulation. This produces a small amount of corrosion that never becomes worse. It stops shortly after it begins and no harm is done.

Note that any system with improperly installed dielectric couplings are subject to severe corrosion. As noted previously, galvanic corrosion does not depend on dissolved oxygen. Coupled iron (or steel) and copper equipment set up an electric current flow between them. The ferrous metal rapidly deteriorates. Use proper isolation between dissimilar metals to prevent this corrosion.

Corrosion and biological control

Hydronic systems with open water supplies need special treatment. Besides filtration to remove dirt, the water requires chemical treatment to kill biological growth (bacteria and mold) and to prevent the accumulation of minerals and salts. A variety of chemicals are available to properly treat water. Some combination chemicals are available that perform more than one function. They can prevent corrosion, halt biological growth, and limit scale buildup.

Many open water systems operate at moderate temperatures too low to kill bacteria. These include cooling towers (open flow and sump systems) and spray humidifiers. Left unchecked, biological contamination of a water system can cause serious problems. Systems operating at temperatures below 130°F are potential breeding grounds for *Legionella* bacteria and other pathogens.

Bacteria and mold contamination can cause a variety of problems. They produce a slime layer over most of the wetted surfaces in the system. This reduces the heat transfer ability of coils and heat exchangers, as well as clogs strainers and filters. Many biological contaminants are dangerous to people and can drift far from their source to infect people downwind. The original outbreak of Legionnaire's disease occurred from cooling tower drift (mist) that migrated into an HVAC system's outdoor air intake.

Algae can grow on wetted surfaces exposed to sunlight. Common on untreated cooling towers, it forms thick clumps that break off and flow into the system, quickly clogging the system's strainers.

Biological treatments include chlorine disinfectants. Like the chemicals used to treat drinking water and swimming pools, these attack and kill germs that can breed in moderate temperature water. Disinfectants must be used constantly to prevent biological growth. Although easy to add (usually by pouring some concentrate into the system's sump or fill point), it must not be neglected.

Water open to the atmosphere evaporates. As it leaves the system as a vapor, it escapes without any of the minerals that it originally came with. These are left behind. As more water evaporates, more minerals accumulate in the remaining liquid. Mineral concentrations increase. Make-up water added to the system just adds more minerals.

Consider a full glass of water that contains 1 g of dissolved minerals. Now, allow half the water to evaporate from the glass. The remaining half glass of water still holds 1 g of minerals. Add half a glass of fresh water to make up for the evaporated water and it brings in another ½ g of minerals. The glass now contains 1½ g of minerals.

If this evaporation and make-up continues, the mineral content in the glass constantly increases. Water leaving the glass takes no minerals with it, while the fresh water adds more. This effect leads to the idea of *cycles of concentration*. As the water in the glass evaporates and is replenished, the mineral concentration rises. The concentration in the glass at a given time, compared with the concentration in fresh water, gives the cycles of concentration. If there were 1½ g of minerals in the glass of water after it was refilled, the water has had 1½ cycles of concentration. The longer the evaporation and make-up process continues, the higher the water's cycles of concentration become. Figure 5.14 illustrates this effect.

Eventually, the water is saturated with minerals. The amount of minerals the water can hold depends on its temperature. Cold water can hold more minerals than warm water. Water saturated with minerals at a low temperature loses its ability to hold all of them when heated. The minerals form a layer of scale on the surface where the heating takes place.

For example, if water was absorbing heat in an air conditioner's condenser, the scale would form on the outside surfaces of the condenser's tubes. This is the hottest location in the system, and the water's ability to hold minerals is lowest at that point.

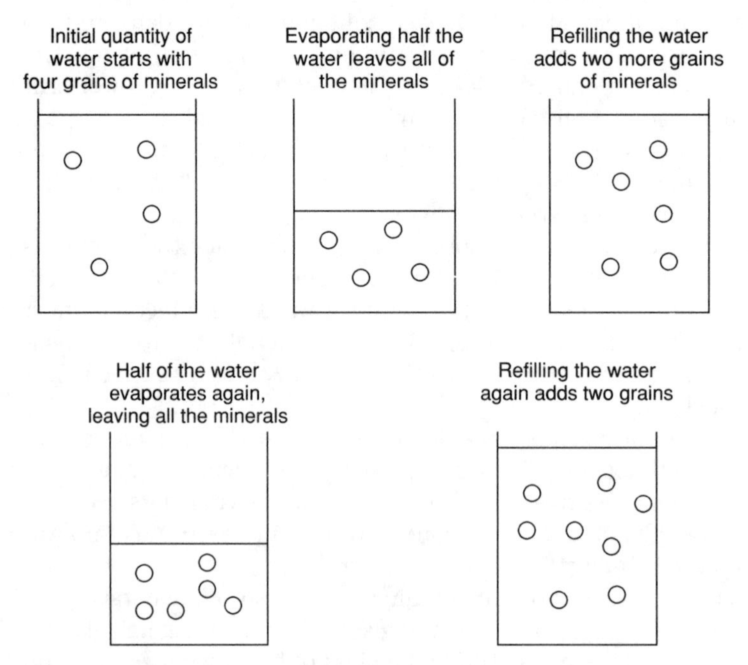

Figure 5.14. Cycles of concentration illustration. Minerals and salts will become saturated and cause deposits if the process continues long enough.

Scale also accumulates where the water evaporates. As it vaporizes, the mineral concentration momentarily increases. This can be enough to deposit a small amount of solid minerals on the evaporation surface. After a short while, a layer of scale accumulates.

Scale contamination must be controlled. Scale fouling can combine with biological contamination to rapidly damage an open hydronic system. Scale has a rough, complex surface perfect for slime to adhere to. The constantly renewed scale surface allows fresh biological growth to take place continuously.

Mineral buildup is not as easy to control as biological growth. Systems require a combination of chemical treatment and fresh water flushing to prevent scale. Chemicals can be used to prevent minerals from dropping out of solution (or soften them into a "mush") and forming a hard crust on the system's surfaces. Flushing the system with water washes out minerals and keeps the cycles of concentration from becoming too high.

Most water treatments use sulfites or sulfates to control scale. Sulfur attracts calcium and prevents it from forming hard scale that will adhere to the system's surfaces. Instead, the calcium circulates in the water longer (and at higher temperatures) and can be flushed out.

Ozone is another scale, corrosion, and biological treatment chemical. Small amounts of gaseous ozone are dissolved in the water where it flows through the system. It must be constantly replenished because it migrates out of the water wherever it comes in contact with air. Also, ozone is an unstable chemical. It breaks down into ordinary oxygen after a short time, but this does not cause a problem.

Ozone is three bonded oxygen atoms, while ordinary atmospheric oxygen has two atoms. The ozone molecule is a *super oxidizer* that destroys all mold and bacterial cells on contact. This action can limit corrosion.

Much of the corrosion that occurs in water systems is caused indirectly by biological growths. The slime layer set up by mold and bacteria can be extremely acidic and can seriously damage the metal underneath. This layer can be difficult to break up once it is established, and its presence prevents many chemicals and disinfectants from killing the underlying infection. Ozone, however, can cut through the slime layer because it oxidizes it on contact. In effect, it "burns" the slime like a fuel. Once the metal is cleaned, and the bacteria and molds are killed, ozone does not readily attack it.

Ozone is supplied by a generator attached to the water system. Electricity or ultraviolet light produces ozone from a flow of air or oxygen. It is injected under low pressure into the water as a mist of very fine bubbles and allowed to flow into the piping system. Because the amount of gas that water can hold drops with increasing temperature, ozone injection is done where the water is coolest. This allows the most gas to be absorbed.

Most scale control systems use fresh water to maintain low cycles of concentration. More water is added to the system than evaporates. The "extra" water runs out the system's overflow drains to the sewer, taking the dissolved or suspended minerals with it. Simple systems use a continuous water feed to maintain a flow of fresh water. However, there are significant disadvantages to this type of system.

As water is constantly added and drained, treatment chemicals are being lost and diluted. Biological and scale controls are going down the drain along with the water used to limit the cycles of concentration. This is very expensive. An automatic feed and drain control prevents this. These controls use a *conductivity sensor* to monitor the condition of the water. As the mineral concentration in the water increases, its electrical conductivity rises. The controller's sensor tracks this.

When the controller detects excessive minerals in the water, it opens the hydronic system's drain. Most systems have an automatic make-up water system that adds fresh water as the controller drains it away. If make-up water is not automatically added, the controller opens a solenoid valve connected to the fresh water supply. Excess water drains out of the system through the overflow outlet. Minerals are carried out with the water. Figure 5.15 illustrates a typical controller system with a fresh water supply meter. Either way, the control reduces the system's cycles of concentration just

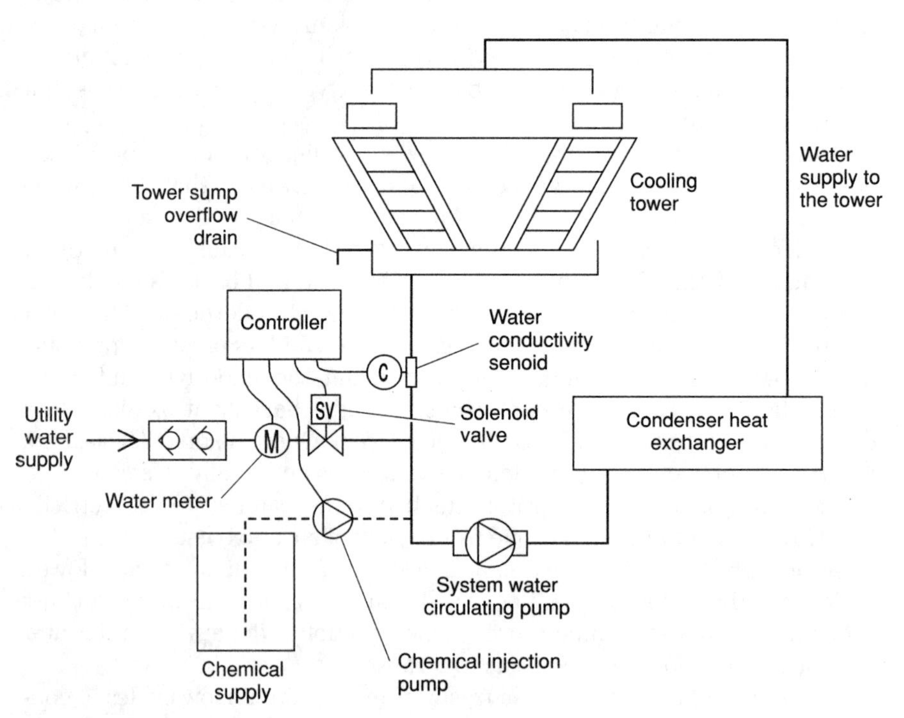

Figure 5.15. Conductivity controller system. Mineral loaded water discharges out the tower drain as fresh water is added.

enough to prevent scale accumulation. No more water treatment chemicals are lost than necessary.

Most controllers also operate a pump that adds replacement chemicals to the water. The control also decides the amount of chemical to add based on how long the drain has been open or how much make-up water has been added. A special water meter on the fresh water line tells the controller how many gallons have been added. This information allows the controller to add the proper amount of chemicals. It pumps in the proper amount to make up for what was lost when the drain or fresh water supply was open.

Most chemical pumps are peristaltic types. A restriction designed to pinch the chemical's delivery tube closed is rolled from the pump's inlet toward the outlet. This action forces a slug of chemical out of the pump and into the hydronic system. The controller controls the amount of time the pump is operated and can tell very precisely the amount of chemical delivered.

Automatic controllers are a positive addition to open hydronic systems. Their initial costs are quickly covered by reduced maintenance and energy costs, as well as less chemical use. Although some clients might not appreciate the benefits of automatic water treatment, and will resist paying for it, they should seriously consider it.

Antifreeze

Another class of chemical treatment used with hydronic systems is freeze protection. Many systems, open and closed, can be exposed to freezing conditions. Using chemical antifreeze is a way to prevent damage.

Most of the freeze protection for HVAC systems relies on ethylene glycol or propylene glycol. These are similar to the chemicals used in automobile antifreeze, but there are important differences. Propylene glycol is essentially nontoxic, while ethylene glycol is highly poisonous. Many codes require the use of nontoxic antifreeze solutions where a chance for potable water contamination might exist.

Most codes consider these conditions to exist where a single-wall heat exchanger is used between the utility water connection and the freeze-protected system. Some codes also call for nontoxic antifreeze when make-up water is directly connected to the freeze-protected system. A backflow preventer in the make-up water line might be required to permit the use of toxic antifreeze.

The antifreeze used in most HVAC systems is ethylene glycol based. This material is acceptable when there is no way for hydronic system water to come in contact with potable water. Heating and cooling system antifreeze is specially blended with corrosion inhibitors for use in building systems.

Don't use automobile antifreeze in an HVAC system, however. Auto coolant is designed to operate with a large variety of metals. To help in preventing corrosion (some of it galvanic), vehicle antifreeze contains silicates. These inert silicone compounds coat the surfaces of the passages they flow through, building a barrier to corrosion. This is not needed or desirable in commercial heating or cooling systems. Automobile cooling systems are designed to compensate for the silicate buildup that occurs. Passages are wide and open to help prevent clogging. An automobile radiator can become clogged, however, if the coolant is not changed often enough. The silicates clog the small tubes after a while if the coolant remains in the system too long. Coolant in commercial HVAC systems does not require regular flushing. Use antifreeze specifically designed for heating and cooling systems.

All antifreezes affect the flow characteristics and heat transfer abilities of water. The water's specific gravity and viscosity increase, causing the system's pumping losses to increase. Less flow occurs at the same pressure, or more pressure is needed to provide the same flow as pure water.

Antifreeze and water solutions don't transfer heat as readily as pure water. The molecules of antifreeze act as an insulator that reduces heat transfer. This means that heat exchangers (including cooling towers, coils, and other exchangers) do not operate as efficiently with an antifreeze mixture as with plain water.

Both effects mean that antifreeze solutions should be used with care. Designs that have little spare capacity for flow or heat exchange can suffer with the addition of antifreeze. Systems that are sufficiently overdesigned will function acceptably with this lower efficiency.

Typical antifreeze solutions transfer heat with approximately 10% less capacity than a system using plain water. The pump head is also approximately 10% higher to get the same flow. These values vary depending on the concentration of antifreeze used, with higher concentrations causing more losses.

Deciding on the proper concentration of antifreeze is generally straightforward. Consult the manufacturer for the concentration that provides freeze protection sufficient for the expected exposure. Because of the extra costs and losses involved in using too much antifreeze, it is usually best to err a bit on the low side. Too much antifreeze provides no benefits.

Putting in a bit too little antifreeze rarely causes problems. That is because, even when moderate freezing takes place in the system, damage does not occur. Mixtures of water and antifreeze don't freeze the same way pure water does. Plain water freezes solid over a small temperature range. Once it begins to freeze it rapidly continues until all of the water is solidified. This does not happen with antifreeze.

An antifreeze solution freezes slowly over a range of temperatures. When it is lowered to its published freezing point, suspended ice crystals form in the liquid. As the temperature continues to fall, more of the liquid solidifies by crystallizing additional liquid into slush. The mixture must be chilled well below its specified freezing point to allow a solid block to form.

While the solution is a thin slush it continues to flow through the system. This prevents freeze damage by allowing the mixture to swell and expand evenly. The system's expansion tank absorbs the volume increase without causing excessive system pressure.

Discuss this with the antifreeze dealer before settling on a concentration to use. It is very likely that the system will work well and be sufficiently protected with less antifreeze than expected.

Installing antifreeze in a closed system should be done with care. Mix a solution of water and antifreeze in a drum and fill the system with the mixture. Don't fill the system with pure antifreeze and top it off with plain water. The materials will take a long time to blend. A slug of pure antifreeze will circulate through the system, significantly altering flow and heat transfer characteristics as it moves through the piping. Also, the plain water running through the system is prone to freezing if it exposed to cold conditions before it has mixed with the antifreeze. It is better to premix the solution before adding it to the system.

Chapter

6

Commissioning, Testing, and Balancing

Starting an HVAC system is usually simple. It is best, however, to follow a few procedures to ensure that the system runs correctly from the start. Besides throwing a switch, the system needs to be checked out and tested to be certain there are no problems that might cause trouble. System startup is also the time to perform a preliminary flow balance. This ensures that the system operates within the design specifications and that any necessary corrections are made right away.

Waiting until later for system balancing can be a mistake. Often the building's ceiling and walls are closed in after the HVAC ducts and pipes are installed. It is better to do a quick check for proper basic operation before the system becomes difficult to work on.

Balancing HVAC systems is necessary when heating or cooling is distributed over any moderately sized area. Balancing allows the system to operate the way it was designed, with all areas receiving the intended amounts of heating and cooling.

As described in Chapter 2, transfer of heat usually depends on movement of a fluid. The only exception is radiant heat transfer. All other systems use air or water to carry heat, and they must be distributed in the proper proportions to allow energy transfer to happen as planned. Areas that receive too much heating or cooling become uncomfortable, while those receiving too little conditioning are even worse. Because the system can deliver only a fixed amount of energy, imbalance robs one area to feed another.

Even VAV air systems require balancing. Each zone controls the volume of its air supply by adjusting its box. However, if the total system is not balanced correctly, the maximum amount of air each zone needs isn't available.

Besides properly distributing energy, proper balance can be essential for ventilation. Air systems providing outdoor air are often designed to deliver portions of the air to various locations. Proper balance is required to ensure that all of the building's occupants receive the correct amount of fresh air.

Most water systems (but not all) require balancing. Those with automatic flow control valves, reverse returns, nonadjustable diversion tees, or secondary pumps (covered in Chapter 7) do not require balancing. These systems make up the minority of installations, however, and most hydronic systems do need balancing after installation.

The following sections provide guidelines for starting and performing simple system balancing. Complex systems, including VAV designs, require special tuning and adjustment dependent upon the specific equipment installed. These systems are beyond the scope of this work and are not covered here. Call special test and balance contractors to address these systems.

Air Systems

Much of the equipment that should be installed in air supply systems was described in Chapter 4. This includes dampers in noncritical branches to limit air flow to the design levels. After the system is installed, set all of the noncritical branch's duct dampers to approximately 75% open and the supply diffuser's dampers 100% open. Be sure that the correct air filters are in place.

Start the system fan with a clamp-on ammeter on its motor leads. The rotational direction of three-phase motors should be confirmed. Check that the fan is turning the right way in its housing. If incorrect, reverse any two connections from the power source to the motor. That will make the motor spin in the other direction.

Check the motor's current draw to ensure that it is not overloaded. If it draws more than its rated amperage, shut off the fan and find the cause. Confirm that the proper sheaves (pulleys) are in place and that nothing is binding the drive. A fan overload can also occur if it is handling too much air. If there is nothing wrong with the drive, try partly restricting the fan's inlet air. If that corrects the problem, set the restriction to allow the motor to draw about 80% of its rated current.

Assuming the current draw is correct, inspect the system for proper operation at all air outlets. Check for noise and degree of air movement throughout the system. If there is no ceiling (or other obstructions) the ducts should be checked for leaks at joints, seams, and couplings. Excessive vibration and duct noise should be corrected. Find the causes and make any necessary repairs.

The actual balancing process can now begin. First, figure out the total air flow through the system. There are several instruments available to help with this, including velometers, anemometers, and pitot tubes. Other instruments, such as low-differential pressure gauges, are also available but are not used as often.

Velometers and anemometers measure the speed of moving air. Some use a miniature propeller that rotates as air moves past it. This moves a needle on a meter that is calibrated in air speed. Other meters use a small wire that carries an electric current. These "hot wire" anemometers determine the air speed by the amount the wire is cooled by the passing air. Its resistance drops as more air flow passes over it and cools it. While both types of instruments can do the job, the hot wire type is smaller and can probe air speeds inside an almost closed duct. Propeller types must be held in the air stream and are most suitable for use in open flows. They work well if you want to find the speed of air discharging from a wall or ceiling register.

Pitot tubes do not measure air speed. Instead, they directly measure the air velocity pressure. They can be used when the air velocity is at least 700 fpm. The moving air is directed into the face of the instrument where it enters a set of tubes. This lifts a column of fluid in the instrument. The height that the fluid lifts is proportional to the velocity pressure of the moving air. Because velocity pressure depends only on the air's speed (at normal densities), the scale indicating the fluid's lift may be calibrated in feet per minute. The air's total pressure (velocity plus static) is higher than its static pressure alone. Two openings on the tube separately pick up both the static and total pressures. The pressures are mechanically subtracted and the difference is the air's velocity pressure. Figure 6.1 illustrates a pitot tube measuring instrument.

Many pitot tubes use water or an alcohol solution as the sensing fluid. The lower the density of the fluid, the more easily the instrument can read low air flows. Some of them have the pressure or velocity tube and scale mounted on an incline. This amplifies the vertical lift of the liquid by forcing it to flow further in the tube to give the needed vertical movement. The scale expands (with more room between calibration points) to allow more precise measurement.

Once the air's velocity is found, its volume can be determined. Air speeds vary within a duct. Air usually flows fastest in the center and more slowly near the duct's walls. Air flows most slowly at the corners of rectangular ducts. Friction from the boundary layer clinging to the duct walls causes this effect. Areas farthest from the walls carry the air at a higher speed. This all means that no single air speed reading in a duct provides the average speed. Several readings (the more, the better) must be taken within the duct to find the average speed. These should be done systematically to get a good estimate.

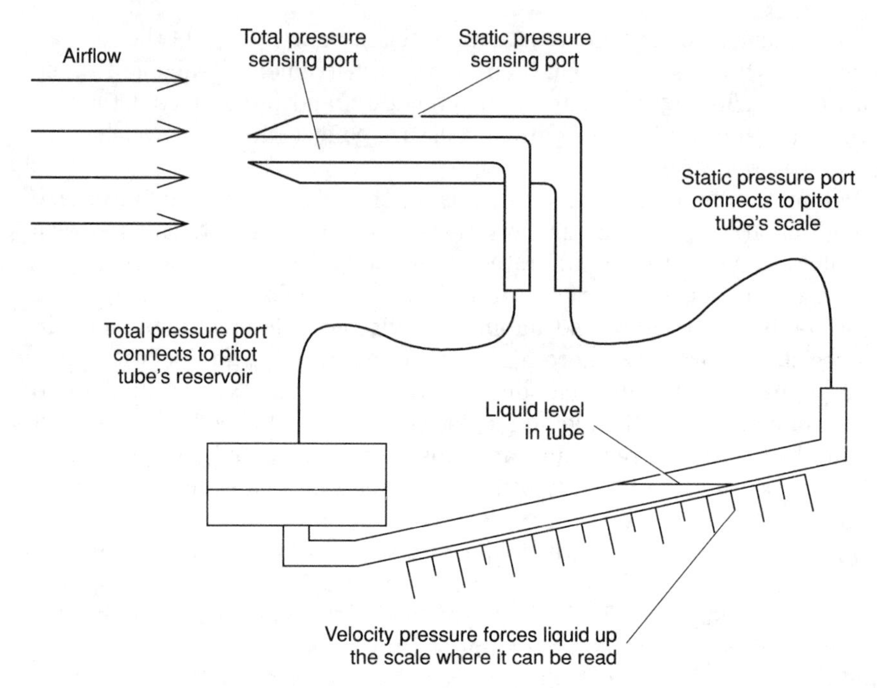

Figure 6.1. Pitot tube operation. The sloped scale amplifies the change in the fluid's height. A lower density fluid also responds to smaller pressure changes with large changes in fluid height.

There is a technique to use when figuring out air velocity in a duct. A set of air speed readings must be taken to find the average speed. This gives the duct's total velocity. Plan to take 12 or 16 readings in any rectangular duct. Rectangular ducts require more readings than oval or round ducts because their airflow is asymmetric. Round ducts should have a minimum of eight readings to arrive at an accurate average.

It is important to measure the airflow where it is least likely to be disturbed by upstream or downstream conditions. Do not measure too close to duct turns, dampers, reducers, or branches. Each of these causes extra turbulence for several feet downstream. Avoid these areas. Try to take measurements in an area of straight duct. At least 5 ft of straight duct is recommended. If long, straight runs are not available, select an area with as long a straight upstream run as possible. Turbulence from upstream disturbances carry downstream farther than disturbances beyond the measuring point affect upstream flow.

Taking the measurements is simple. Drill a small hole in the side of the duct, just large enough to admit the anemometer or pitot tube's probe. Round ducts need a single hole and rectangular ducts require three holes. The three holes must be spaced to sense air in equal areas from the duct's top and bottom walls. Measure the distance from the duct's top to bottom and divide this distance by six. One hole should be spaced this distance from the duct's bottom, another the same distance from the top surface. The last hole should be placed halfway from the bottom to top. Figure 6.2 shows the proper hole locations.

Specific procedures for rectangular ducts and round ducts are slightly different. To read rectangular ducts, insert the probe inside the duct far enough to place it seven-eighths the distance between the near and far walls. Be certain that it is facing directly into the airflow. Read and record the air speed. Withdraw the probe from the duct a distance equal to ¼ the duct's width. It is important to get this correct, so measure the probe's insertion length in the duct. Measure and record the velocity reading at this point.

Repeat this procedure two more times, moving the probe back out of the duct the same ¼ of the duct's width each time. A total of four readings are obtained from the hole, evenly spaced from the far wall to the near one.

Repeat the entire set of four readings for the other two holes. Record all of the readings and add all of them together. Divide the sum by 12 to find the duct's average velocity.

Round ducts must be measured a bit differently. Because the area of a circle increases with the square of the diameter, eight unevenly spaced readings are needed to survey the airflow. This gives two readings from four "same area" zones inside the duct.

Insert the probe into the duct the distances shown in Table 6.1. The distances in the table show how far into the duct (measured from the hole) the probe should be placed. Some probes have a distance scale on their side that can be used to directly read the insertion depth.

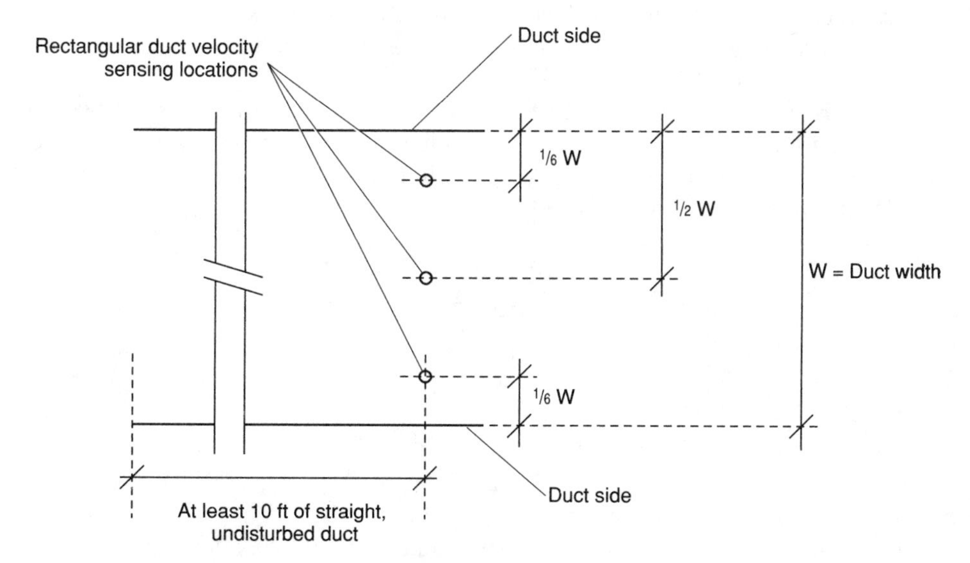

Figure 6.2. Rectangular duct probe locations for checking air velocity.

TABLE 6.1 Insertion Depths (Inches) for Round Duct Velocity Traverses

Duct diameter (in.)	Probe/test number 1	2	3	4	5	6	7	8
10	0.33	1.07	1.98	3.75	6.25	8.02	8.93	9.67
12	0.40	1.28	2.38	4.50	7.50	9.62	10.72	11.60
14	0.47	1.49	2.78	5.25	8.75	11.22	12.51	13.53
16	0.54	1.71	3.17	6.00	10.00	12.83	14.29	15.46
18	0.60	1.92	3.57	6.75	11.25	14.43	16.08	17.40
20	0.67	2.13	3.96	7.50	12.50	16.04	17.87	19.33
22	0.74	2.35	4.36	8.25	13.75	17.64	19.65	21.26
24	0.80	2.56	4.76	9.00	15.00	19.24	21.44	23.20
28	0.94	2.99	5.55	10.50	17.50	22.45	25.01	27.06
32	1.07	3.41	6.34	12.00	20.00	25.66	28.59	30.93
36	1.21	3.84	7.14	13.50	22.50	28.86	32.16	34.79

Record the velocity at each distance, add them together, and divide the sum by eight. The result is the duct's average air velocity. Don't neglect readings where the probe is inserted only a short distance into the duct. These provide a true picture of the flow in the duct, accounting for the slow mass of air near the walls.

Table 6.1 shows round duct insertion depths for various common duct diameters. Use these figures if doing velocity traverses on round ducts whose diameters match those in the table. Multiply the distance factors given in the first column of the table by the duct's diameter if the duct being tested is not in the table.

Example: Find the average air speed and volume in a 14-in. diameter round duct. First, find the insertion depth of the probe for the eight required readings. The 14-in. diameter duct requires readings be taken at the depths shown in Table 6.1.

Start with the smallest insertion depth (½ in.), and orient the probe directly into the airflow. Record the reading and repeat the process seven more times, once for each insertion depth. For the example, assume the following velocities were recorded:

Velocity	Feet/minute	Depth (inches)
1	275	½
2	368	1½
3	496	2¾
4	553	5¼
5	577	8¾
6	488	11¼
7	401	12½
8	266	13½

The average of these speeds is their sum divided by eight. With the data above, the average is 428 fpm. Chapter 4 gave the formula for finding the air speed in a duct given the air volume and the duct diameter.

$$\text{fpm} = \frac{144 \times \text{cfm}}{(0.785 \times \text{Diameter}^2)}$$

Rearranged to show cubic feet per minute as a function of diameter and velocity, the formula is

$$\text{cfm} = \frac{\text{fpm} \times \text{Diameter}^2}{183}$$

Using the example's values, the equation becomes

$$\text{cfm} = 428 \times \frac{14^2}{183} = 458 \text{ cfm}$$

For reference, rectangular duct volume is found from its velocity, width, and depth by the formula

$$\text{cfm} = \frac{\text{fpm} \times \text{Width} \times \text{Depth}}{144}$$

where width and depth are both in inches.

After finding the volume in the system's main duct, compare it to the design requirement. It can be adjusted up or down by varying the fan's speed according to the fan law: The volume of air varies directly with the fan speed. Adjust the sheave diameters (while watching the current draw) or winding tap and establish the proper air flow.

Fans without sheaves or multiple-speed taps can be adjusted by controlling their inlet air flow. Avoid excessively restricting the air flow, if possible, to keep the fan operating efficiently.

Once the main duct's air flow is set, adjust the major branches. First, check the air delivered to the critical path's outlet diffuser. This is most easily done with a hood-type velometer. These mount to the diffuser's frame and capture all of the air discharged. Various hood kits are available to use with different diffuser sizes and types.

If the critical diffuser's output is slightly high, adjust its internal damper. If it's acceptable, then the total pressure drop in the system is at its design point. Too little output shows that the pressure drop in the system is too high. It is likely that too much air is being delivered to noncritical branches.

Adjust the major branches' dampers so they are more closed (starting with branches closest to the fan) and recheck the critical diffuser's output. You should be able to set it to the proper volume with some tuning of the branch dampers.

Finally, if the main duct and critical diffuser's volumes are correct, balance the main branches. Do volume traverses of them to confirm that they carry their design requirements. Adjust them with their dampers to deliver the correct amount of air.

Some give and take between branch airflows exists. As air is reduced in a branch by closing its damper, the pressure in the main duct increases. This causes more air to be delivered to the other branches. The opposite condition can also happen: opening a damper in one branch can cause less air to be delivered to other branches. This effect can cause some branches that were already set to become out of balance.

Because of this, it is a good idea to go back and confirm that branch air flow remains correct after adjusting other branches. The main duct and critical branch are the minimums to retest as the work proceeds. Also, regularly recheck the motor's current draw to make sure it is not becoming overloaded.

A balance report should include the final air delivery volumes from every diffuser in the system and the air volumes drawn through all return and outside air inlets. Use a diffuser velometer to check the diffuser and register inlets and outlets, and use traverse measurements to establish the outside air supply duct's volume.

Water Systems

Proper water flow in hydronic systems provides the correct amount of heating or cooling. The idea is the same as air balancing: Adjust the flows to deliver the correct amount of water to all parts of the system. Most systems should provide a way to check and adjust the main flow through the pump. Unlike air systems, however, not all hydronic circuits require balancing. Some circuits are self-balancing, allowing the water to adjust its flow to conditions in the system.

Specific water flow circuits are shown in Chapter 7. While they vary in connections and details, all systems that require balancing follow similar procedures.

Water flow in pipes is tested with several types of instruments. Devices commonly used for HVAC system flow measurements include ultrasonic flowmeters, pressure drop orifices, and rotameters. Each has its advantages and disadvantages. Figure 6.3 shows each type of flowmeter.

Ultrasonic flowmeters are the easiest to use. They clamp onto the outside of the pipe and can read the speed of the water flowing inside. The reading is taken through the pipe's wall. The hydronic system remains closed because the meter reads the flow by measuring the way the flow affects sound transmitted inside. While they are expensive, their ability to read a closed pipe's flow is a tremendous advantage. Ultrasonic flowmeters don't work well if the flow has just run past an elbow or other fitting. Use them on straight pipes several feet past a fitting to get the most accurate reading.

Flow restriction gauges are also available. To use them, install a flow restriction orifice in the pipe to be measured. The system's water runs through the orifice and its pressure is reduced by the orifice's restriction.

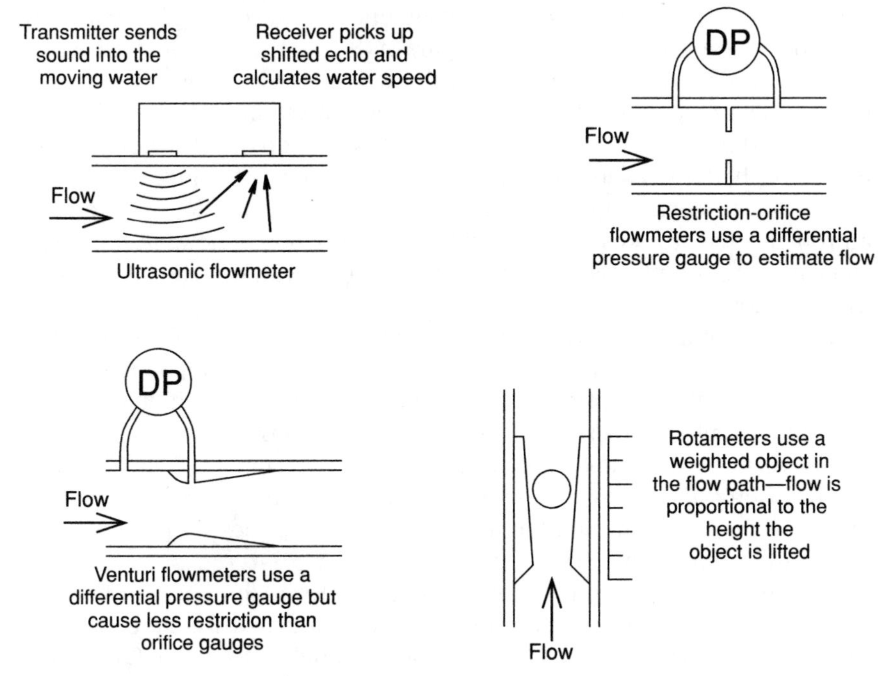

Figure 6.3. Various flowmeter types.

The pressure drop is related to the speed of the water flowing in the pipe. Orifices are available with different openings to allow measuring various ranges of flow without excess restriction.

Related to orifice flow restriction gauges are venturi devices. Water flows through an area with a diameter reduction (venturi section) along its length. The smaller diameter causes the water speed to increase, but does not significantly increase the pressure drop along the pipe. The water's speed increase causes its static pressure to decrease compared to its pressure within the main part of the pipe. Higher water speeds, associated with greater flow rates, cause more of a pressure drop within the venturi. The instrument compares the static pressure before and within the venturi to find the flow rate.

Another pressure loss test is done at balancing valves fitted with pressure taps. Pressure taps are small openings in the valve's body provided just up and downstream of its plug. By placing a pressure gauge in each opening (or a differential pressure gauge in both openings), the pressure drop across the valve can be read. While a tapped valve's pressure drop usually won't provide the actual gallons per minute flowing through the pipe, knowing the valve's pressure drop can be useful. The higher the pressure drop across the valve, the lower the flow through it.

Rotameters are direct-reading instruments that use the momentum of the moving water to lift or displace a part of the instrument. Most use a plug that fits inside a tapered tube. Water flows from the bottom of the tube to the top. The plug, normally resting near the bottom of the tube, lifts from the upward flow of water. The tube is larger at the top than the bottom. The plug is pushed up to a level where its weight balances the flow.

Other displacement-type flowmeters are also available. Some meters use a propeller that rotates in a housing to measure water flow. Others use a vane and pointer, pivoted to move past a scale as water pushes the vane in the direction of flow.

All these units, except the ultrasonic flowmeter, require a direct connection to the piping system. This can be a disadvantage. It is impossible to find the flow through an existing system without shutting it off, draining it, cutting, and installing the proper fittings. Once the readings are done, the piping needs to be disturbed again to remove the instrument. This can make the ultrasonic flowmeter the wisest choice. Although more expensive to purchase than the other units, they are easy to use and can pay for themselves after only a few jobs.

Another way of confirming proper water flow through a boiler or chiller is to measure the water's temperature change as it runs through the unit. If the heat added or taken out is close to the nameplate rating, the flow can be found with the formula

$$\text{gpm} = \frac{\text{Btu/hr}}{(T_d \times 498)}$$

The value calculated for the water flow using this method is quite accurate. If the heater or chiller is working properly the water's temperature change is proportional to the flow.

The requirements for balancing a hydronic system are similar to those for an air system. All gate and balancing valves should be set to the wide-open position. Automatic control valves should be fully opened. Check that all strainers have the correct screens installed and that all drains are closed. If using a flowmeter that must be installed into the piping, place any fittings that are needed into the piping before proceeding.

Completely fill the system with water. To do this properly, fill it from the bottom and allow air to escape at the top. The system's pump will remove some of the remaining air out of the system, but don't expect it to remove large amounts trapped far above it.

Most systems can be filled with plain tap water. Those that must use an antifreeze solution should have it blended before filling the system. Mix it to the proper proportion with water in a drum and use the system's pump to draw it into the piping. This can be done with a hose temporarily connected to the pump's inlet.

Some designers specify that a one-time corrosion inhibitor must be added to the initial fill. Treat this like antifreeze: mix it with the water before filling the system.

Once the system is completely filled, check that the expansion tank has its proper charge of water and air. Confirm the pressure in precharged diaphragm tanks before and after filling. If more pressure is needed, adjust it with a compressor or hand pump.

Next, start the pump. Confirm that a three-phase pump is turning in the right direction and that the current draw is within the motor's full-load amperage limit. Shut it off if the current draw is excessive. This probably means that there is a serious restriction in the system or a shutoff valve has been left closed. If a three-phase motor is turning the wrong way, swap any two of the three connections to the power source.

Once the pump is running properly, check the entire system for leaks. Inspect every pipe and fitting for integrity and correct any problems. Completely drain copper pipe before soldering. If a copper system does need some minor repairs, mark all of the fittings that must be corrected before draining the system. This allows you to fix all the problems with one drain-down.

Disinfect any systems that will carry potable water after leak testing. This is often done with a chlorine solution according to specific codes.

After the system has been proven to be free of leaks, test the boiler or chiller. The boiler can be used to help remove any air trapped in the piping. The hotter the water is, the less air it can hold. Therefore, heating the water to approximately 200°F drives much of the air dissolved in the water out of solution. It will form bubbles and travel to the system's air eliminator. Let the hot water circulate through the system for at least 4 hours to remove the air.

While the system is running, test all of the coils and fin tubes to see if they are receiving water. Check for noises, vibration, and pipe surge (shaking) in the entire system. Parts of the system that should receive water might be air blocked. Open the fitting there to remove ("bleed") any air. Once water comes out of the bleed, close it, and check the system again.

A properly operating system runs smoothly and quietly, with no surging or gurgling sounds. Either condition suggests something is wrong that should be corrected. Loose supports, water hammer, or excessive velocity can cause pipe surge and shaking. It is important to correct pipe surge. This condition eventually weakens the pipe's joints and causes them to fail.

Gurgling sounds suggest air is traveling through the piping. That condition might correct itself after several hours as the air is caught by the eliminator. If the air does not leave after 4 hours, loosen some fittings high in the system, bleed any trapped air, and close the fitting once no more air is discharged.

Now the system flow can be checked. Using a flowmeter or the water's temperature change, check the flow through the boiler or chiller. Adjust it, if necessary, by closing a balancing valve. If you are using the temperature change method to check flow, allow enough time for the inlet and outlet temperatures to stabilize before figuring out the flow rate.

Remember that although the chiller or boiler might be working fine, the rest of the system might not be putting a significant load on it. But watch that the water is not excessively heated or cooled. While the chances are the equipment's safety systems won't allow dangerous system temperatures or pressures, don't depend on them. Monitor the temperature, and shut off the boiler if the temperature gets over 210°F.

Also, be sure that heated water does not go to a chiller. Even a shut-down chiller can be seriously damaged by hot water. The refrigerant's pressure becomes extreme and can rupture the safety vents. If there is any connection between the chilled water and heated water systems, double check that the shutoff valves to the chiller are closed when using the boiler.

Very cold water can also damage an operating boiler. If testing the chiller, the system's water is likely to be cooled to between 40°F and 50°F. Close the boiler's isolation valves and do not send cold water to it. The sudden temperature shock can rupture the boiler's tubes.

Also, keep in mind that chilled water can cause condensation. While cold water won't directly hurt an idle boiler, the condensation dripping into its firebox and wetting the tubes or coils can cause corrosion damage. Always keep chilled water out of the boiler.

Once the main pump's flow is confirmed, verify branch flows. Adjust the balancing valves until the specified water flow is established in each circuit. Recheck the main system flow and the pump motor's current draw, and fine tune the system to get all of the flows correct. Lock all of the balancing valve's plugs in place.

If the system uses a nondiaphragm expansion tank to collect air, confirm that the water level is correct. It might be necessary to remove some air that came out of the piping to be sure the correct amount of water is in place. Allow the system to cool to room temperature before rechecking the expansion tank's level. A hot system naturally forces more water into the tank, giving it a higher water level than its room temperature level.

Air can usually be removed from nondiaphragm tanks by opening the special bleed fitting at the tank's bottom. Water should enter from the system's make-up supply and fill the tank to the top of the fitting's fill tube (approximately two-thirds to the tank's top).

Chapter 7

Heating Systems

Heating systems are available in many styles and types to match almost any need or situation. They use a wide range of fuels and their distribution methods fit many requirements. The cost and efficiency of heating systems have both improved as advanced technology has been adapted to these systems.

This chapter provides an outline of the heating systems that are available and which are best for particular applications. System delivery types (forced hot air, hydronic, and radiant) are described, including the advantages and drawbacks of each.

Heat Sources

Many fuels can provide heat. While solar, coal, and wood systems (including pelletized cellulose) are available, these are not used for most HVAC systems. Normal residential and commercial systems operate with either a fossil fuel or an electric source.

Fossil fuels

The most popular heat sources are fossil fuel based. These fuels include natural gas, liquefied petroleum (propane and butane), and oil or kerosene. Fossil fuel systems are often more economical to operate than electric systems and require less expensive building wiring.

Fossil fuel systems burn fuel at a controlled rate in a combustion area. Combustion demands a constant supply of air from either the ambient space or a dedicated outside air supply duct.

Burning fuel is placed nearly in contact with one wall of a heat exchanger. Heat exchangers carry the energy into the air or water used to provide the building's heat. Burning fuel and hot combustion products on one side send heat through the heat exchanger's wall to the air or water on the other side.

Exchanger walls should be thin enough to provide the highest efficiency, but thick enough to resist the severe combustion environment. They need a large surface area to collect as much heat as possible.

Natural gas. Natural gas is one of the more popular fossil fuels. It consists primarily of methane, a molecule with one carbon and four hydrogen atoms. If the heater is operating properly, the gas oxidizes into carbon dioxide and water vapor. The low toxicity of its combustion products allows unvented natural gas cooking to be used indoors.

This does not mean that the flue discharge from a gas fired heating system is not harmful. Heaters can produce enough carbon monoxide to kill a building's occupants. Vent the flue according to local building codes or the requirements of the National Fire Protection Agency (NFPA). The NFPA's Standard 54 describes in detail the specific requirements for properly venting gas fired equipment. These must be followed exactly. A mistake in this area could be fatal.

The production of water vapor as a by-product of combustion means that flue gases are humid. Since the gases are hot, the water remains in a vapor form and passes out of the flue with the carbon dioxide (and some carbon monoxide). If the gases are sufficiently cooled the water vapor condenses into liquid water.

Natural gas is sold by the hundreds of cubic feet, abbreviated CCF. Most fuel gas is blended with nitrogen (an inert, nonflammable gas) to provide a heat value of about 1050 Btu/ft^3. One CCF contains the energy equivalent of 105,000 Btu. For most estimating, designers round this off to 100,000 Btu/CCF.

Another natural gas measure sometimes used is the *therm*. One therm is 100,000 Btu, but it is often used in place of CCF when specifying natural gas systems.

Gas is supplied to HVAC heating systems at low pressure. Most units work correctly with a gas pressure between 5 and 10 in. water. Utilities supply a regulator at the meter to reduce the gas pressure from the street distribution level of 30 psi. Most gas is distributed in buildings at between 7 and 10 in. water. Because of the low distribution pressure, gas piping must be installed with low pressure losses. NFPA Standard 54 includes tables showing the proper size pipes to use for various system capacities and run lengths.

Heating systems do not get the full Btu potential out of the gas. Some heat is lost out the flue with the exhaust gases. The amount of heat retained and used by the heater is a measure of its efficiency. Efficiency of fossil fueled heating devices is measured by their annual fuel utilization efficiency (AFUE). This rating considers the unit's operation during an entire year's operation. The test includes times of heavy load, light load, long and short cycles, starts and stops. It is a well developed standard that helps purchasers select the most efficient systems and units.

Higher efficiency units cost more than lower efficiency ones. This makes it tempting to install the lower efficiency units to keep job costs as low as possible. Because many building owners are interested in saving money, the option of installing a high-efficiency system should be presented. When the lower operating costs over the system's life are considered, the more efficient heating unit might be the lower total cost alternative.

Natural gas heating systems have become very efficient. Condensing heat exchangers cool the exhaust gas below its dew point, allowing the water vapor formed by combustion to liquefy. This process extracts the additional heat of water vaporization from the exhaust. Every pound of water condensed contributes over 1000 Btu to the building. The cooled flue gas is usually discharged outdoors through a PVC pipe. No metal or ceramic flue or chimney is required.

Starting most gas furnaces requires several steps. Because most do not use a standing pilot light, and use powered combustion gas venting, start-up can be involved. Most systems use a series of relays and proving switches to begin heat delivery.

When the thermostat (or other control) requires heat, it closes a relay in the heater. This usually energizes a flue gas discharge fan. This fan forces the combustion gases out of the flue or exhaust pipe. A pressure switch connected to a pitot tube in the flue confirms two things: the air from the flue fan is flowing, and it is not blocked. A failure of either halts the start-up.

Once the flue discharge fan has been proven to be running properly, the gas is lit. Several options are available. Some systems allow a small amount of gas to flow to a pilot light burner where it ignites from a high-voltage spark or a heated wire. Again, the unit has sensing devices to prove that the ignition occurs within a specified time. If the flame does not start in the allotted time, the gas is shut off and the heater stops its starting sequence.

Once the pilot burner is lit, the heater allows the main flow of gas to start. The gas lights easily from the large pilot flame, ensuring a quiet combustion start.

Some heaters do not use a pilot gas burner. Instead, the main gas is turned on once the flue discharge fan's operation has been proven. A spark or hot wire ignitor located at one or more of the main flame tubes starts. Gas reaching these tubes lights and a distributor sends burning gas to the other tubes.

This type of system is much more sensitive to a slow start than pilot types. Because the full flow of gas starts to all burners simultaneously, before any of it has ignited, some unburned gas might accumulate around the flame tubes. Once ignition is started, this gas can light noisily. If the flame distributor doesn't work properly, an entire flame tube might not light. It discharges gas into the combustion area while adjacent tubes are at full fire. This can cause an explosion. Install and set up nonpilot heaters with care to ensure that they start reliably.

If anything goes wrong in the starting sequence all heaters go into *lockup*. When this happens the thermostat or other control must be shut off to reset the starting sequence. Resetting from the thermostat is usually done by setting it to a low temperature. It must be set below the ambient air temperature so that it stops calling for heat. It's usually only necessary to lower the thermostat's temperature for a few seconds to reset the furnace or boiler. After resetting, the system can attempt to start again.

A common failure point with gas heaters is the ignitor. The ignitor usually has some constant contact with part of the flame while the unit is firing. This contact point might serve as a flame sensor to shut the system off in the event the flame goes out. Constant exposure of the ignitor to the flame can cause it to wear and erode. High temperature and plentiful oxygen in the flame can cause the ignitor to oxidize. The ignitor should be the first thing tested when servicing a "won't start" complaint.

Very efficient gas heating plants use pulse combustion technology. Gas is not supplied in a steady stream with a standard gas valve. Instead, the gas blends with air in a special mixing chamber and enters the combustion chamber in a series of pulses. Each pulse of gas burns individually, and the momentum of one pulse discharging out of the exhaust sucks in the next.

Pulse combustion burners get their high efficiency from the combustion chamber's very large surface area. This also serves as the heat exchanger, and the burned fuel is cooled almost immediately to below 100°F. Efficiencies of over 90% are common with pulse combustion systems.

Many high-efficiency furnaces and boilers have combustion air inlets piped from the outside. Make these connections per the manufacturer's instructions to allow the unit to burn only outside air. This helps to keep combustion air clean and system efficiency high. Condensing heat exchangers can be quickly damaged if the heater burns contaminated indoor air. Extremely minute quantities of chlorine gas in the combustion air can greatly accelerate the heat exchanger's corrosion. Chlorine in the air from running chlorinated water or doing laundry nearby with chlorine bleach can cause damage.

Natural gas heating systems are very popular. They are moderately priced and have gained widespread acceptance. Many clients request gas heat if it is available.

Liquefied petroleum. Propane and butane are gases provided in liquid form for heating systems. Refined from oil, they are sold by the pound, stored in high-pressure tanks, and have a large heat value. Liquefied petroleum (LP) gas contains more Btu per pound than natural gas. Most systems use propane, but a blend of propane and butane is sometimes considered LP.

Propane has a heat value of 21,700 Btu/lb, or 2480 Btu/ft^3. Butane's values are 21,300 Btu/lb and 3220 Btu/ft^3. The additional energy (compared with natural gas) comes from their more complex structure. Their molecules have more carbon and hydrogen, so more oxidation occurs when they are burned.

Most of the heating systems available for natural gas can be fitted to run on LP. The orifices that feed gas into the manifold must be calibrated for LP gas, and pressure regulators are usually set to a lower output. These changes allow the heater to keep its original Btu per hour output value with the higher heat content fuel.

Bulk propane and butane is supplied at a much higher pressure than natural gas. Use care when piping these fuels before a pressure reducing regulator. At 120°F, propane's pressure reaches 225 psi. While butane's pressure is only 56 psi, both can cause serious leaks if the piping is not installed with care. Follow local codes and NFPA regulations carefully when piping LP gas.

Oil. Oil heat is commonly used where natural gas is not available and electricity costs are too high. While new technology has improved the efficiency of oil fueled systems, they are not quite as efficient as the best natural gas units. The problem is due to oil's complexity.

Fuel oil is a complicated mix of many compounds. When burned, they form gaseous products that include sulfur, ash, and other residues. These must be kept from condensing on the heat exchanger. If the exhaust gases are cooled too far, they form a residue layer on the exchanger's surface that can reduce the exchanger's efficiency. Flue gases must be kept hot enough to prevent their condensation.

This hot exhaust takes heat out of the building. While the efficiency of oil heaters can be enhanced with special burners and combustion air controls, the need for a high-temperature flue discharge is a limitation.

Oil is available in several grades, referred to by numbers. Higher grade numbers indicate thicker, more viscous oil, and lower number oils correspond to more highly refined oils. Lower numbered fuels cost more than the higher numbered ones. Most commercial and residential heating systems use #2 oil, while industrial units handle less expensive #4 or #6 oil.

Kerosene is similar to fuel oil, and most oil burners can handle it in limited quantities. It is much more expensive than fuel oil and is usually not used in HVAC systems.

Number 2 oil has a heat content of 143,000 Btu/gal. At approximately 7.3 lb/gal, this equates to 19,600 Btu/lb. While this heat value is comparable to liquefied petroleum, its combustion is not as efficient.

Oil's contaminants cause a soot coating to form when the flame impinges upon the heat exchanger surface. While a well-tuned burner can minimize this, the ash content of oil makes some buildup unavoidable. Oil fired systems must be cleaned regularly to maintain their efficiency. A very thin layer of soot insulates the heat exchanger's surface, seriously reducing heat transfer.

Most oil burners operate with a high-pressure pump. If the burner is located close to the storage tank, the burner's pump draws fuel out of the storage tank. No extra pump is required. If the tank is located far away from the burner, or well below its elevation, an auxiliary pump might be needed at the tank.

Auxiliary pumps deliver oil from the main tank to a small *day tank* placed near the burner. These small tanks usually hold about 75 gal of oil, enough for at least several hours of operation. When a float in the day tank drops, it closes a switch and automatically turns on the auxiliary pump. Oil transfers from the main tank to the day tank until the float turns the pump off. The burner's pump draws its fuel from the day tank as needed.

The burner's pump boosts the oil to high pressure. Different burner designs operate at different specific pressures, but most provide over 100 psi. The high-pressure oil goes to a burner orifice and sprays into the combustion chamber. As the oil leaves the orifice it atomizes into a spray of very fine droplets. Oil in this form is easily ignited.

The fine mist of atomized oil droplets passes through an electric arc ignitor. A high-voltage spark passing through the oil mist fires the fuel. Unlike gas heater ignitors that only operate at the start of burner operation, oil ignitors constantly run while the burner is on.

Newer burners, called *flame retention* types, provide more efficiency than previous designs. Oil is exposed to more turbulence to atomize it better, and the flame is controlled to improve its impingement on the heat exchanger.

Most oil burners are reliable. The pump operates with meshing, toothed wheels (resembling a pair of gears) to push the oil to the orifice. The ignitor electrodes are large-diameter rods that resist flame erosion. But a common problem with oil burners is clogging.

All burners have a filter to clean the incoming oil before it reaches the pump. Because oil can accumulate debris (from the delivery truck or water condensation in the tank), the filter is likely to become clogged at some point. It is usually no problem to remove the filter and clean or replace it to restore heater operation.

Water can also cause oil system contamination. Tanks located outdoors or underground are prone to collect water from rain or condensation. This water reacts with the oil, forming a nonflammable foam or scum that clogs filters and lines. If a creamy, brown foam (that resembles dirty whipped cream) is found in the filter, it is a sure sign of water contamination. The tank must be cleaned.

Small amounts of water can be removed with chemicals added to the tank, but significant amounts must be physically removed. Pump the dirty oil out and dispose of it, flush the tank with kerosene, and load new oil.

Regulations dealing with oil tank installation and oil disposal are stringent. The U.S. Environmental Protection Agency (EPA) requires that underground tanks, or above ground tanks with a surface touching the ground that cannot be inspected, be registered. Tanks must be leak tested and oil inventory and delivery records must be maintained.

No *significant spills* of oil can be disposed of into the ground or sewers. Oil must be disposed of with a registered disposal company licensed to handle it. If an oil spill is found, or evidence of a past spill is noted (including contaminated soil), notify the EPA or your state environmental office. Failure to do so can subject the party to significant fines and the possibility of further legal action.

Oil burning heaters cost more than gas fired units. The burners are more complicated, raising equipment costs. If a client insists on using oil heat, or if you are replacing an existing oil unit, this is the way to go. Otherwise, consider gas, heat pumps, or electric heat as an alternative.

Electric resistance

The simplest, most efficient heat source is electric resistance. Systems do not get any easier than these.

Electric resistance heat operates at perfect, 100% efficiency. All of the energy used by the heater becomes heat. The only waste is energy lost through the furnace or boiler jacket to the surrounding space. If the heater is located in the heated space it works without any losses.

Electric heaters are often installed where there is no fossil fuel available, or where electricity costs are low. Electric heat is the least expensive to purchase and install. The lack of combustion eliminates the need for any special ventilation or air supply.

Some electric heaters are installed to supplement fossil fuel systems. Others provide spot heat to a small area where it might be difficult to deliver heat from another source. Their ease of installation and control make them the top choice for warming a chilly area where it is not practical to install other heat. For example, it is usually less expensive to use an electric heater if a remote area far from hydronic piping needs to be warmed. The possibly higher operating costs of the electric heat might be justified to avoid high installation costs.

Electric heat wiring must be able to carry the required power draw. The heater might be the largest electricity consuming device on the premises. Be sure of the supply capacity when installing a new electric heater where one was not used before.

Many electric resistance heaters are baseboard units. Because no central combustion unit is needed with electric heat, the heat can be generated directly in the area where it is needed. Each room or zone can have an individual thermostat to operate its heater to whatever set point the occupant wants. Baseboard heaters are quiet, inexpensive, and are installed along the outside walls where heat is needed.

Other spot electric heaters include ceiling- and wall-mounted forced air units and radiant panels. These give the installer flexibility in placing the heat in the exact location where it is needed. Without ducts or pipes to run, the installation is fairly simple. Adequate electric wiring can usually be run with flexible conduit or cables, making the supply adaptable to tight-fitting locations. There are also central forced air and hydronic electric heating systems.

There are few parts inside most central electric heaters. The heating elements are long bars approximately ⅜ in. in diameter. Current passes through a wire embedded in the center of the bar and heat passes out through the bar's surface. The bars are covered by an insulating material, usually ceramic or glass. This material must be able to withstand the severe thermal expansion and contraction that occurs when the heater turns on and off.

Heat transfers directly from the electric elements with no heat exchanger. The heating elements are directly in contact with the air or water flow to be heated. Designed like the tubes in a coil, the fluid passes around and through the elements. Many electric air heaters use fins between the elements to enhance the heat transfer. Therefore, duct-mounted electric heating assemblies are often called *coils*, even though no fluid circulates inside them.

Many large duct-mounted or water immersed electric heaters operate with more than one stage. Heating elements are energized as the demand for heat increases. Less demand causes fewer heaters to operate. The control is usually a set of thermostats or aquastats installed in the path of the incoming fluid. The greater the difference between the incoming temperature and the desired space temperature, the more heating elements operate.

Some space thermostats have built-in two-stage heating controls. The first stage energizes if the space temperature falls one or two degrees below the set point. This is the first, low-stage heat setting. If the temperature continues to fall more than three or four degrees below the set point the second stage energizes. This turns on all of the heating elements, allowing a fast warmup.

These thermostats help a system that has been set back during unoccupied hours to quickly recover by using all of the available heating capacity. Once the space is warmed to nearly the set point, the low-power heat economically maintains the setting.

Heaters powered with three-phase electric service can be staged with *wye-delta* switching. The heating elements operate either in series with each other (wye) and generate less heat, or in parallel (delta) for full heat output. Although all the elements constantly operate, the wye connection produces about ⅓ the heat of the delta connection. All of the heating elements are turned off when no heat is required.

The heating elements are usually directly controlled with high-power contactors, or relays. These allow the low-voltage control circuits to turn on the system. Inspect the relay contacts when servicing electric heat systems. They can become worn or burned from switching the elements off and on.

Some heaters work properly only when the air or water flows at a specific minimum speed. If the flow stops or slows, the elements can overheat. They might warp, sagging into the walls of the heating unit, or the wires might burn out and open. These heaters have a safety switch to prevent them from operating if flow stops. Air handlers use a *sail* switch (with a paddle hanging in the air stream), and boilers use a water flow or pressure switch. In either case, check the switches if the heater doesn't operate.

Air-to-air heat pumps

A special form of electric heater is the heat pump. There are two types of heat pumps: air to air and water source. Water source heat pumps are described in Chapter 9.

Many air conditioners offer heat pump circuits as an option. They are often chosen as package terminal air conditioners (PTACs) used in hotels, nursing homes, and some offices. There are also central system heat pumps available for homes and moderately sized commercial buildings.

Heat pumps are specialized air conditioners that move heat energy into or away from a space. The system can be reversed to cause heat to flow from the space to the outside (like a regular air conditioner) or from the outside to the inside space. This is considered its heat pump mode.

Air-to-air heat pumps have two refrigerant coils and a reversing valve. When cooling is needed, the indoor coil functions as the evaporator and heat is removed from the space. The outdoor coil functions as the condenser and heat removed from the space is discharged outside.

Reversing the flow of refrigerant swaps the coils' functions. The outdoor coil becomes the evaporator and absorbs heat from the outside air. The coil absorbs heat from air only when its surface is colder than the air temperature. Air at 50°F can deliver significant amounts of heat to a coil at 25°F. As the outside air passes over the coil its temperature drops. Outdoor air discharged from an operating heat pump is often well below freezing because of its loss of heat. Heat removed from the outside air enters the interior space. The indoor coil functions as the system's condenser, discharging the system's heat inside. Indoor air warms as it passes over the hot coils.

It might be helpful to think of the heat pump as a *heat amplifier*. Heat absorbed at the low outdoor temperature is intensified in the heat pump to a higher temperature. This higher temperature heat is discharged inside.

The heat delivery capacity in the heat pump mode is approximately as high as the system's air conditioning capacity. A heat pump rated for 1 ton of air conditioning delivers about 12,000 Btu of heat per hour.

Air-to-air heat pumps do not always operate in the heat pump mode. If outdoor temperatures fall below freezing the system cannot extract heat from the air. When the air's temperature approaches the refrigerant's temperature, heat transfer decreases. Backup heaters run to satisfy space requirements. These backup heaters are usually electric resistance heating elements installed in the air handler cabinet. They take over when the system cannot operate in the heat pump mode.

Other electric heaters on the outdoor coils control frost. When cool (but above freezing), moist outside air is further cooled by the evaporator, frost can form. Water condensed out of the air freezes on the cold surface of the coil. This insulates the coils and can block their air flow. Electric resistance heaters thaw the coils' surfaces.

Defrost heater controls operate from several *frost sensing* systems. Outside temperature sensors are sometimes used with a timer to start the heaters when frost is possible. The heaters run for several minutes each hour when the outside temperature makes frost likely. Other systems monitor the outside temperature and refrigerant pressure. A blocked coil causes less evaporation to occur, reducing the system's pressure. A sensor on the compressor's suction line starts the defrost heaters whenever the pressure falls below a set point and the outside temperature is low enough. When the pressure returns to normal the heaters are shut off. These systems are more efficient than timer-controlled ones because the heaters are only used when required.

Energy transfers very efficiently when the system runs as a heat pump. Outside heat is the main source of energy delivered into the space, so these systems deliver more heat energy than is drawn from the power source. Heat pumps typically use about ⅓ the energy of an electric resistance heater of comparable capacity. Energy and operating cost savings are significant.

Depending on the climate and fuel costs, heat pumps can operate at lower annual costs than fossil fuel systems. They always cost less to operate than electric resistance heaters. The more they operate in the heat pump mode, the higher the efficiency and savings. There are few savings, however, if a heat pump is using its backup resistance heat most of the time. The climate should have relatively long periods of above freezing weather. Much of the eastern United States has a climate compatible with air-to-air heat pumps.

Like electric resistance heaters, air-to-air heat pumps make ideal supplemental or spot heat units. Their operation as an air conditioner can be presented as a bonus for the client. They might never have considered installing air conditioning, and the idea of getting it included with an energy efficient heater might be very appealing. Air-to-air heat pumps cost more than straight electric heat units, but the extra purchase cost can usually be recovered in a few years with lower electric costs.

The efficiency of heat pumps operating as air conditioners is rated by a factor called the heating system performance factor (HSPF). The HSPF is found by operating the system at specific inside and outside conditions. These tests ensure that all manufacturers use realistic conditions for the rating. The higher a unit's HSPF, the greater its efficiency and the lower its operating costs.

Most air-to-air heat pumps are rated with an HSPF, but their air conditioning mode efficiency might also be important for the application. These ratings are covered in Chapter 8, but include the energy efficiency ratio and the coefficient of performance (EER and COP). Like the HSPF, higher EER and COP ratings show that a unit is more efficient.

Heat recovery

Some systems use energy wasted from other processes for comfort heating. These are grouped together as heat recovery systems. Most of these systems use refrigeration condensers or manufacturing processes as a heat source.

Heat recovery systems are usually used as a backup source of heat, installed to save energy costs for the primary heating system. They are usually connected to the primary heating system with controls to allow the recovery unit to automatically start when it is economical to use it.

Most heat recovery systems are custom designed for the application they serve. While HVAC technicians usually do not become involved in their design, some familiarity with the operation of these systems is helpful.

Flue. Heat recovery units are available for residential and small commercial heaters. They are installed into the main heater's flue pipe. Hot furnace or boiler exhaust gases rise past the surfaces of a set of steel tubes. The flue gases surround the tubes' walls, but cannot get inside. Room air is blown into one end of the tubes and discharges out the other end.

Many of these heat exchangers are used with wood burning stoves and fireplaces, but models are available for oil and gas fueled heaters as well. They are not compatible, however, with some of the newer, more efficient heaters.

Units that cool the flue gases below or close to the condensation point should not use external heat recovery exchangers. Corrosive liquids that form quickly destroy the integrity of the steel tubes, leading to a carbon monoxide leak into the space.

Refrigerant. Air conditioning and other refrigeration systems might operate in one area while another requires heat. Computer rooms, large refrigerated areas, and core areas of large buildings might all require constant cooling, no matter the season.

Air conditioning and refrigeration heat recovery systems are usually built in one of two ways: direct air heating or water heating. Direct air systems are the simplest. The system's condenser is installed in an air duct connected to the building's warm air supply. Some systems use dampers on the duct to allow the condenser to operate all the time. The dampers vent the condenser heated air out of the system when it isn't needed. They close when recovered heat is wanted, diverting it into the building. Building air passes through the condenser and all of its discharged heat helps warm the building.

New direct air systems might not be permitted by some codes. Concern over the possibility of refrigerant leaks into the ventilation air has caused some code officials to disallow these new systems. Refrigerant leaks in or near the condenser would be inducted directly into the building. Because the refrigerant might be toxic, many building codes don't allow systems that might expose occupants to leaks. Technicians should check with the local code enforcement officials before repairing direct air heat recovery systems. Even repairing one of these systems might be illegal. The system might have to be removed.

Water recovery systems use an auxiliary condenser built like a shell and tube heat exchanger. Refrigerant runs through the tubes and water passes by them in the surrounding jacket. Waste heat passes from the refrigerant into the water through the tubes' walls.

Air conditioning systems that use water heat recovery condensers also require a standard heat rejection device. This might be a dry cooler or cooling tower (used with a water-cooled condenser). There might not always be a demand for hot water, but the system needs to continue operating.

Use safety controls to prevent the water from overheating. Condensing refrigerant can heat the water to its boiling point, causing the pressure to drastically increase. Steam can be injected in the piping and the resulting water hammer might damage it. Use water flow and temperature safety controls to divert the refrigerant to the nonheat recovery condenser, if necessary. Any time water stops flowing through the heat exchanger or its temperature becomes too high, the controls shut off the heat recovery condenser. Some systems use a second-stage control as a backup. It stops the refrigeration system if the recovery tank's water continues to rise beyond the first shutdown point.

Some water heat recovery systems heat domestic (potable) water. Check applicable codes to see if specific requirements must be followed before installing or replacing these units. Many codes require that double-wall heat exchangers be used, with the space between the walls vented to the atmosphere. This prevents the possibility of refrigerant leaking into the water supply where it could be ingested.

Also, the water temperature of some potable systems must be closely controlled. Health care facilities, for example, must supply water no hotter than 110°F to the resident's faucets. Be certain that the system will not cause the water to overheat.

Process. Many manufacturing plants produce waste heat. Some plants use it for comfort heating when it is needed. Process heat might be available in gases, steam, hot water, or heated air. Some manufacturing plants use heat recovery for almost all of their comfort heating needs. Comfort heating systems in some facilities are designed around the available waste heat from the manufacturing process. If the process has leftover steam or hot water, the building's heating system uses it. The "free" heat is the system's primary fuel. These facilities then integrate a backup heating unit into the primary heat recovery system.

It is essential that any manufacturing process heat recovery system not impair the process in any way. It cannot cause a system pressure buildup (or loss) or prevent the flow of the fluid. The HVAC system must operate "invisibly" to the process, and get off-line quickly if a problem develops that could hurt the manufacturing process.

Some process waste gases are very hot. Flues used for incinerators, waste gas burning, or kilns carry hot gases out to the atmosphere or to industrial scrubbers. These are sometimes tapped as energy sources for heat recovery.

Water can be sprayed directly into the gases to absorb heat as they are driven out of the stack. The water cools and cleans the gases, and a steady supply of hot water suitable for heating is provided. Use materials resistant to the corrosive compounds picked up from the gases to filter and handle the water. Water-to-water heat exchangers might be used to prevent contamination of the HVAC system. The contaminated water stays on one side and the hydronic system water stays on the other.

Some process gas heat recovery systems use a heat exchanger in the gas stack. A set of corrosion resistant pipes or tubes is installed in the stack's wall. These form the heat exchanger. Water flowing through the tubes collects heat from the gases. The water then delivers this heat to the heating system. Because the gases never directly touch the water, special cleaning is usually not necessary. These systems allow the building's hydronic heating water to flow through the stack's heat exchanger, or use a second heat exchanger to couple the stack exchanger's hot water into the building's. The choice depends on the degree of risk associated with a possible leak.

Some gases are more toxic and corrosive (and might cause more leaks) than others. With a single stack exchanger, a leak can cause the entire building's heating water to be lost into the stack. This can ruin the manufacturing system. A separate heat exchanger system, however, can be designed with little water from the stack system. A leak in this case would not be as damaging.

Waste process steam can be directly used in coils and shell and tube exchangers for HVAC systems. The system must accept the steam at the delivered pressure and be able to handle surges that might accompany machine operation.

Heated water is often a by-product of manufacturing processes. It can be used directly in hydronic systems if it is not contaminated. A heat exchanger is usually used to prevent the process water from contaminating the HVAC system and to allow the systems to operate at independent pressures.

Heat amplifiers. Heat recovery can often be enhanced with a refrigeration-type heat amplifier. Where large quantities of low-temperature, low-grade waste heat are available, an amplifier booster can economically increase its temperature enough for comfort heating.

Waste heat boosters are refrigeration chillers. The process fluid is cooled (from, for example, 80°F to 50°F) with the booster's evaporator. Its condenser operates at a higher temperature, usually approximately 160°F. This is sufficient to provide hydronic heating. Direct air heating from the amplifier with an air-cooled condenser might not be allowed. As noted earlier, concerns about refrigerant leaks into occupied areas have made many of these systems illegal.

Forced Hot Air Systems

Most heating systems use forced hot air. Because most air conditioning systems require forced air, the air handler can be used for both heating and cooling. It is more economical to install a single air delivery system than two, so many clients and installers make forced hot air their first choice.

Hot air systems are available in a wide range of system styles and capacities. Technically, an air delivery system that includes a hot water heating coil can be considered a forced hot air system. Most technicians, however, consider true forced hot air systems to be only those that include a furnace or electric heating coil. This is the type of forced hot air system considered here.

Most forced hot air systems are fossil fueled. Because the combustion takes place at one location, the energy must be distributed from there to the space. Forced air lends itself to this well. Electric furnaces are available, but are not as common. Forced hot air systems operate as a cabinet-mounted furnace or a rooftop integrated system. Furnaces are available for horizontal, vertical upflow, and vertical downflow applications.

Furnaces

The central forced air heating device used to heat a flow of air is the furnace. Most furnaces are controlled by the building's space thermostat. When heat is needed in the building, the burner starts immediately. Air does not flow at first to allow the heat exchanger's temperature to rise over about 110°F. This prevents the unit from delivering an uncomfortable blast of cool air when it starts its heating cycle.

The blower starts when an internal thermostat (called a *limit* control) senses that the heat exchanger is properly heated. Once the heat exchanger is warm enough, the furnace's blower starts and delivers heat to the space.

When the space temperature reaches the thermostat's set point, it turns off the furnace. Heat production stops, but the fan runs until the heat exchanger's temperature falls below the limit control's set point. This helps extract every bit of heat out of the furnace for greater efficiency.

Furnaces receive the heating system's return air in a plenum. A filter cleans the air and it passes into the fan's inlet. The fan forces the air through the heat exchanger and out to the supply ducts. It returns to the inlet and the cycle repeats.

The fan is installed upstream of the heat exchanger for an important reason. Air pressure just past the fan is the highest anywhere in the system. If the heat exchanger leaks, the air's higher pressure prevents combustion gases from entering the air side of the exchanger. If the fan were downstream of the heat exchanger, the opposite would occur. Low air pressure would draw flue gases into the heat exchanger, allowing a possible carbon monoxide hazard.

Many furnaces use indoor air for combustion and to help the flue draft. No outside air inlet is needed. With these units it is also important to avoid leaks in the return air system, or any return air openings, in the same room as the furnace. Under some conditions the fan can draw air down the chimney and into the nearby return air inlet. This prevents the furnace from properly venting and can cause all of the flue gases to enter the building.

Constant improvements in furnace heat exchangers have been made over the years. Once always made of cast iron, most new heat exchangers are now made of thin, ceramic-coated steel. They intercept as much of the flame and heated combustion air as possible to maximize their efficiency. As noted earlier, some heat exchangers cool the flue gases below their dew point. This causes the water produced by combustion to condense, liberating the water's heat of vaporization for additional heating of the space.

Condensing furnace heat exchangers quickly drain the water away from their surfaces. Plastic, corrosion-proof trays and tubes carry the water out of the furnace where it is disposed of. Most codes require that the condensate discharge into a sanitary sewer, not a storm drain or sump. Use corrosion-proof pipe for the drain line. The condensate can contain sulfuric, nitric, and hydrochloric acids from the combustion process.

Many modern furnaces have combustion air inlets that receive outside air. Using outside combustion air helps keep the space from having a negative pressure. Consuming indoor air for combustion and flue gas removal causes a net air loss from the building. The furnace's outside air inlet should admit enough air to feed both the combustion and the discharge of the flue gases. Thus, combustion and flue gas removal can occur without using any indoor air.

Electric heat storage. Some central electric furnaces are built with an integral thermal storage mass. These are bricks of dense, rock-like material that absorb and store heat. Heat storage furnaces are popular where the utility offers different electric rates depending on the time of day.

To trim peak demands, some utilities offer customers cheaper electric power when the total demand is lowest. This is usually at night. Energy storage furnaces allow a building to use this cheaper power.

Because electric furnaces do not need special heat exchangers, burners, or flues, they can be more compact than fossil fuel units. Heat storage furnaces use this extra space by filling it with hundreds of pounds of heat storage bricks (see Figure 7.1).

The furnace operates all night, while the electric rates are low, heating the bricks to a very high temperature. Air is drawn through the heating elements, around the bricks, and back to the elements. If the building needs heat while the bricks are warming, some hot air leaves the furnace and goes to the space.

After the bricks are hot, and the electric rates have gone back up, building heat is supplied entirely by the bricks. The heating elements are left off as long as the bricks stay hot. When the building needs heat, the fan sends air around the bricks and into the building. If the storage load is well matched to the building's heating requirements, the electric elements can remain off until the utility rate drops again.

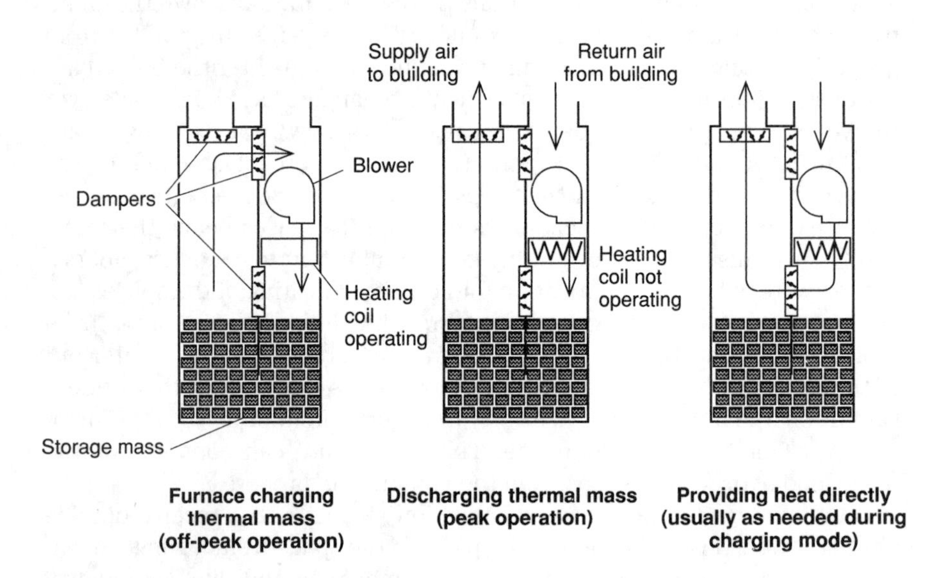

Figure 7.1. Heat storage electric furnace. Capacity of storage mass is selected based on the number of Btus that must be delivered during peak periods.

This can save a lot of money. The off-peak electric rate is usually much lower than the regular rate, and storage systems can make electric resistance heat very economical.

Hydronic Heating Systems

Hydronic heat is often the preferred system for clients that do not need to integrate heating and air conditioning. Hot water heat is quiet, it raises no dust, and it is easily broken into multiple zones.

Hydronic heating systems operate with fin tube radiators, convectors installed near the ceiling, coils in air handlers, and radiant tubing in floors, ceilings, or walls. The greater heat carrying ability of water compared with air means that small pipes can run through narrow openings instead of the large chases required by ducts.

The versatility of hot water heating systems makes them the first choice for many "tricky" applications. Many technicians do not have a good understanding of them, however, and might prefer to use forced air, even if a hydronic system would be easier. There is no need to be puzzled over hydronic systems. Their operation is as simple as a hot air system, and their versatility allows their use in many more applications.

System pressure

Hydronic systems use a flow of water to deliver heat to a building. The entire system must be under positive pressure to prevent pump cavitation, water boiling, and excess air formation.

System pressure decreases with increasing height. Every foot of system height decreases the pressure by 0.435 psi. Because all parts of the system should be at a positive pressure (including the highest parts), the system's pressure must accommodate the highest point.

Minimum pressure at every point in the system should be no lower than 10 psi. As a practical working minimum, the lowest pressure on even the simplest system should be at least 12 psi. While the highest parts of the system should have adequate pressure, the lowest area's pressure must not be too high.

Pressure at the expansion tank sets the level for the entire system. It is the "reference point" that other parts of the system use when establishing their pressure. Set the expansion tank's pressure with the make-up water regulator to the desired room temperature value. It must be high enough to provide 10 psi at the system's highest point, but not be so high that low points are pressurized beyond their safety valve's operating point.

A good room-temperature expansion tank pressure is the minimum required to provide 10 psi to the highest locations. Running the pump and heating the water adds additional pressure.

An example: A three-story building has hydronic coils in the ceiling at each floor. The boiler, pump, and expansion tank are in the basement. The boiler is on the floor and the tank is 8 ft above it. The third-floor ceiling is 48 ft above the basement floor. Estimate the expansion tank's room-temperature pressure needed to properly pressurize the system.

First, note that the coils in the third-floor ceiling should have 10 psi. They are 48 ft above the basement floor, but a little closer to the expansion tank. We have to find the distance from the third-floor coils to the expansion tank. This distance is found by subtracting the height of the tank from the coils' height. The equation is

$$\text{Distance} = 48\text{ ft} - 8\text{ ft} = 40\text{ ft}$$

The pressure loss caused by the 40-ft height is found by multiplying the pressure/elevation factor by the height difference:

$$\text{psi} = 0.435\text{ psi/ft} \times 40\text{ ft} = 17.4\text{ psi}$$

The pressure needed at the tank is the sum of the pressure loss and the 10 psi desired at the coils.

$$\text{Desired pressure} = 17.4\text{ psi} + 10\text{ psi} = 27.4\text{ psi at the expansion tank}$$

If a closed, diaphragm-type expansion tank is being used, precharge it with its air fitting to about 25 psi. If a nondiaphragm tank is used, its pressure will take care of itself as the system is filled with water. In either case the make-up water's pressure regulator sets the final pressure.

What is the pressure at the boiler? Because the boiler is lower than the tank, its pressure is higher. The additional pressure is found by multiplying the pressure/elevation factor by the height difference.

$$\text{psi} = 0.435\text{ psi/ft} \times 8\text{ ft} = 3.48\text{ psi}$$

The pressure at the boiler is the sum of the tank pressure and the pressure gained from the height difference.

$$\text{psi at the boiler} = 17.4\text{ psi} + 3.48\text{ psi} = 20.9\text{ psi}$$

If adjusting the make-up water regulator from a gauge at the boiler, set it to the 20.9-psi reading.

Boilers

Hydronic system boilers are similar to furnaces. Water circulates through a heat exchanger (or around a set of electric heating elements) and absorbs heat. A pump circulates the water after it is heated and delivers it to the building.

Operation of the burner or heating elements is controlled by the zone's space thermostat. The boiler starts and runs whenever heat is required. Fossil fuel boilers start with the same systems that furnaces use (as described earlier). Once the flue draft and burner flames are present the water begins to heat.

An internal aquastat senses the boiler's water jacket temperature. When the water temperature reaches the set point, a set of contacts closes and starts the pump. Hot water is circulated through the building's pipe loop by the pump to any zone that calls for it.

Once the space thermostat is satisfied, the boiler turns off. With single- and some multiple-zone systems the pump runs until the boiler's heat exchanger cools below the aquastat's set point. After the heat exchanger has cooled sufficiently, the pump stops.

It is important to have enough water flow while the boiler is running to prevent overheating. Insufficient flow could allow a stationary or very slow-moving pocket of water to linger in the unit. If it absorbs enough heat it might reach its boiling point. The steam could cause water hammer, noisy operation, or completely choke off the water flow. It could also lead to more serious problems if it forces water out of the boiler. Water volume in the boiler must be sufficient to ensure that all of the heat exchanger surfaces are constantly wetted.

Having too low a water volume can be dangerous. Operating a boiler without enough water causes portions of the heat exchanger to overheat. The surfaces heat well past the boiling point, and any water that comes into contact with them instantly flashes to steam. This can cause an explosion.

The mass of water suddenly brought to the boiling point expands its volume by greater than 25 times. This leads to an instantaneous, huge pressure increase in the boiler. It might be far too much for the safety relief valve to handle. Excessive pressure can cause the boiler or its connected piping to burst, spraying the area with flying metal and scalding water.

The potential for boilers to explode has been decreased with modern designs. Engineers have reduced the amount of water in the boiler with smaller heat exchanger tubes, decreasing the available mass that can explode.

Besides explosions, however, overheating can cause other damage. A heat exchanger with poor flow or one that has run dry can warp and split from excessive temperatures. While not as disastrous as an explosion, the unit will be ruined and must be replaced.

Lack of water or flow is not the only problem to avoid with a boiler. The unit has to handle the system's water temperature swings. Boilers made for hydronic heating are designed to do this. Others (for example, storage hot water tanks usually used for domestic hot water heating) are not.

Hydronic heating system boilers heat circulating water from relatively cold conditions up to the full discharge temperature. While the entire temperature increase does not take place in the first pass of water through the boiler, this can still put severe stress on the unit. In addition, fossil fueled boilers might have to deal with flue gas condensation. At the start of the heating season, or anytime the system has been shut down for an extended period, the system's water is below 70°F. When this water circulates into the boiler at the start of its heating cycle, it can chill the heat exchanger sufficiently to cause exhaust gas condensation. This condition usually is short lived (until the system water heats).

Radiators

Hydronic heating units transfer heat from the hot water into the surrounding space. Fin tube radiators or air handler coils are usually used for this. They are both water to air heat exchangers.

Many details of air handler coils are described in Chapter 4. Air handler coils are the "bridge" between hydronic heating systems and air systems. They are usually used in larger systems, where a hot water system provides both perimeter radiation heat and air heating.

Most hydronic heating systems use fin tube radiators, often mounted near the floor. These are very simple devices. Fin tube heaters consist of a copper pipe with thin metal plates (fins) bonded or crimped onto the pipe's outside surface. The assembly resembles a stack of metal plates with the tube extending through a hole in the center of the stack.

As hot water runs inside the tube, heat conducts through its wall to the fins. The large surface area of the fins heats the nearby air, causing it to rise and convect out of the heater's housing. Air enters the housing from below to replace the air that was warmed and escaped. This air heats up and the cycle repeats.

Note that baseboard "radiators" are actually convectors. Little heat is transferred to the space by radiation. They are called radiators because many of them were installed to replace old steam radiators when steam systems were converted to hot water. The name stuck. Despite the possible confusion, keep in mind that fin tube radiators work by convection. Air is heated and rises into the space.

Fin tubes are available in a wide range of styles and capacities. Most use standard copper pipe for the tube, making it easy to connect lengths of fin tube with standard components. Different tube diameters and fin types (size and spacing) give different amounts of heat output per foot of tube length at a given average water temperature.

Various cabinets covering the fin tubes can be used to protect them and keep them clean. In fact, some very stylish heating units are available from Europe for applications where conventional types are unacceptable. Hydronic system convectors include equipment beyond fin tube radiators. Elegant towel warming racks, boot drying stands, benches, and even stairway banisters are available to carry heating water. While they might act more like true radiators than convectors, they are good for systems where the need for spot heating or unobtrusive heating units is essential.

Piping layouts

Water travels in a loop of pipe with various accessories and fittings. The basic components were covered in Chapter 5, and some of the ways they connect to form systems are described in this chapter. Refer to Figure 7.2 for a basic schematic of a boiler heating source. There are other ways to make the connections, but this method is typical. There are a few things to remember about the "best ways" to make hydronic system connections.

The fundamental components needed for any hydronic heating system are the boiler, expansion tank, pump, and the heating units. Water flows through the boiler, receives heat from the hot surfaces of the heat exchanger, and is drawn into the pump. The pump increases the water's pressure and sends it into the building. Heating units there (including fin tube radiators, air handler hot water coils, and radiant loops) deliver the heat to the space. The water returns to the boiler and pump, is heated, and is carried to the space again.

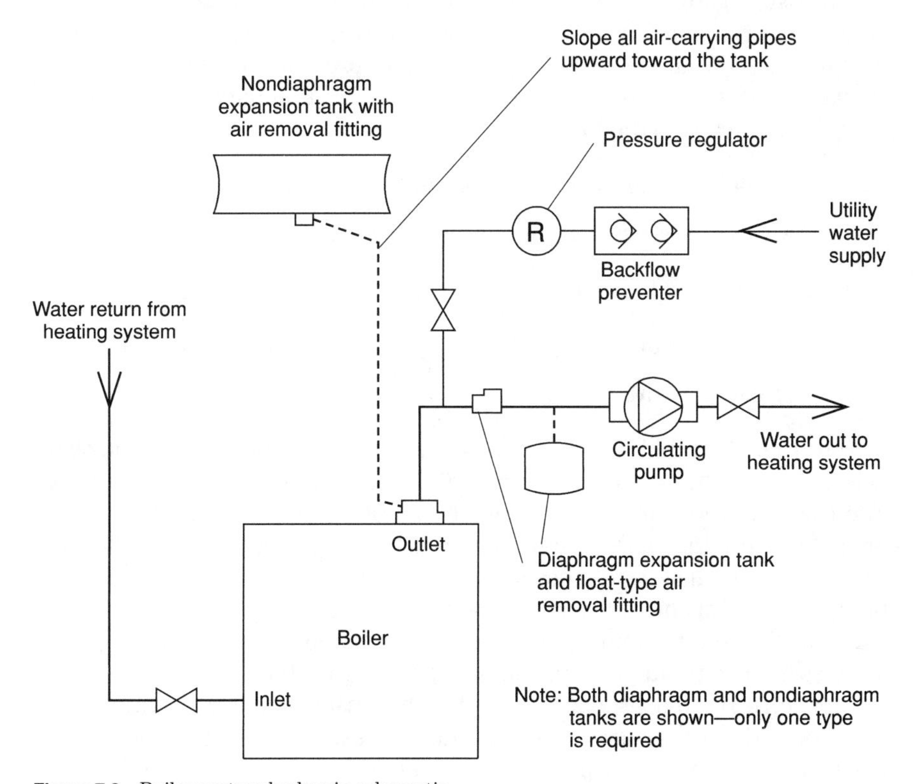

Figure 7.2. Boiler system hydronic schematic.

There are a variety of piping systems that the designer can choose from. First, consider the connections of the boiler, expansion tank, and pump. Then, review the various connections for the heating units to the boiler and pump.

The connection of the pump to the expansion tank is important. One type works much better than the other. By placing the expansion tank on the pump inlet (as shown), the building's supply piping pressure increases. The supply and return pressures are higher than the expansion tank's pressure. If, however, the pump's outlet is connected to the expansion tank, the building's system pressure will be less. Supply and return pressures will be less than the expansion tank's pressure.

The expansion tank is the point of constant pressure in hydronic systems. It never varies (with the system at a constant temperature). Having the pump's inlet connected to the tank forces the inlet pressure to exactly equal the tank pressure and the pump's outlet pressure to increase above the tank's pressure. Having the pump's outlet connected to the tank forces the outlet pressure to equal the tank pressure. The pump's inlet pressure drops below the tank's pressure. Consider this example:

A pressure gauge on an expansion tank reads 15 psi when the system's pump is not running. The pump produces a pressure increase of 4 psi from its inlet to outlet. What would the inlet and outlet pressures be if the pump's inlet was connected to the tank? What if the outlet was connected to the tank?

Inlet connected to the tank:

Inlet pressure = 15 psi (it assumes the tank pressure)

Outlet pressure = 15 psi + 4 psi = 19 psi

Outlet connected to the tank:

Outlet pressure = 15 psi (now it equals the tank's pressure)

Inlet pressure = 15 psi – 4 psi = 11 psi

Note that the tank's pressure does not change. Its pressure is the same whether the pump is running or not. Since the pump does not send water into (or take water out of) the tank, it cannot influence the tank's internal pressure. Only the water's thermal expansion can change the tank's pressure. The general system pressure is higher with the tank connected to the pump's inlet. This provides several benefits.

Higher system pressure helps keep dissolved air from forming bubbles in the system's piping. When the water is carried high into the building, its pressure decreases because of the elevation change. Having a few more pounds per square inch of residual pressure can keep the system clear of air. Higher pressure water can keep the air in solution so fewer bubbles form.

Higher pump inlet pressure also helps prevent cavitation. The ability of water to vaporize from the pump's impeller turbulence depends on its pressure. More pressure helps prevent turbulence from starting.

The best location for an air remover depends on the type of remover used, and the type of expansion tank (diaphragm or open type) in the system. Air removers that feed bubbles into a pressurized tank work best when located near the open expansion tank. The bubbles can float up into the tank where they do no harm.

Closed diaphragm-type tanks do not act as a collecting station for air. Use a remover with an internal float and valve to allow the air to discharge to the atmosphere without a tank. The best place for this type of air remover is high in the system on the main water return pipe. Pressure is lowest there and any bubbles formed in the piping probably pass by that point. Be certain that the air remover is accessible for future maintenance.

If the air remover is installed in a high location, put an open pipe on its air outlet. The pipe should extend upward at least 12 in. This catches any water that might bubble out of the remover when it vents air, and prevents dripping below.

Other options are available for connecting the heating units to the pump and boiler. The choices might not be quite as easy, but each has advantages.

Series and parallel piping. Water can be delivered to heating units (fin tube radiators, air handler coils, radiant coils, etc.) in a variety of ways. All units must ultimately be connected to the boiler and pump's supply and return connections. Series or parallel connections can be made, depending on the situation.

Series piping connections are the simplest (see Figure 7.3). The unit closest to the boiler and pump receives all of the water. Water then flows to the next unit in line, with all of the flow going through it. This process continues from one to another, until all of the units are connected. The last heater in the chain is hooked to the boiler and pump's return line.

Simple series loops like this are fine for a small hydronic system serving a single area. All of the heat goes into the area anyway, so it doesn't matter if some heaters deliver more heat than others.

It is difficult to balance heat delivery with a simple series loop. All of the units handle the full flow of water. They must be sized to handle the total water flow even when they need to deliver a small amount of heat. If the size is reduced (to match the heat requirement), pressure losses for the entire system will increase. Flow capacity will suffer.

Connecting units in parallel (as shown in Figure 7.4) has several advantages. Each unit has access to the same temperature water, and their flows can be individually controlled with zone valves. Flow through one unit does not depend on flow through any other. If a parallel system is con-

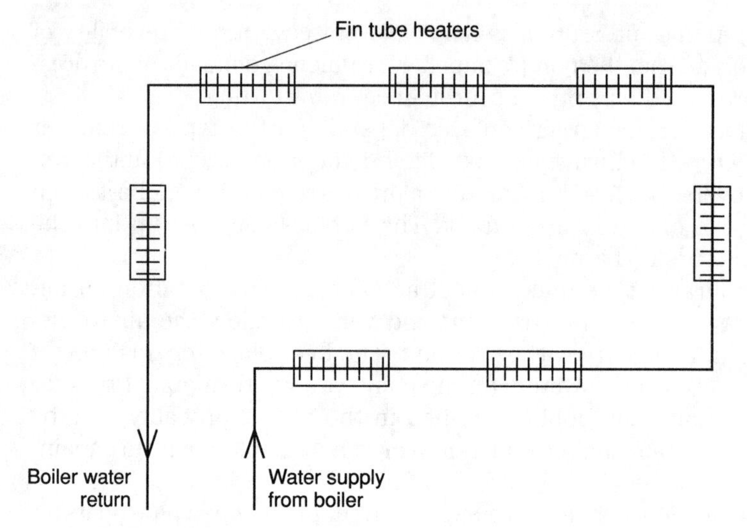

Figure 7.3. Series connected hydronic loop. Water temperature changes as it travels through the loop.

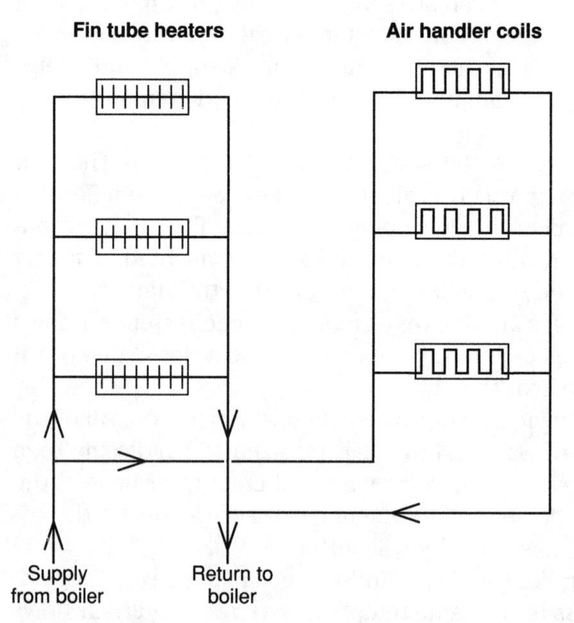

Figure 7.4. Parallel connected hydronic loop. All units receive the water at the same temperature.

nected in a reverse return (described later) configuration, it is completely self-balancing.

Disadvantages of parallel connections usually include their cost. More pipes are needed to provide each unit with its own separate supply and return connection. If only a few heating units are needed and they are to be zoned together, it is usually better to use series connections.

Many piping systems are combinations of parallel and series hookups. Each zone has its individual heaters connected in series, and each set of series connected heaters is connected in parallel to the others. This allows you to use a single balancing valve for each zone and permits each zone to operate independently.

Straight and reverse return piping. Parallel hydronic heating pipe loops can take several configurations. All heating devices (radiators, coils, tubing loops, etc.) need both supply and return connections. The way these are connected can make a big difference in the system's performance.

Where fan coil radiators, air handler coils, or radiant loops are parallel connected to common supply and return main lines, each has its own connections to the pump's supply and return. Usually the supply and return mains follow each other from the boiler room to the most distant heating unit. Both pipes take the same basic path. Along the route, individual zones or units connect to both lines wherever it is convenient.

Units closest to the boiler room are the first to connect to both the supply and return lines. The unit at the farthest location is the last to connect to both lines. This type of loop is a *straight return* system. Water runs along the supply main to a unit's connection and feeds the unit. The return water leaves the zone and connects to the return main. From there it goes straight back to the pump and boiler.

This can lead to serious pressure imbalances. The first heating unit or zone is closer to the pump and boiler than the others. Its supply and return connections are the first ones in the line, with the supply and return connection points closest to the pump and boiler. At this position, the supply line has the greatest pump pressure, and the return has the greatest suction. The first zone has the largest pressure difference from supply to return and the greatest amount of water flows through it.

On the other hand, the heater or zone connected to the far end of the main supply and return lines gets very little pressure. Flow resistance is highest on the long supply and return lines and less water flows to the end. Water prefers the path of least resistance, and the last heater or zone in line starves for flow. This situation causes overheating at the unit closest to the boiler and serious underheating in the farthest unit that must be corrected.

Automatic or manual balancing valves set the proper system water flows in each branch. Balancing valves are needed to add pressure drop to the units close to the pump. A *reverse return* system, however, is self-balancing and needs no balancing valves. Reverse return connections allow every unit to have the same pressure drop no matter where in the system it is. Each unit is hydraulically the same distance from the pump, no matter how physically far away they really are.

Figure 7.5 shows the differences between a straight and reverse return system. Note that water always takes the same distance path through the reverse return system, no matter which heating unit it flows through. This similarity of distance gives it equal flow characteristics.

The very first unit to receive water from the pump is at the far end of the return side of the system. Similarly, the last unit on the supply is closest to the return side. Units between them have "in between" flows on the supply and return sides, but there is no easiest path for the water to take through any one unit.

Reverse return paths can take several forms. Figure 7.6 shows a reverse return system of heaters installed around the outside walls of a building. While its piping might not look as obviously "reversed" as the schematic's, water can flow equally through all units.

Multiple-zone secondary systems. All standard series and parallel hydronic loops (as discussed so far) are *primary* systems. Secondary systems are a bit more complex, but provide more flexibility. They allow some of the system's water to bypass each heater or zone. The amount of water permit-

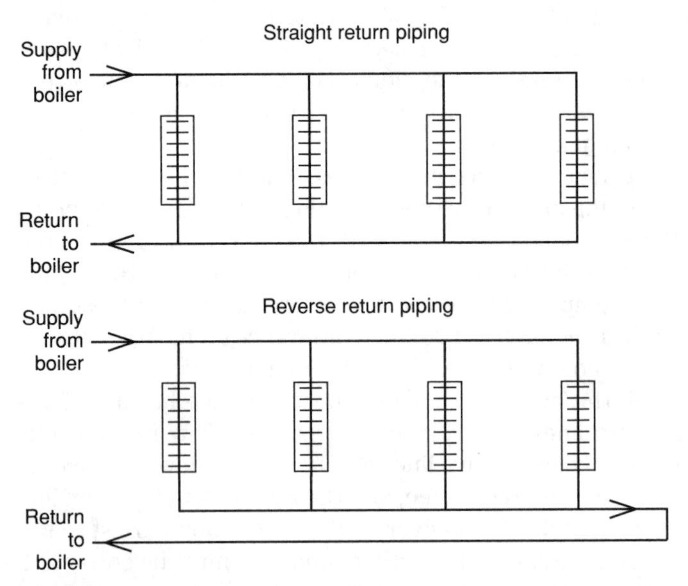

Figure 7.5. Straight and reverse return parallel loops. The reverse return arrangement equalizes pressure differences between all units regardless of their distance from the supply.

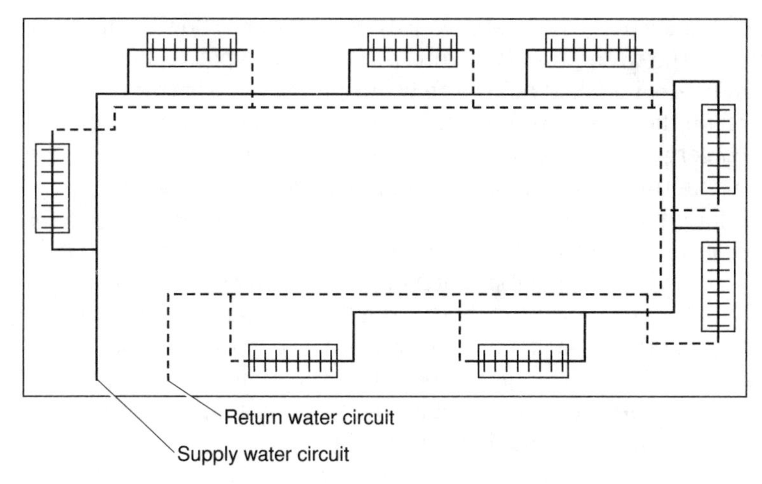

Figure 7.6. Perimeter reverse return loop.

ted to flow through the zone matches the heat delivery needed there. All of the system water not needed by a zone bypasses it. Secondary systems work for most applications, both simple and complex.

Simple systems use flow diversion tees to send most of the water past an individual fin tube radiator. Note the typical applications shown in Figure 7.7. There are two ways to use these tees.

If the fin tube unit is physically above the pipe carrying water, only one tee is usually used. Install it on the heater's outlet side, with a standard plumbing tee used on the inlet. The pressure drop caused by the diversion tee and

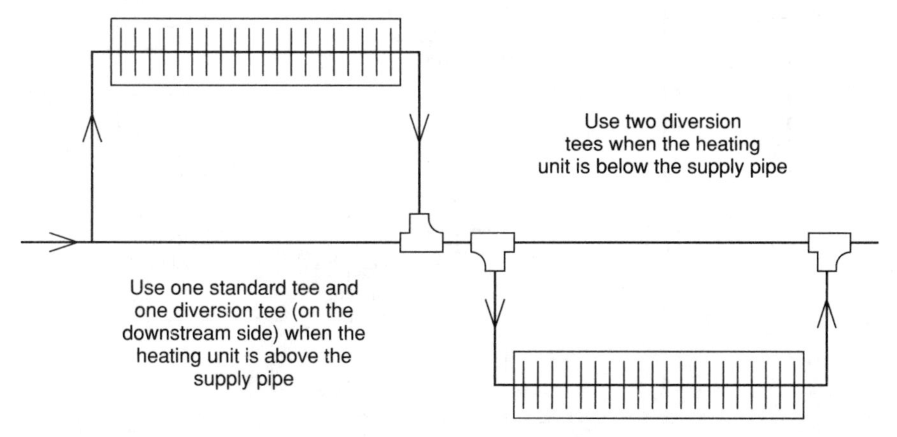

Figure 7.7. Use of diversion tees. Each tee causes a higher pressure drop in the "straight through" flow to divert water to its branch leg.

the natural buoyancy of the hot water causes some water to flow through the radiator. Most of the water, however, runs through the bypass tube.

Fin tube radiators installed below the water piping require two diversion tees, one each on the radiator's inlet and outlet. Because hot water wants to flow up (as cooler water flows down), two tees are needed to cause enough flow through a fin tube unit below the supply pipe.

Unless it is found to be necessary, don't use two diversion tees on systems that have radiators located above the supply pipes. The flow through the fin tube will be excessive, causing too much pressure drop and heat loss.

The main difference in operation between a flow diversion secondary system and a simple series loop is temperature. Because most of the water bypasses each radiator, its average temperature stays higher in the earlier parts of the loop. More heat is available for units later in the system, so there is less likelihood of them being starved for heat.

Beyond diversion tee secondary piping are pumped secondary systems. These are generally used for commercial systems where total flows are higher than residential systems and multiple zones are used. While pumped secondary systems might look confusing at first, their operation and design are simple.

Figure 7.8 shows the essentials of a pumped secondary system. The system has two separate zones to supply heat to the space. Note that there are three pumps: the main boiler pump that circulates water through the building and one pump serving each zone. Further, each of the zone's pumps is bypassed by the main supply pipe. This is the key to their operation.

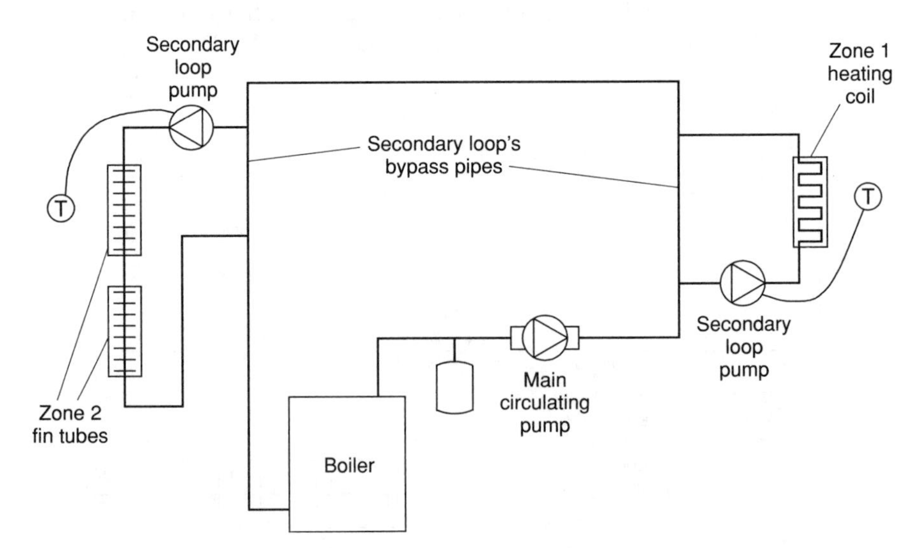

Figure 7.8. Two-zone pumped secondary system. When the secondary pumps operate some flow is diverted out of the bypass pipes.

The main pump's job is to send water around the main loop (including the boiler). It does not flow water through any of the coils, radiators, or other heating units. If it does cause water to run through these units it means that the system is not piped correctly. The main pump should only send water through the main pipe loop. As the water passes through the loop, it picks up heat in the boiler and carries it out into the system.

Each zone's pump is responsible for passing water through its heating units only. These might be series or parallel connected coils or fin tube radiators, all fed from the single zone pump. When the pump is not on, no water flows through the zone's loop. Operating the zone pump switches the zone's heat off and on.

The zones and the main pipe loop share a short length of pipe called the zone's *bypass*. This is where the zone receives its heat.

While each pump is only responsible for its own flow (main pump for the main loop, secondary pumps for their zones), water can pass freely between them. Zone pumps really do not push water backwards through the bypass pipe as the diagram suggests. Instead, they steal some of the main loop's flow and send it into the zone. The main pump is not affected by this, however. The same amount of water is removed by the zone at its inlet connection as is delivered at its outlet connection. As far as the main pump is concerned, its flow is the same whether the bypass pump is on or off.

Zone pumps also are not affected by the main pump's operation or by the operation of the other zone pumps. A constant supply of water is available whether flow is already moving through the bypass pipe or not. The only difference is that, when the main pump is running, the water it pumps originates from the main loop. Because the same amount of water returns to the main loop at the other connection, there is no effect.

Designing a pumped secondary system is very easy. Because each pump has its own specific job to do, the pressure losses are easy to calculate. There is no need to make the main pump handle the pressure losses at each heating unit, so smaller pumps are usually used with these systems.

Balancing is automatic. Flow through the main loop is constant everywhere. When a zone needs flow, its pump starts and flow is delivered. When heat is no longer needed at that zone, its pump shuts off and the flow stops. Main system flow is unchanged.

One thing that must be considered when designing any secondary system is the temperature loss of the water as it runs through the main loop. The temperature gradually declines, providing less heat to units farther along in the loop. Plan for this so that the downstream units have sufficient capacity at the lower water temperature.

It is easy to estimate the temperature at any point in the system. The temperature is found when figuring out the system's heat delivery requirements. Consider this example (see Figure 7.9).

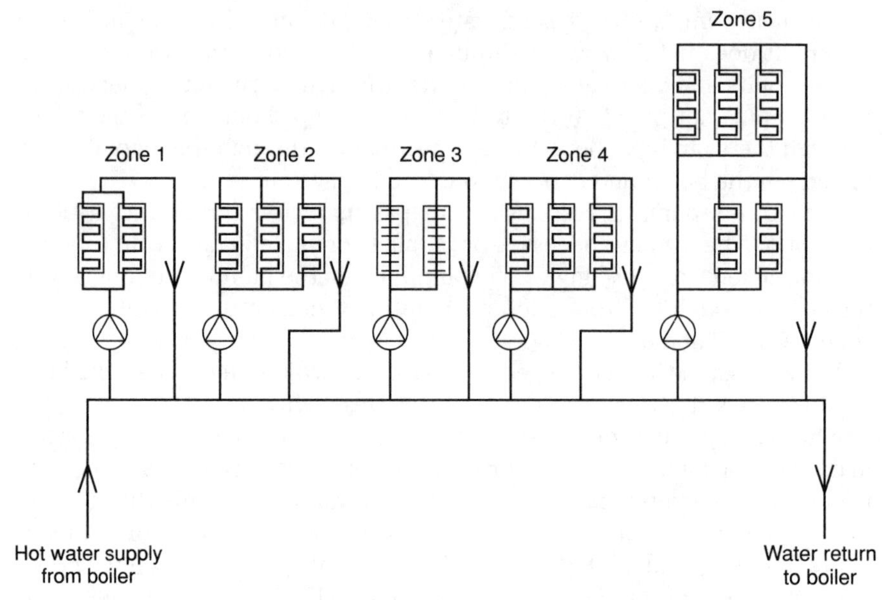

Figure 7.9. Example five-zone pumped secondary system.

A building requires a 1,800,000-Btu/hr boiler. It is to deliver 190°F water with a planned 20°F temperature drop through the main loop and each zone. Five secondary zones are connected to the loop as shown, with heat losses estimated to be

Zone	Btu/hr
1	260,000
2	320,000
3	190,000
4	420,000
5	610,000

Find the required water flow through the main loop and each zone, and the temperature of the main water flow as it leaves each zone. From Chapter 2 the formulas for waterborne heat transfer are

$$\text{gpm} = \frac{\text{Btu/hr}}{(T_d \times 498)}$$

$$T_d = \frac{\text{Btu/hr}}{(\text{gpm} \times 498)}$$

First, find the main loop flow required to carry 1,800,000 Btu/hr with a 20°F temperature change.

$$\text{Main flow gpm} = \frac{1{,}800{,}000 \text{ Btu/hr}}{(20 \text{ gpm} \times 498)} = 181 \text{ gpm}$$

This is the flow required for the main system pump. Next, find the flow required for the zone 1 pump. Since we know the zone uses 260,000 Btu/hr and the water loses 20°F, the equation becomes

$$\text{Zone 1 flow gpm} = \frac{260{,}000 \text{ Btu/hr}}{(20 \text{ gpm} \times 498)} = 26 \text{ gpm}$$

The zone 1 output temperature would be

$$190°\text{F} - 20°\text{F} = 170°\text{F}$$

The 26-gpm flow of 170°F water from zone 1 blends with the rest of the main system's flow (that bypassed zone 1) of 190°F water. The temperature after the two streams blend is between 170°F and 190°F. Its temperature is found with the above equation using the zone 1 heat loss of 260,000 Btu/hr and the main system flow of 181 gpm.

$$T_d = \frac{260{,}000 \text{ Btu/hr}}{(181 \text{ gpm} \times 498)} = 2.9°\text{F}$$

This is the temperature change of the main flow. Subtract the change from the initial 190°F to find the final temperature. The main flow's temperature after zone 1 is

$$190°\text{F} - 2.9°\text{F} = 187.1°\text{F}$$

Now find the required flow for the remaining zones and the temperature change of the main flow as it leaves each zone's outlet.

$$\text{Zone 2 flow} = \frac{320{,}000 \text{ Btu/hr}}{(20 \text{ gpm} \times 498)} = 32 \text{ gpm}$$

The main flow temperature difference after zone 2 is

$$T_d = \frac{320{,}000 \text{ Btu/hr}}{(181 \text{ gpm} \times 498)} = 3.6°\text{F}$$

The main flow temperature after zone 2 is

$$187.1°\text{F} - 3.6°\text{F} = 183.5°\text{F}$$

The zone 3 flow is

$$\frac{190{,}000 \text{ Btu/hr}}{(20 \text{ gpm} \times 498)} = 19 \text{ gpm}$$

The main flow temperature change after zone 3 is

$$T_d = \frac{190{,}000 \text{ Btu/hr}}{(181 \text{ gpm} \times 498)} = 2.1°\text{F}$$

The main flow temperature after zone 3

$$183.5°\text{F} - 2.1°\text{F} = 181.4°\text{F}$$

The zone 4 flow is

$$\frac{420{,}000 \text{ Btu/hr}}{(20 \times 498)} = 42 \text{ gpm}$$

The main flow temperature after zone 4 is

$$181.4°\text{F} - \left(\frac{420{,}000 \text{ Btu/hr}}{(181 \times 498)}\right) = 176.7°\text{F}$$

The zone 5 flow is

$$\frac{610{,}000 \text{ Btu/hr}}{(20 \times 498)} = 61 \text{ gpm}$$

The main flow temperature is

$$176.7°\text{F} - \left(\frac{610{,}000 \text{ Btu/hr}}{(181 \times 498)}\right) = 169.9°\text{F}$$

This final main flow temperature, 169.9°F, confirms the original goal of the system's 20°F temperature drop. The water started out at 190°F and did drop to about 170°F. When designing a secondary system, be certain to check that the heat output of the coils or radiators will be sufficient at the lower water input temperatures.

Zone 5 of the previous example receives water at 176.7°F and discharges it at 156.7°F. Its average temperature of 166.7°F is well below the zone 1 average temperature of 180°F, so zone 5 will need larger capacity equipment to give an equivalent output to its space. This is not a problem, but should be expected during operation. Check the unit's specifications or with the manufacturer to confirm that heat output will be sufficient. If information is not available, Table 2.4 gives factors showing the amount of heat transfer available from typical hydronic devices at different water system temperatures.

Other multiple-zone systems. An advantage of hydronic heat is the ease with which the system can be divided into more than one zone. There are many advantages to multiple zones, including individual temperature controls to suit specific needs and lower operating costs. Hydronic systems can be split into zones in a variety of ways.

The secondary system described previously is intended for multiple zones. It effectively delivers heat to multiple independent locations. While secondary systems do an excellent job of dividing and distributing heat to multiple zones, they might not be suitable for smaller, simpler systems. Residential and small commercial hydronic systems are easy to zone with a combination of automatic valves and pumps.

Zoned systems can use a single pump to deliver water to the entire system. Multiple automatic valves provide flow as needed for each zone. A thermostat in each zone's space controls the operation of the single boiler and the valve feeding water to its zone. See Figure 7.10 for a schematic of this type of system.

The boiler starts when any zone needs heat. Its full capacity is available for any and all zones that call for heat. When the boiler's heat exchanger heats sufficiently, the pump starts.

Zones requiring heat open their electrically operated valves. This allows water flow through the zone. The pump might have to operate with valves open for one or all. Use a low-speed pump with a 1750-rpm motor on this type of

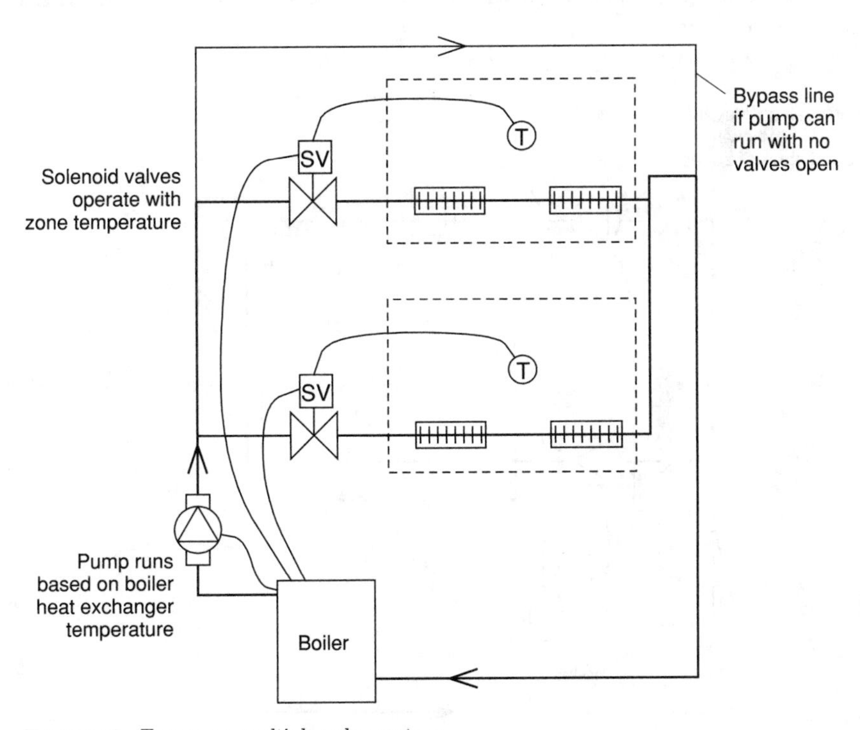

Figure 7.10. Two-zone, multiple-valve system.

system. Low-speed pumps have a fairly "flat" pressure delivery curve. Output pressure is reasonably unchanged no matter how many zone valves are open.

Some systems with zone valves close the valve when the thermostat is satisfied. The boiler shuts down (unless another zone needs heat), but the pump continues to run. If the pump runs with no water flow, it (and the water inside it) might overheat. Systems with these controls must have a bypass line from the supply to the return pipes to always allow water to flow. A ½ in. pipe allows enough water to run through the pump to prevent overheating, but does not bypass so much water that the zones starve.

Other multiple-zone systems avoid dead-heading the pump by allowing it to run only if at least one zone valve is open. The controls work like this: When a zone thermostat calls for heat it starts the boiler and opens its valve. After the boiler's heat exchanger warms up, the pump runs and water flows to the zone. When the thermostat is satisfied the boiler shuts off, the pump is turned off, and the zone valve closes. This arrangement traps more heat in the boiler, but avoids any chance of pump damage.

Multiple-pump systems do not use electrically operated valves. Instead, each zone has its own circulating pump controlled by the zone thermostat. When a zone needs heat it turns on the boiler. The zone's pump starts when the heat exchanger reaches the proper temperature. Figure 7.11 illustrates this system.

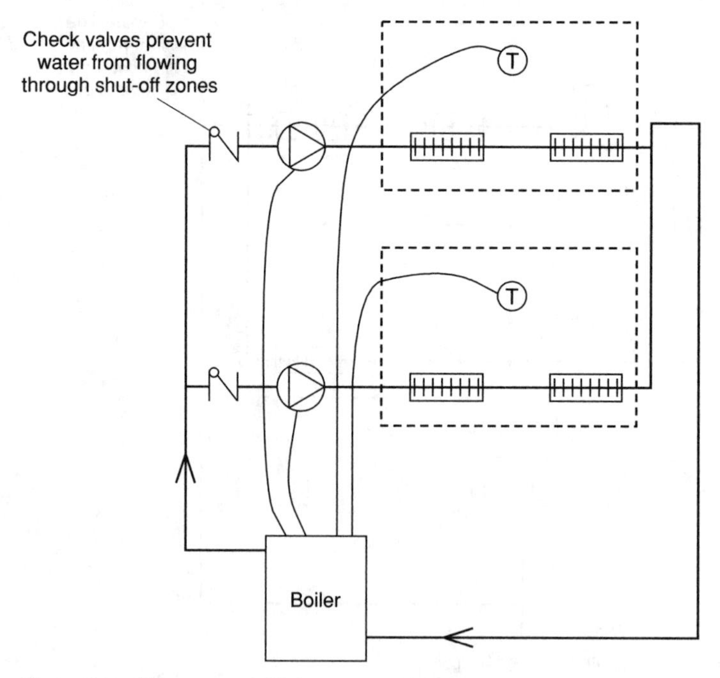

Figure 7.11. Two-zone, multiple-pump system.

Other zones that need heat can start their pumps if the boiler's heat exchanger is hot enough. Each zone starts the boiler, or keeps it running, as long as it needs heat. When the zone's thermostat is satisfied, it turns off its pump and turns the boiler off. The boiler keeps running, however, if another zone is using heat.

Multiple-pump zoned systems can also cause some unusual flow conditions that can mimic unrelated problems. Nonoperating pumps can allow water to flow through them in either direction. Water always flows from areas of high pressure to areas of lower pressure. If a turned-off pump and its hydronic circuit represent a path for water to take, it will flow through it. If it is a very low resistance path, most of the water will use it instead of the intended route through the proper zone.

Water improperly backing through a zone can cause several problems. Hot water is diverted from the zone it is supposed to go to. This can make the zone that needs heat be too cold and the zone the water leaks through too hot. The supposedly nonoperating zone receives some heat whenever the operating pump runs. These effects are often confused with a system air lock or clog. If the cold zone is bled (causing water to flow out of the vent), it might begin to heat. Once the bleeding is stopped, flow ceases, and it cools again.

Many circuits use check valves to prevent water backflow, and these can become hung up in the open position. If servicing a multiple-pump system that is acting erratically, with the wrong zones heating when a pump runs, inspect the check valves. One or more is probably stuck open.

Radiant heat piping. True radiant heat systems do not use wall- or floor-mounted fin tube radiator heaters. Instead, a large heated surface or panel delivers heat energy from infrared energy to objects in the space. Very little convection takes place. Because of this, there are some differences between convective and radiant hydronic systems. Water temperatures are usually limited to 130°F to 140°F with radiant systems, where they might reach close to 200°F in convective circuits.

There is usually no problem with piping systems that serve only one type of heat or the other. Adjust the operating limit of the boiler to the appropriate setting. Combining radiant and convective circuits in one heating system can, however, be a challenge. Because the convective system must have high-temperature water, the boiler should be set to deliver it. This is the highest temperature requirement. However, the radiant system should not use very hot water. Not only would the heat losses be excessive, but the heated surfaces could become dangerous. Many radiant systems use heated concrete floor slabs. Heating them to over 140°F could cause burns.

Many combined systems use a modified secondary system to feed the radiant loop (see Figure 7.12). High-temperature water from the main system feeds the radiant loop, but is first blended with some of its own cooled re-

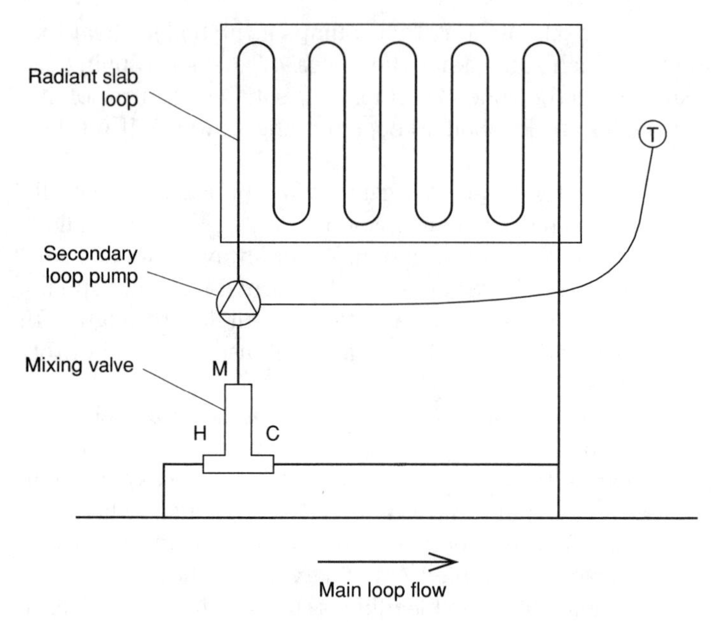

Figure 7.12. Secondary radiant loop with mixing valve. Set the mixing valve to reduce the radiant loop's temperature to its required maximum.

turn water. The result is that the water feeding the radiant loop is cooled to the required temperature. The system uses a three-way tempering valve to regulate the radiant loop's temperature independent of the main system.

Tempering valves are independent temperature regulators. They mix hot and cold streams of water and produce a single stream of warm water. The temperature of the output water is set with a knob or screwdriver adjustment. As water circulates through the radiant loop and the tempering valve, it cools below the valve's set point. This opens the hot water inlet, drawing more high-temperature system water into the loop. As it enters, an equal volume of the cool loop water leaves the radiant loop and returns to the main system and back to the boiler for heating. The tempering valve continually regulates the amount of hot water admitted to the system to keep its heat output constant.

The radiant system's thermostat starts and stops the heat the same way that any other secondary system's control does. It starts and stops the secondary pump. There is no effect on the tempering valve. The tempering valve only provides the proper water temperature to the loop.

System diversity

Larger heating systems are more efficient if they have a degree of *diversity*. This means that the system can deliver a range of heat output levels.

Most heating plants have two output levels: off and on. Adding diversity to a system allows it to reduce its heat output when demands are lower. This provides real fuel savings, because starting and stopping a large heating system causes some inefficiencies.

Losses occur when the heat exchanger and other parts must be warmed every time the system cycles on. Allowing it to run longer with less heat output prevents frequent starts and stops, eliminating much of the warm-up losses.

Many boilers are available in, or can be connected in, multiple stages. Large multiple-stage boilers have sets of electric heating elements or heat exchangers that can operate individually or together to provide a range of heat outputs. The broader the heat delivery range the boiler can provide, the wider the diversity.

Wide diversity is helpful for applications where the expected heat demands can vary over a wide range. For example, some climates have mild heating seasons with only a few extreme swings each season. The heating plant must be capable of handling the extremes, but the system's efficiency suffers when the oversized heating system carries the normally small load. Cost savings are available if the heating system delivers only a fraction of its full capacity. To do this it must prevent all of the burners or elements from running when less heat output is needed. Multiple-stage boilers are usually available in larger capacities (more than 1,000,000 Btu), but similar systems can work and provide efficiency improvements in almost any application.

Secondary boiler systems. While very large multiple-stage boilers are available, it's easy to connect individual boilers for multiple-stage operation. When full heat is needed, all the boilers feed heat to the system. When less heat is needed one or more boilers are shut down and the rest carry the load. Controls make starting and stopping the individual boilers completely automatic.

Most multiple-boiler systems divide the total load into two or three units. Each unit should be the same size to equally share the load. A multiple-stage boiler system also provides a measure of system backup. With several units available to provide heat to the building, the failure of any one is not a catastrophe. If the building has partial heat it won't freeze. While it might not be comfortable, as long as at least half of the building heat is available it should avoid freezing.

The simplest method of piping a multiple-stage boiler system is the secondary piping loop described earlier. Instead of placing heat delivery units (for example, fin tube radiators) on the secondary circuit, install one boiler on each circuit. Refer to Figure 7.13 for a two-boiler example. Its operation is the same as the secondary hydronic systems described earlier. The only difference is the direction of heat flow: the multiple boilers deliver heat into the loop instead of taking it out. All the other principles are the same.

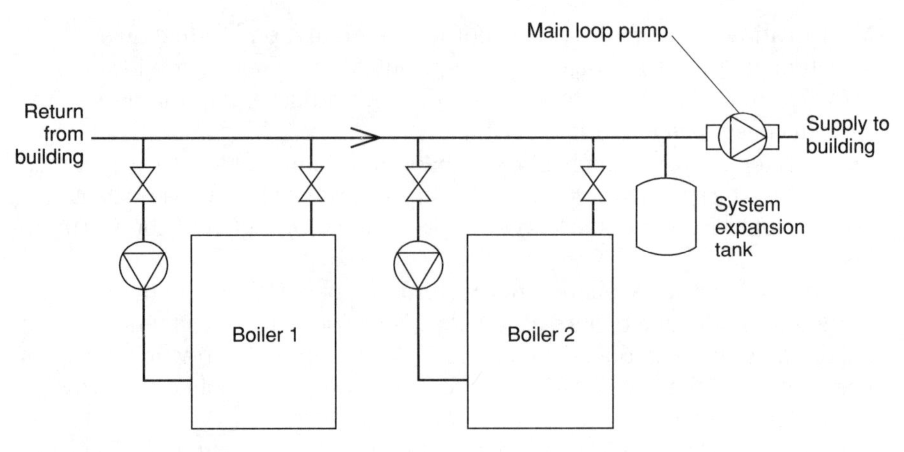

Figure 7.13. Two-boiler secondary pumped system. Each boiler receives water flow only when its secondary pump is running. Otherwise, all water passes through the bypass pipe.

Another reason for adding a second-stage boiler is to provide additional heating capacity. When additional boiler capacity is needed because of a building addition it might be more economical to add a small boiler than to completely replace an existing unit.

Converting the boiler system to secondary flow allows the original unit to be used with a new one. Total capacity is increased and the extra flexibility of the secondary system is a benefit. Both the original and the new boilers operate on their own secondary loops. The client saves the difference in cost between a new small boiler and the much larger one, and keeps the original investment in the older unit.

Most multiple-stage systems operate with one boiler as *lead* and the other operates as *lag*. The lead boiler starts first and finishes last, while the lag unit takes care of the peak heating requirements. The space thermostat starts the lead boiler and its secondary loop pump, and the temperature of the loop's water starts the main pump.

Automatic controls start the lag boiler based on the temperature in the main loop's return water. If the lead boiler can keep it sufficiently hot, the lag boiler remains off. If the return water is too cold, and the space thermostat still requires heat, the lag unit starts. A time delay prevents the lag boiler from starting too soon. This allows water heated by the lead boiler to circulate through the system and return to the boiler and temperature sensor.

If the return water becomes hot enough, the lag boiler turns off and the system again operates on a single boiler. Finally, if there is no demand for heat from the building thermostat the lead boiler is shut down.

Temperature scheduling. Hydronic system diversity can be improved by *scheduling* the loop's water temperature. This means that the loop's maximum temperature is decided in part by the outdoor temperature. As the outside temperature falls, controls allow the loop water temperature to rise. Warmer outside temperatures cause the loop's maximum temperature to fall.

Scheduling works with both primary and secondary systems to improve efficiency. The natural effect of less heat delivery from a system at lower water temperatures matches the lower demands with higher outside temperatures.

Scheduling controls require two temperature probes. One is installed in the boiler's heat exchanger or outlet pipe and the other senses outdoor temperature. The control starts the system's pump when the boiler heats the loop temperature to the schedule point. The boiler can continue heating the water an additional 5°F to 10°F. If the water heats to that point, the control shuts off the boiler even if the space thermostat continues to call for heat. The boiler starts again only when the water temperature falls below the schedule point.

On days when the outside temperature is relatively high, lower temperature water circulates through the system. The boiler does not run long enough to heat the water past the point needed to keep the building comfortable.

Radiant Heating

Radiant heating systems are beginning to become very popular in this country. They work very well in almost any climate, save energy, and are completely silent. While there is no possibility of delivering air conditioning radiantly, these systems are installed in many new homes as an "invisible" heating system.

The actual heat delivery with radiant systems is effected by warming the floor, wall, or ceiling surface. Electric or hydronic heat sources merely serve to warm the actual radiant surfaces to an appropriate operating temperature.

Radiant heating systems typically operate at lower temperatures than space heating systems. Hydronic loops might run at full capacity with only 110°F to 130°F water compared to the 200°F temperatures often used for convective systems. The low temperature requirement is related to the sensitivity of radiant heat transfer to temperature differences. As noted in Chapter 2, increasing radiant temperature differences drastically increases heat flow. This effect allows low-temperature systems to provide full heat output.

Radiant heat transfer from a large, warmed panel causes solid objects in the space to warm. After a few hours, most objects within view of the heated panel assume a surface temperature quite close to that of the panel. This uniformity helps to reduce convective drafts and stratification. Properly installed systems that use heated ceiling panels heat the floor to within 5°F of the ceiling's temperature. This small difference in temperature prevents significant stratification.

Even the inside surfaces of windows are heated by the radiant panel. Heating the windows helps prevent convective drafts that usually come from large glazed areas during cold weather. It also helps prevent humidity from condensing on the glass, reducing the window's tendency to fog.

Space air temperatures in buildings that are radiantly heated are often lower than with convective systems. Air can be cooler because radiant heat warms the building's occupants directly and it is not involved in the heat transfer process. Occupants feel as comfortable in radiant heated systems with air at 60°F as they would in ordinary systems with 75°F air. Because air is essentially transparent to infrared radiation it absorbs none of it. The fact that it warms at all is because the room's contents heat enough to convect some of their surface heat into the air.

Cooler air temperatures with radiant heating translate into significant energy savings. Heat losses due to infiltration are related to the difference in temperature between the outside and inside air. Less difference in temperature means that less heat is lost. Many radiant systems use 10% to 20% less energy than convective heating designs.

A secondary benefit of cool interior air temperature is better humidity control. Air's relative humidity drops as it is heated. Keeping the inside air cool helps maintain the humidity without using humidifiers. This can keep the occupants more comfortable.

These energy savings mean that hydronic system designs can be done differently. Most climates, even those with very cold conditions, need less heating capacity from a radiant heating system than they do with a standard system. It is reasonable to estimate the required capacity based on the total square feet of floor area to be heated. Because the air in the space is only incidentally warmed, the total area is a better indicator of the amount of heat required.

Most residential systems need approximately 30 to 50 Btu/hr of radiant heat per square foot of floor area. This does not include heat supplied to any ventilation air. To be conservative, it is usually safest to allow 50 Btu/hr per square foot for radiant heating and estimate the infiltration losses separately.

For example, assume that a 4500-ft^2 building has a total volume of 58,000 ft^3. Find the required radiant and infiltration air heating capacities. The indoor air temperature is expected to be 65°F and the outdoor design air temperature is –20°F. First, the radiant portion is

$$\text{Btu/hr} = 50\ \text{Btu/hr/ft}^2 \times 4500\ \text{ft}^2 = 225{,}000\ \text{Btu/hr}$$

Infiltration losses, based on 1 ACH, can be found with the equation:

$$\text{cfm} = \frac{\text{ACH} \times \text{Space volume}}{60}$$

$$\text{cfm} = \frac{1 \times 58{,}000\ \text{ft}^3}{60} = 967\ \text{cfm}$$

The heat loss related to infiltration is found with the formula

$$\text{Heat loss} = 1.08 \times \text{cfm} \times T_d$$

$$\text{Heat loss} = 1.08 \times 967\ \text{cfm} \times (65°\text{F} - -20°\text{F}) = 88{,}770\ \text{Btu/hr}$$

The total heat loss is

$$225{,}000\ \text{Btu/hr} + 88{,}770\ \text{Btu/hr} = 313{,}770\ \text{Btu/hr}$$

This is very conservative, and results in a moderately oversized heating system. It works acceptably, however, for most installations.

Estimate the total heat loss for the entire building and the losses for specific areas inside. Proportion the total heat loss of the building into each outside room by comparing the size of the room to the total building size. For example, if a room represents 20% of the square feet of the building, then approximately 20% of the heat should be delivered to it.

Radiant tubing or electric heat tapes provide a given amount of heat per foot. Consult the product specification manual or ask the manufacturer for information on specific products and applications. They can help estimate the amount of tubing or tape needed to provide the heat that each room and the entire building will require.

Most radiant systems are installed in the floor. As noted in Chapter 2, radiant heat transfer depends on the "view" that the radiator has of the objects in the space. The floor is an excellent radiator because the entire room is constantly in view of it.

Other radiant systems are installed into ceilings or walls. These work the same way as any radiant system, but can provide more controlled spot heating for problem areas. It is easy to add electric radiant panels to a surface to solve problems that might be difficult for conventional units.

In order for a radiant heating system to be successful, it is helpful to have the building tight. All radiant heating systems rely on the occupants being warmed directly by heat energy radiated from the warmed surface. Cool drafts in the room can cause a chilling effect if they blow over someone being warmed with a radiant system. While cold drafts are uncomfortable with any heating system, they can be a greater problem with radiant systems.

Make sure that the clients understand this and seal drafty windows or doors that might cause trouble. If a new building is receiving radiant heat, discuss this need with the general contractor. Everyone should understand that a perfectly working radiant heating system might not be satisfactory in a drafty building.

Thermostatic controls in radiant heated spaces should be installed within the radiant surface's view. It is important to expose the thermostat to the radiant energy to properly control the heating system. If it only senses the cool interior air it can't properly control the system.

Slab hydronic and electric

Embedding hot water piping or electric heating elements into a concrete floor slab is the most common means of installing radiant heat. The technique has been well developed and successfully used in many applications. Radiant floors are effective for heating most commercial and residential spaces.

Heating tubes or low-wattage elements are placed approximately 1 in. below the floor's surface. They are usually placed in a serpentine pattern to distribute heat evenly over the area, with closer spacing near outside walls.

The ground should be well insulated below the radiant floor to confine the heat to the slab. Uninsulated soil conducts much of the heat away from the concrete, so 2 in. of rigid foam insulation is usually sufficient. Insulate the outside perimeter of the radiant slab, too. Use 4 in. of thick rigid foam extending at least 2 ft below the surface to prevent the slab's heat from escaping to the outside.

It is important to provide radiant heat over the entire floor surface and not confine it to the perimeter areas. Because radiant heating systems do not heat the room's air, the room might feel uncomfortable if no heat is available near the interior areas.

Most interior areas do well with approximately 50 Btu/hr per square foot of floor area. The outside perimeter of the floor (2 to 3 ft from outside walls) should receive up to 75 Btu/hr per square foot of floor area. Generally, areas at outside walls should receive 50% more heat than interior surfaces. This provides extra energy to offset the colder surfaces that must be warmed there.

Note that these rules of thumb are maximums. Warmer climates might not require large amounts of heat. Smaller radiant systems work well if matched to total heat loss estimates as described in Chapter 3. These estimates provide a generously sized radiant heating system. Proportion the total heat energy in the space to provide the correct amount of energy delivery for the need.

For example, a room measuring 32 ft by 27 ft requires 32,000 Btu/hr. The heating requirement was estimated with the methods outlined in Chapter 3. One of the 32-ft-long walls has an outside exposure. The hydronic tubing being used delivers 40 Btu/ft when operated at the planned 120°F average temperature and buried under concrete. How much tubing should be used and how should it be arranged?

The total length of tubing needed is equal to the total Btu requirement divided by the tubing's output per foot.

$$\text{Length} = \frac{32{,}000\ \text{Btu/hr}}{40\ \text{Btu/ft}} = 800\ \text{ft}$$

The tube should be looped in a serpentine pattern, with 50% more tubing installed along the outside wall. This perimeter area requiring additional heat is the area of a 2-ft wide "strip" along the 32-ft-long wall.

$$32\ \text{ft} \times 2\ \text{ft} = 64\ \text{ft}^2$$

The interior area requiring less heat is the total area of the room less the perimeter area calculated previously.

$$32\ \text{ft} \times 27\ \text{ft} = 864\ \text{ft}^2 - 64\ \text{ft}^2 = 800\ \text{ft}^2$$

The heat density (Btu/hr/ft^2) for the interior space is found with the equation

$$\text{Heat density} = \frac{\text{Total Btu/hr}}{(\text{Inside area} + (1.5 \times \text{Perimeter area}))}$$

$$\text{Heat density} = \frac{32{,}000\ \text{Btu/hr}}{(800\ \text{ft}^2 + (1.5 \times 64\ \text{ft}^2))} = 35.7\ \text{Btu/hr/ft}^2$$

The heat density of the perimeter area is 1.5 times greater than this, or

$$\text{Heat density} = 1.5 \times 35.7\ \text{Btu/hr/ft}^2 = 53.6\ \text{Btu/hr/ft}^2$$

Find the tubing's spacing by dividing the tube's output per foot by the required density.

$$\text{Spacing (interior area)} = \frac{40\ \text{Btu/ft}}{35.7\ \text{Btu/hr/ft}^2} = 1.12\ \text{ft or } 13.5\ \text{in.}$$

$$\text{Spacing (perimeter area)} = \frac{40\ \text{Btu/ft}}{53.6\ \text{Btu/hr/ft}^2} = 0.746\ \text{ft or } 9\ \text{in.}$$

Therefore, the tubing loops should be 9 in. apart within 2 ft of the outside wall and 13.5 in. apart elsewhere. This delivers the design amount of heat to the space.

Carpeting or other floor coverings can affect the amount of heat transfer available from a slab system. The covering's surface becomes the radiant panel and heat must flow from the slab through the covering to warm it. It is easy to compensate for the insulating effects of floor coverings. Raise the water temperature (with hydronic systems) or use closer heat tape spacings (for electric systems) when heavy carpeting is anticipated. Consult with the radiant tubing or heat tape dealer to find out how much additional heat flow is needed when carpeting is used.

In the past, radiant hydronic tubing installed in concrete was always made of copper. While bare copper is an excellent material for general piping use, it does not perform well over time when encased in concrete. It is subject to corrosion from the acids and galvanic action of soil and the alkali reactions of concrete. Most modern systems use plastic tubing to carry the water.

The best plastic radiant tubing has an exterior oxygen barrier to keep this gas out of the system's water. Ordinary plastic tubing is permeable to oxygen, and it can enter the system and cause corrosion. Most closed hydronic systems do not use chemical corrosion inhibitors, but rely on the early absorption of minor amounts of dissolved oxygen from the system water. Permeable tubing, however, can provide a constant supply of oxygen to continue damaging the system. Tubing with an oxygen barrier costs more than tubing without it, but it will pay for itself in the long run.

Electric heating wire (or tape) is available in several stock wattages and is sold by the foot. Most of these use parallel circuits that allow the wire to generate heat anywhere along its entire length. The heat tape itself resembles a flat wire television antenna lead. It is wider than it is thick and covered with a tough heat, moisture, and wear resistant jacket. Copper wires at the tape's edges carry full line voltage through its entire length. Between the conductors the heat generating material resists the flow of electricity. Heat is generated in the tape as current passes between the two conductors, crossing its width and passing through the heating material.

Some tapes can self-regulate their temperature. The microscopic structure of the heat generating material changes as its temperature increases. This increases its resistance to electric current, decreasing the heat generating capability of that area of the tape.

Self-regulation takes place independently all along the length of the tape. If one part of the tape is heavily insulated or overlaps itself it will overheat. Self-regulating tapes automatically sense the higher temperature in that area and reduce the heat output there. Thus, overheating is avoided. Once the current is turned off the internal structure of the tape returns to its original form and it heats normally again. This automatic temperature limiting can take place repeatedly without damaging the tape.

Wall and ceiling systems

Most of the functional descriptions for floor-installed radiant systems are true of wall and ceiling systems. They differ only in the installation details required for their applications.

Ceiling-mounted radiant heating systems do not cause excessive air stratification. While the idea of placing a heat source at the highest location in a space might sound wrong, radiant systems work very well with this setup. They operate so differently from convective hot water or forced air systems that ceiling source radiant heat works with almost no stratification.

Convective and forced air systems heat the space's air to warm the room and its occupants. This happens best if the air is uniformly mixed and a layer of cool air does not lie near the bottom of the space. Cool air feels chilly to the room's occupants, and higher thermostat settings are used to compensate. Meanwhile, a layer of warm air lies near the ceiling, too high to provide comfort for the room.

Radiant heat does not warm the space's air. Instead, it transmits energy to the objects in the room, warming them. Because ceiling-type radiant systems have a perfect view of the floor, much of the energy from the ceiling increases the floor's temperature. Temperatures on the floor surface in rooms with ceiling radiant systems become very close to that of the heated ceiling temperature.

The warm floor spreads heat via convection to the room's air. The total effect is the best possible convective heat source: a large, moderate-temperature heat source at the bottom of the space. Floor and ceiling temperatures become very uniform and stratification is not a problem.

Most wall or ceiling radiant systems are electrically operated, but there are also some unusual hydronic floor and ceiling systems using buried tubing or large metal pans. The warm surfaces are exposed and allowed to radiate heat, but these hydronic systems are very uncommon.

Ceiling or wall radiant systems can be used to heat the upper floors in buildings that have slab radiant heat for the first floor. They work the same as floor-installed systems, with the same advantages and requirements. Because wall systems might not have a full view of the room and its contents, more than one wall might require heating.

Electric radiant systems are simple. A heating wire or tape is buried under or within the plaster or gypsum board used for the surface, or an independent panel can be used that hangs from the wall or ceiling. The former systems are intended as permanent installations, while the latter are often used for supplemental heating.

Tubing or heat tapes buried under plaster or gypsum board must be secured to the framing or lathe holding the plaster or board in place. Thin (⅜-in.-thick) gypsum board, or a similar thickness of plaster, applied to the heater's surface ensures that sufficient heat transfer can take place. The material should not be overheated, so hydronic systems should be limited to about 130°F to 140°F. Heat tapes should be self-limiting to prevent them from exceeding these operating temperatures.

Insulate the back surface of the heated panel to limit heat losses (if they face an unheated space). It is possible, however, to heat two rooms with one warmed wall located between them. Install the heat tubing or tape in the middle of the partition, with alternating runs on each side of the studs. The tape or tube runs back and forth in the wall cavity, secured to both faces on alternate passes.

While the heat is on the interior space of the studs gets quite hot, so special electrical wiring might be required if it runs through these spaces. Heat tapes or tubes should not be installed within 6 in. of electric switches or outlets.

Electric radiant panels are very easy to install. They clip to the wall or ceiling and require a simple connection to a line voltage thermostat and power source. Panel radiant heaters can supplement any type of heat (forced air, convective, or radiant) if the main system doesn't work well.

Chapter

8

Air Conditioning Systems

Air conditioners cool and dehumidify occupied spaces. As described in Chapter 3, there are many ways that buildings can gain heat. The most common way of removing this heat is with an air conditioning system.

Operation

Air conditioners use a fluid (refrigerant) to carry heat from one place to another. Because heat naturally flows from higher temperature areas to lower temperature areas, the fluid must work against the natural direction of flow. This requires energy.

Air conditioners use a vapor cycle refrigeration system. This system includes four major components: compressor, condensing heat exchanger, thermal expansion valve, and evaporating heat exchanger. Figure 8.1 illustrates schematically these major components. The compressor drives fluid through the rest of the system. As it travels it changes phase from gas to liquid and back to gas. Heat is absorbed and rejected at the desired locations, moving it from the lower temperature area to the higher temperature one. The compressor provides the energy required to drive heat against its natural flow.

Compressors

Refrigerant starts at the compressor inlet as a low-temperature, low-pressure gas. The compressor raises the refrigerant's pressure (and temperature). Air conditioning compressors are usually positive displacement devices. These

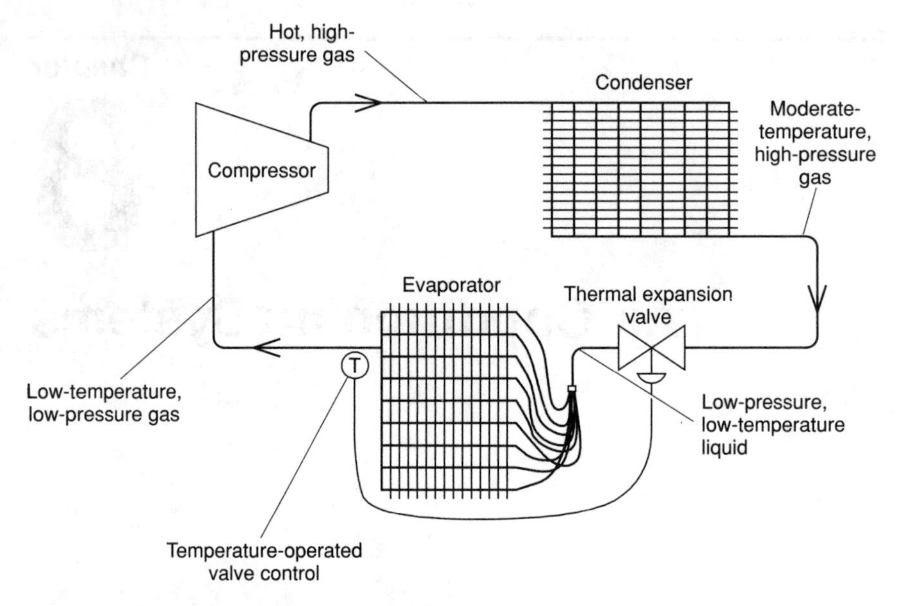

Figure 8.1. Basic refrigeration circuit with direct expansion coil.

include reciprocating (piston), screw, and scroll units that physically squeeze the refrigerant from the low-pressure intake port to the output. While the compressor pushes the gas to the output port, the gas' volume decreases. This increases its pressure and temperature.

Positive displacement compressors cannot handle liquids. A liquid is incompressible and can cause serious damage. Liquid that enters a compressor inlet can quickly destroy the compressor.

Compressors are usually classified into two families: hermetic and semihermetic. Hermetic units are supplied sealed in a steel case (often described as a "tin can"), while semihermetic units have separate motors and compressors. Hermetic units cannot be serviced. The drive motor and compressor are sealed together in the case to avoid the need for a seal resistant to refrigerant leaks. Semihermetic units are serviceable. They can be disassembled and the internal components (valves, pistons, bearings, etc.) can be replaced.

Larger compressors are usually semihermetic designs. Their replacement costs are usually so high that it is worth disassembling and servicing them when parts wear out. Semihermetic compressors can be remanufactured, and rebuilt units are available as replacements.

Smaller units are often hermetic types. It is less expensive to swap out the entire assembly than to try to repair it. Hermetic units are less likely to leak, and the lack of external gaskets makes them more resistant to contamination.

Centrifugal refrigerant compressors are not positive displacement devices. An impeller, built like a specialized fan, accepts gas at its hub and spins it at high speed out to its periphery. The spinning action increases the gas' pressure enough for it to be used in the air conditioning system.

Because of the high output to input pressure change required, most centrifugal compressors use more than one stage to boost pressure several times. This distributes the work to several impellers and each can run more slowly than a very high-speed, single-stage unit.

While centrifugal chillers are not positive displacement compressors, they also should not handle liquid refrigerant. The impact of liquid striking the high-speed turbine blades can break the blades.

Surge can also damage centrifugal compressors. Similar to the surge that an overrestricted fan experiences, compressor surge occurs when it attempts to push the gas into a too high pressure outlet. The outlet pressure exceeds the capability of the impeller, and some gas slips backward through the rotating blades. Avoid conditions that encourage compressor surge. It can damage the impeller and housing, requiring that the entire unit be torn down and rebuilt. Proper control of the system head pressure prevents surge.

Most centrifugal compressors are used for water chillers. Their large size, and the temperature and pressure ranges of the refrigerants they use, make them unsuitable for most direct air cooling systems.

Refrigerant temperature increases as it is compressed. Depending on the operating conditions and the type of refrigerant, it might leave the compressor at more than 300°F. This hot gas can be used for refrigerant heat recovery systems. By sending it through a heat exchanger, water can be heated from the otherwise wasted heat of the gas.

Condensers

The hot, high-pressure gaseous refrigerant leaving the compressor next goes to a heat exchanger called the *condenser*. This unit cools the gas below its high-pressure dew point. As the refrigerant passes through the condenser it changes from a gas to a liquid. Air conditioning systems use high-capacity condensers to completely liquefy the refrigerant.

Condensers are usually air or water cooled. Air-cooled condensers (also called *dry coolers*) use a coil for the gas and one or more propeller-type fans to drive air through the coil. This removes heat from the refrigerant inside the tubes. Some larger condensers operate several fans sequentially, depending on the air conditioner's load.

More heat rejection requires more fans. These staged fans usually operate with a pressure control. Additional heat handled by the system causes higher gas pressure in the condenser. The pressure sensors detect the higher refrigerant pressures and start additional fans.

The condenser pressure, also called the system's *head pressure* (because it equals the compressor's cylinder head discharge pressure), is an excellent gauge of system operation. This is covered in further detail later.

Water-cooled condensers are usually shell and tube heat exchangers. Refrigerant flows through the tubes and rejects heat to the surrounding water-filled shell. These systems might include cooling towers or recycling systems so that the warmed water can be recovered. If the system dumps the water to the sewer it usually includes an automatic valve to shut off the water flow when the air conditioner is not running.

Other water-cooled condensers include shell and coil, tube in tube, and plate types. Shell and coil condensers reverse the arrangement of the shell and tube type. Water flows through the inner coil and the refrigerant condenses inside the outer shell.

Tube-in-tube designs flow refrigerant through one of two tubes. A larger diameter tube encloses a smaller sized one. Refrigerant or water flows either inside the smaller tube or in the space between the two tubes. Often they are piped in a counterflow arrangement, where the refrigerant flows in the opposite direction of the water. This provides more efficient total heat transfer.

Plate-type exchangers are usually used for smaller systems. Alternating flows of refrigerant and water make up a "sandwich" of fluids, with each separated by a stainless steel plate. These are inexpensive, but are impossible to clean or repair.

Another water-cooled condenser is the evaporative condenser (a type of cooling tower). Water remains at the tower (other than the drain lines) and cools the refrigerant enclosed in a set of tubes. Evaporative condensers can have water discharge temperatures close to the ambient air's wet bulb temperature. In many climates this is well below the dry bulb temperature for most of the year. Cooling the refrigerant to the ambient air's wet bulb temperature allows faster and more thorough condensation. System head pressure drops and the system operates more efficiently.

Usually, the lower the condenser's operating temperature and pressure, the less energy the system uses. Lower condenser temperatures produce lower system pressures. Lower pressures allow the compressor to do less work.

In cases where there are long runs of refrigerant lines, the liquid's pressure might drop excessively due to flow and friction losses. This might allow it to become low enough to start to boil within the line. This blocks flow through the line and significantly reduces its capacity. To prevent this, the liquid refrigerant must be cooled below its condensation temperature. This process is called *subcooling*. Subcooling prevents the pressure loss caused when the refrigerant boils.

Thermal expansion valves

After the condenser, the refrigerant is a high-pressure, moderate-temperature liquid. The heat energy it rejected in the condenser causes it to change from a gas to a liquid and reduce its temperature. It flows in this state to the thermal expansion valve or metering orifice. It undergoes a significant pressure drop as it flows through the partly opened valve or restriction.

Flowing a fluid through a restriction reduces its pressure. This is exactly the same as the pressure loss water experiences as it flows through a pipe. More restrictive pipes cause higher pressure losses. The thermal expansion valve or metering orifice acts like a piping network. Refrigerant pressure is lost as it is pushed through the restriction.

Changing the pressure on a liquid changes its boiling temperature. Higher pressures cause higher boiling points and lower pressures cause lower boiling points. If water is contained in a vessel at 1.0 in. mercury vacuum, its boiling point is 80°F. Conversely, the higher pressure in an automotive cooling system contributes to its resistance to boiling. Pure water won't boil until 250°F under the system's 15 psi working pressure. Refrigerants behave the same way. Reducing their pressure lowers their boiling point.

The refrigerant's pressure is reduced so much as it flows through the air conditioning system's expansion valve that it begins to vaporize (boil) immediately afterward. This pressure loss and vaporization cause the low-pressure liquid's temperature to drop to its new, lower boiling point. This lower boiling point is the design cooling point for the system. For most nonchiller air conditioning systems it is usually about 35°F to 40°F.

For example, an air conditioning system using refrigerant R-22 is usually designed with a suction pressure near 65 psi. This sets its boiling point to approximately 37°F, allowing it to effectively cool air or water.

Note that no heat is added to the refrigerant at this point to cause it to boil. The loss of pressure and the heat it brought from the condenser have caused the liquid refrigerant to start boiling. Boiling drops the refrigerant's temperature. Once it reaches that point, boiling stops. The refrigerant is now a cool, low-pressure mixture of liquid and gas. The refrigerant's new low temperature allows it to easily absorb heat from its surroundings.

Evaporators

After the thermal expansion valve, the low-pressure, low-temperature liquid and vapor mixture enters another heat exchanger called the *evaporator*. There it absorbs heat from the passing air or water. This additional heat causes the rest of the liquid refrigerant to boil. Boiling is a heat absorbing process, so the refrigerant accepts the passing fluid's heat energy.

Evaporators used for water chillers are, like condensers, built as shell and tube heat exchangers. Nonchiller systems, usually called direct expansion (DX) types, use refrigerant coils as described in Chapter 4.

As the gas leaves the evaporator, its temperature is sensed by a probe connected to the thermal expansion valve. The probe causes the valve to open or close just enough to keep the evaporator's discharge temperature a few degrees above the refrigerant's boiling point. The difference between the evaporator discharge temperature and the refrigerant's boiling point is its *superheat*. If, for example, the discharge temperature was 38°F, and the refrigerant's boiling point (at its pressure in the evaporator) was 33°F, it has 5°F of superheat.

It is important for the system to maintain superheat. Most air conditioning systems should maintain about 10°F of superheat during normal operation. If they don't, some refrigerant leaving the evaporator might still be in liquid form. This liquid entering the compressor can damage it. Small amounts of liquid increase internal wear of reciprocating compressors, and large amounts can destroy them. A properly adjusted and working thermal expansion valve prevents the compressor from receiving a slug of liquid.

Once the refrigerant leaves the evaporator it is ready to enter the compressor and repeat the cycle. It has absorbed heat from the evaporator and rejected it to the condenser. This is the desired heat transfer.

Some refrigeration systems, most notably centrifugal water chillers, operate with internal pressures below atmospheric. Normal suction pressure for these systems is a partial vacuum. Refrigerants used in these low-pressure systems include CFC-11 and HCFC-123. The partial vacuum causes these systems to absorb air.

Air in a refrigeration system can cause capacity problems. It uses system volume and increases pressure, making the compressor work harder to deliver the rated cooling. In addition, water absorbed with the air can become acidic when mixed with the refrigerant and lubricating oil. For these reasons, low-pressure systems usually include a purge device.

Purge systems allow the machine to dispose of air and water vapor that has mixed with the refrigerant, but to retain the system's charge. Because most low-pressure systems use ozone depleting refrigerants, purge systems must capture the refrigerant that escapes with the air.

These purge units include small direct expansion coolers that condense gaseous refrigerant out of the vapor exhaust stream. Once liquefied, it is pumped back to the system and the remaining air is vented out.

Capacity Control

Just as heating systems are more efficient with a wide diversity range, cooling systems benefit from the ability to run under partial load. In fact, this is usually essential for comfort air conditioning systems.

Most cooling systems have two functions: reduce the air's temperature and reduce its humidity. While both functions are heat removal, usually only one is controlled. Almost all systems operate with a space temperature thermostat. Humidity reduction is expected to take care of itself when sensible heat is removed.

During times when both sensible and latent heat loads are at their highest design points, the air conditioner should be able to meet the temperature and dehumidification demands. While it runs to keep the space temperature down, it also maintains low humidity. Moist inside air passing over the cold chilled water or evaporator coil is continuously dried.

However, this might not happen correctly at times of low sensible heat and high latent heat loads. This condition is common when the outdoor temperatures or internal loads are not as high as the design extremes. If the system is only able to run at full capacity, it will quickly cool the space to the thermostat's setting and shut itself off. While it is stopped, no dehumidification takes place and the space humidity rises. This is the reason that simple air conditioners with no capacity controls (often used for residential and light commercial systems) should be installed on the "small" side. It is better to have the unit smaller than peak loads demand than to have it run insufficiently when loads are more normal.

Most larger commercial systems have one or more means of limiting their output so that they can run continuously at less than full load. These systems usually include staged compressors, unloaders, and hot gas bypass controls.

Staged compressors are set up to divide the cooling load into two or more portions. Multiple compressors feed their own condenser and evaporator coils installed in the air handler. Each compressor operates independently from the others (except their common controls).

Multiple staged air conditioners run only one system at times of low load. A head pressure sensor on the lead system (the first one to run when cooling is needed) decides when it is loaded to full capacity. At that point the control starts the lag compressor (the second stage) to bring it on line. This brings the full system capacity on to cool the space. The lag unit starts and stops in response to changing load conditions.

Multiple compressors also act as a partial backup to each other, which is helpful in case of a failure. One system can be serviced while the other continues to operate. While the single operating system might not be adequate to completely cool the space, it provides dehumidified air and can keep a system failure from becoming an emergency.

Pressure unloading provides another way of limiting system capacity. Unloading reduces the pressure that the compressor can achieve, preventing it from working as effectively. The pressure output drops as the unloader opens, preventing the full charge of refrigerant in the system from condensing. In effect, the system partly disables itself, allowing it to reach only part of its capacity.

Unloading controls are very efficient. Often used with larger systems, the unloaded compressor runs more easily and consumes less power than when it is fully loaded. This can save a significant amount of energy over a cooling season.

Air conditioning manufacturers use a variety of ways to unload a compressor. All of them, however, limit the volume of gas that is compressed.

Some reciprocating systems open a valve on the compressor head or crankcase that allows some gas to escape during compression. Screw compressors often include a slide valve unloader. When opened, only part of the screw is sealed and it can achieve only partial compression. Centrifugal chillers unload by setting the compressor's static (nonrotating) inlet vanes to less than optimal positions. Less gas can enter the impeller, so the impeller develops less pressure.

Unloading controls, like staged compressors, usually operate automatically depending on the system head pressure. When it increases beyond a set point, the unloader closes and the system increases its capacity. As the head pressure drops, the control reduces the system capacity.

Capacity regulators include external manual and automatic limit setpoint controls. Many screw and centrifugal water chillers include these controls, which prevent them from running more than a fraction of their full capacity. This can be useful in larger systems where there is a possibility of a temperature overshoot. Chilled water temperatures might become too low if the machine applied its full cooling capacity at times of low load.

This can happen when a system initially cools room temperature water. The comparatively high starting water temperature can cause the chiller to run at full capacity. If there is very little cooling load on the system, the freshly cooled water will return to the chiller with little added heat. When the machine cools the water again, ice might form before the head pressure control has a chance to reduce the machine's output. Manually limiting the capacity at system startup helps avoid this.

Automatic external unloading control is also used with some energy management systems. Many electric utilities penalize customers for the monthly peak electrical demand from the entire system. Energy management systems can unload the chiller to reduce the air conditioning system's power at times of peak demand. It might only need to reduce the power for 15 minutes to prevent a significant increase in the month's electric bill.

Hot gas bypass is another air conditioning capacity control. These systems are used with smaller units to achieve some capacity limiting. They are simple and effective, but do not provide the energy savings that multiple stages or unloading can give.

Hot gas bypass systems allow some of the compressor's output of hot, high-pressure refrigerant to skip the condensing and evaporation cycle. Instead, it is recycled back into the compressor's inlet and compressed again. This effectively reduces the amount of refrigerant flowing through the system, but the compressor continues to handle the full mass of gas.

Simple automatic solenoid valves are used for hot gas bypass systems. A metering orifice allows a set amount of gas to run from the compressor's discharge port to the solenoid valve. When the valve opens, the gas passes through it back to the compressor's inlet.

Head pressure controls operate the hot gas bypass solenoid. A switch is provided on some systems to allow the valve to be manually energized to run the system at partial capacity. However, because there is no significant energy savings while the system is bypassed, the only benefits are to help prevent transient overcooling and improve space humidity control.

Efficiency Ratings

An air conditioner's efficiency is rated by several factors. Because the amount of heat they remove from the space exceeds the electric power consumed, a classic measure of efficiency does not work. Efficiency is the energy output divided by its input. Air conditioning systems would have efficiencies over 100% by this measure. The reason is that the heat output at the condenser does not all originate from the power supplied by the electric source. Most of it comes from the evaporator coil and this heat makes it appear to operate at more than 100% efficiency. To provide a meaningful efficiency measure, the coefficient of performance (COP) and energy efficiency ratio (EER) are usually used. The higher these ratings are, the better.

The COP for an air conditioning system is the amount of energy removed by the evaporator divided by the energy used in the compressor. All units must be consistent. For example, if a 6-ton air conditioner requires 5500 W to operate, its coefficient of performance is found by

$$\text{Heat removed} = 6 \text{ ton/hr} \times 12{,}000 \text{ Btu/ton} = 72{,}000 \text{ Btu/hr}$$

$$\text{Power used} = 5500 \text{ W} \times 3.41 \text{ Btu/W} = 18{,}755 \text{ Btu}$$

$$\text{COP} = \frac{72{,}000 \text{ Btu/hr}}{18{,}755 \text{ Btu}} = 3.84$$

The energy efficiency ratio is similar, but the unit used for measuring the heat removal is Btu per hour and the electric power unit is watts. The above example's EER is

$$\text{EER} = \frac{72{,}000 \text{ Btu/hr}}{5500 \text{ W}} = 13.1$$

The two ratings are similar, but EER has gained widespread acceptance because of its use by the government and many energy conservation codes. Most equipment sold must meet minimum standards for efficiency, and these standards are described as an EER rating.

Energy efficiency ratios are estimated by the equipment manufacturer under specific test conditions. These might not be ideal conditions for the unit's operation, but all equipment is tested under the same conditions. Comparisons of different systems are meaningful because they all are tested the same way.

Service

Many factors influence an air conditioning system's operation. Interior and exterior temperature, the condition of the condenser and evaporator coils, and the amount of refrigerant charge all affect how the system works.

Generally, the lower the temperature in the system, the lower its internal pressure. Refrigerant, like most substances, expands when heated and contracts when cooled. Running a system at lower temperatures causes the refrigerant to contract and reduce its pressure. The opposite is also true: systems operating at higher temperatures have higher pressures.

Any system that is working poorly with excessive head pressure probably has a heat problem. The most likely cause is that the condenser cannot reject enough heat to satisfy the load. This reduces the system's capacity as it tries to work against the excessive pressure. If the pressure becomes too high, safety controls shut the system down to prevent damage.

Check that the condenser receives a good flow of air, and that it is not exposed to any unusual heat sources. Clean and straighten an air-cooled condenser's fins, and confirm that the blower operates properly. Even a fan shroud left off the case can divert air away from the condenser coil and cause excess head pressure.

Water-cooled condensers and evaporative units usually fail because of fouling of the tubes' exterior surfaces. The tubes should be cleaned thoroughly to remove any mineral scale or biological slime.

Sufficient water flow through the exchanger should be confirmed. Most systems permit the condenser water to warm up to 20°F as it passes through the shell. If the tube's surfaces are clean and the temperature increases no more than 20°F, then the water flow is acceptable. Too much of a temperature increase usually means that the water flow is insufficient.

Very low head pressure might show that the system's condenser is too cold. This can be a problem in any season. The condenser must limit its heat rejection in the winter. A *low ambient* control is supposed to keep head pressure from falling too low. The control shuts off the condenser fan and can close dampers to prevent outside air from reaching the coil. If the low ambient control fails, pressure drops. This can cause the system's refrigerant to liquefy because the thermal expansion valve won't be able to drop the pressure enough to vaporize the refrigerant.

The amount of refrigerant in the system influences its operation. Air conditioners are meant to operate with a specific amount of fluid. Too little and the pressure will be too low, too much and the pressure will be too high.

Symptoms of too little refrigerant can be surprising. If the charge is slightly low, pressure losses are not too great. A minor pressure loss at the compressor's suction fitting reduces the fluid's boiling point below 32°F. Ice forms on the evaporator coil because of the low-pressure refrigerant in it.

One of the first symptoms of insufficient charge is ice on the evaporator coil. The ice blocks the airflow through the coil, and can cause a damaging freeze in the chiller's heat exchanger. If the charge falls much beyond the ice-forming point, the system stops working. Not enough refrigerant remains to be compressed to the point of condensation.

Besides monitoring head pressure, compressor current draw should be examined. A clamp-on ammeter is an excellent tool to use when examining air conditioners. Compressors in properly operating systems draw their rated current. Current draw well above or below this suggests a problem.

Too low current draw, along with a low head pressure reading, means that the compressor is not working as hard as it should. Low current draw and poor performance usually mean an internal compressor problem. Bad valves that do not seal properly are a typical cause of low current draw and poor performance.

Too much current draw with normal head pressure might be caused by a restriction in the compressor's suction (inlet) line. Confirm that the filter and tubing are not dirty or restricted.

A larger than normal current draw, accompanied by too much head pressure, can show that the system is overloaded. Check that the condenser can get rid of its heat and that the condenser's refrigerant discharge line is not clogged or restricted.

Most systems include refrigerant filter dryers. Units are available for the liquid and vapor sides of the system. These hold any contaminating particles or water that enter the system. A ceramic filter element absorbs any water to keep it out of circulation. Change these filters whenever the refrigerant system is opened.

Some filter dryers have removable cores so you can change just the ceramic element without discharging all of the refrigerant from the system. Use removable core filter dryers if the system has become heavily contaminated. The system will need several filter changes to capture all of the dirt circulating inside. It's easier and less expensive to change filter dryer cores than to discharge and evacuate the entire system just to install a new filter.

Handle air conditioning systems carefully when they are opened for service. The section on "Refrigerant regulations" covers some of the rules regarding discharge, recovery, and recycling of refrigerants. Care must be used even after the refrigerant has been removed.

It is important to keep all open parts of an air conditioning system as clean and dry as possible. Dirt and moisture can cause an otherwise good system to fail. Dirt can foul the thermal expansion valve or become hung up on the compressor valve's seats.

Water can cause even worse damage. Water (even humidity in the surrounding air) exposed to the internal lubricants and refrigerant of a system becomes highly acidic. Few surfaces inside an air conditioning system are corrosion resistant, and the acid will quickly damage them. Acid can score the compressor's housing, corrode the cylinder bore, and pit the valve seats, effectively destroying them.

Evacuate an air conditioning system when it is reassembled to remove all traces of air and water. Air contaminates a system by taking up space intended for refrigerant. Its presence reduces the operating capacity of the system.

The compressor's manufacturer can give you specific instructions for properly evacuating a system. Most call for three vacuum purges, with pure, dry nitrogen gas used between evacuations to help mop up any remaining water. The vacuum attained must be high enough to remove over 99.5% of the air and water. This equates to a 0.15 in. mercury vacuum.

Noncondensable gases (like air) in an air conditioning system cause serious efficiency losses. These gases migrate to the condenser and prevent it from transferring heat away from the refrigerant. Also, noncondensable gases increase the system head pressure, causing the power consumption to increase. Avoid leaving any air or other contaminants in the system.

It's easy to test for the presence of air or other noncondensable contamination in a system. Shut the system off and allow the refrigerant to cool to either the ambient air or condenser cooling water temperature. Measure the pressure at the condenser and compare it to the refrigerant's saturation pressure at the system's temperature. If the condenser pressure is higher than the saturation pressure it shows that noncondensable gases are present in the system.

Refrigerant must be added carefully. It is important to add the proper amount, and to do it so that no damage to the system will occur. Most systems need to have refrigerant added as a vapor to the compressor's inlet. Adding liquid at this point can destroy the compressor.

Refrigerant supply bottles should be kept upright when filling a system. There is no dip tube in the bottle, so as long as it is upright vapor can reach its outlet. Tipping or inverting the bottle floods the system with liquid.

It might be helpful to warm the bottle of some refrigerants to fill the system more rapidly. Never use a heat gun or torch. Placing the bottle in a pan of warm water provides enough pressure to quickly fill almost any system. Don't use water over 150°F to prevent excessive compressor inlet pressures.

There are several ways to fill a system with refrigerant: current draw, weight, sight glass, and superheat. Each has its advantages.

Watching the compressor's motor current draw is an excellent way of monitoring refrigerant level in small- and medium-size systems. Many of these systems need only a few ounces of refrigerant for a complete charge, making other methods of filling impractical.

Refrigerant is slowly added to the system while monitoring the current draw of the compressor. Once it increases to just below its rated draw, stop adding fluid. Let the unit run for several minutes and check the ammeter to see if more refrigerant should be added. If the current is still below the rating, slowly add enough refrigerant to bring it up to its proper level.

Adding refrigerant by weight is useful for larger systems where several pounds are needed. Most systems have a specification stating the weight of the refrigerant's charge. While this generally does not include any field supplied interconnecting tubing, it provides a good starting point for charging the system. Add the specified weight for each of the separate components in the system (compressor, condenser, interconnecting lines, and evaporator).

A special scale is used that shows the weight removed from its platform. The refrigerant bottle is placed on the scale and connected to the system. The scale is set to zero and the refrigerant begins charging the air conditioning system. As the fluid runs into the system the scale's reading increases, showing the weight lost from the tank (and, therefore, added to the system). Turn off the flow when the rated amount of refrigerant has been added to the system. This should provide proper operation for systems that do not have very long lengths of interconnecting tubing. Confirm that the system has the correct charge by checking the compressor motor current draw, sight glass, or superheat.

A sight glass provides an effective way of properly charging a system. Sight glasses are fittings with a small window in the side for observing the refrigerant flowing through the system. Usually installed just after the condenser, they can be used when filling the system with refrigerant.

When there is a very small charge, or none at all, the sight glass shows only vapor. It appears clear, possibly streaked with oil. As fluid is added, the glass shows a fog passing through the tubing. This is a suspension of liquid droplets held within the surrounding vapor. When system pressure is high enough to begin condensing most of the refrigerant, the glass shows a bubbly fluid. The system is almost filled at this point. Continue adding fluid until the bubbles disappear from the glass and the fluid runs clear. An occasional small bubble is acceptable if the weather is very hot or the load is unusually high. Check the compressor's current draw to confirm the charge is correct.

A very precise way of charging a system is by checking the refrigerant's superheat as it leaves the evaporator. A special thermometer and a suction pressure gauge are used to check superheat. The pressure gauge fits onto the compressor's inlet fitting and the thermometer is positioned on the compressor's suction line a few inches beyond the evaporator's outlet.

Each type of system and refrigerant has specific temperature/pressure relationships and superheat specifications. As the fluid's temperature rises above its saturation temperature its superheat increases. Monitor the system while filling it to find out when the proper superheat is achieved.

The system's pressure increases as it fills. Once it is high enough for the refrigeration system to begin working, the fluid leaves the evaporator in a superheated condition. It completely vaporizes in the first part of the coil and continues to absorb heat as it runs to the vapor manifold. Its temperature is well above its boiling point from all of the heat it absorbed after vaporizing. Its superheat is excessive for a fully functioning system. More refrigerant added to the system increases the distance it goes into the evaporator coil before it completely vaporizes. Because it does not have as great a distance to travel in the coil as a pure vapor, it cannot pick up as much heat. Its measured superheat decreases.

Superheat falls as the system fills. As noted earlier, one symptom of a slightly low charge is freezing of the evaporator coil. At this point the total system pressure is so low that the evaporation temperature is below freezing. The refrigerant will not, however, be doing much cooling. A low evaporation temperature is different from high capacity. Monitor the superheat by comparing the saturation temperature (as determined by the pressure) with the actual gas temperature.

Stop adding refrigerant when the system's refrigerant superheat is correct. Adding more continues to reduce the superheat, risking possible compressor damage from ingestion of liquid refrigerant.

System Designs

Almost all air conditioning systems are, ultimately, forced air systems. While chillers use water to carry energy away from the cooled air, it is forced air that ultimately cools the space. This means that there are fewer options for air conditioning systems than there are for heating systems. Most commercial and residential air conditioning systems fall into two categories depending upon the physical size and layout of their components: unitary equipment and split systems.

Unitary designs are available in a large range of capacities, with custom rooftop systems available to cool large buildings. Note that package terminal air conditioner (PTAC) units occupy the opposite end of the capacity scale. These are usually used to cool a single room.

Unitary package

Unitary systems have essentially all their components contained in a single package. Fans for the condenser (if it is air cooled) and evaporator are usually included in unitary systems. Essentially they are self-contained air cooling and delivery systems. Many are available with electric heaters or fossil fuel burners to provide a complete summer and winter HVAC package. Unitary systems include rooftop units, vertical and horizontal flow package systems (generally with water-cooled condensers), and PTACs.

Unitary systems just need to be installed into the space and connected to the utilities to provide cooling. Large units connect to a field supplied duct system, but all of the operating controls (except space thermostats) are included in the package.

Most commercial air conditioners installed today are unitary package systems, often rooftop units. Options can be installed to make unitary package systems suitable for most applications.

Rooftop systems are available with vertical or horizontal air intake and discharge. Outside air intakes are optional, as are fully integrated economizers to reduce cooling costs. Gas or electric heat, and hot water or steam coils, are optionally available with the basic unit.

Rooftop systems include an evaporator fan sized to provide 400 cfm/ton of cooling capacity. An adjustable sheave on the motor's shaft allows the technician to match the air delivery to the external duct system's static pressure requirements.

Special equipment, like electrostatic air filters and heat recovery systems are usually not available for package rooftop units. Custom fabricated systems, however, can incorporate almost any option or feature imaginable.

Water-cooled, unitary package systems are usually installed indoors. Most connect to a duct system that distributes the discharged air into the space. Return air can be either ducted to the unit or freely drawn into the integral filter housing. Electric, hot water, or steam coils can be fitted to these units.

Package terminal air conditioners

Package terminal air conditioners represent the low-cost end of unitary systems. Often used for heating and cooling in motels, nursing homes, and schools, they install into an opening in an outside wall. The evaporator section faces into the building and the condenser protrudes outside. Package terminal air conditioners are inexpensive, and operate quietly and efficiently. Several options are available to make them usable in most applications.

Most PTACs are available with heat pump reversing valves and electric resistance heaters to provide space heating. As discussed in Chapter 7, the heat pump mode provides greater efficiency under most operating conditions. Most heat pumps operate economically if the outdoor temperature does not go below freezing.

Other options for PTACs include remote setback control. The thermostat is disabled by closing a switch (usually at the building manager's office) that prevents the air conditioning or the heat from working. The unit provides heat if the room's temperature falls below 50°F, but comfort heating and cooling are disabled. This saves a lot of energy when the space is unoccupied, and the remote control allows the unit to use its thermostat when the room is occupied.

Some PTACs include an integral condensate disposal device. Water condensing on the evaporator coil drains to the outside condenser section. There it collects in a pan, where a wheel attached to the condenser fan slings it onto the condenser coil. The water hits the hot coil and evaporates.

This provides two benefits. Condensate is removed without any plumbing or drips. Units can be installed above doors and sidewalks without the risk of "wetting-down" passersby. Also, no expensive sewer lines are needed to dispose of the water. Second, the water evaporating off the condenser coil increases the unit's efficiency. By cooling the condenser, head pressure drops and less energy is consumed.

This water is free of minerals. Because it condenses on the evaporator coil from humidity in the air, it doesn't leave a mineral buildup as it evaporates on the condenser.

Package terminal air conditioners are an economical way to provide comfort cooling and heating. Ventilation can be supplied with a manually controlled outside air vent, making these units self-contained systems for areas up to several hundred square feet. Extension ducts can be installed on some units to deliver air to a separate nearby area (like a hotel room's bathroom).

Central split

The other major class of air conditioning system is the split system. These include central split direct expansion units and modular split systems. Split systems have their compressor, and usually the condenser, separate from the evaporator. The compressor is often packaged with the condenser and the assembly is installed outdoors. The evaporator coil is installed inside the building in an air duct. Most split systems do not include the evaporator fan. They work with an air handler and duct system installed in the space.

Central split air conditioners are used in residential and commercial buildings where a rooftop or unitary system is not appropriate. Flexibility is their greatest advantage. They can serve air handlers located far inside a building where no outside air is available to cool the condenser. The evaporator coil is the only interior component that must be located in the building, and the noisy compressor and condenser fan remain outdoors.

Central air conditioning systems integrate well with forced air heating units. Common controls (usually a simple heat and cool thermostat) make them easy for clients to operate. The ease of locating the condenser unit almost anywhere outside the building (including on a flat roof) makes these units very popular.

These systems are usually easy to install. The evaporator coil is installed in the air handler (usually just past an existing burner heat exchanger in a furnace), and the condenser is placed on a pad outside. Copper tubing runs between them and connects to fittings on each. Electric power is provided to the condenser unit (to run the compressor and the condenser fan), and a control wire runs from the air handler to the condenser. The building's space thermostat sends a signal through the low-voltage wire to operate the compressor and fan starting relays.

All of the components (including the interconnecting tubing) are available precharged with refrigerant. There is no need to evacuate the system if the tubing is connected quickly and no refrigerant is lost. A small amount of charge might be needed to top off the system.

Install the liquid and suction tubes that run to the evaporator with care. The suction line (the larger diameter tube returning vaporized refrigerant to the compressor) is cold while the unit is running. Insulate it carefully to limit condensation and drips. Carefully seal the outside wall or roof penetration to prevent the insulation from acting like a conduit for water to leak into the building. This insulation is usually polyethylene foam and is porous to moisture.

Humidity in the air tends to condense in the insulation. This can soak the insulation, causing liquid water to migrate along the pipe for long distances. Use closed cell insulation or be sure the insulation's vapor barrier is not broken. See Chapter 2 for more information about insulating refrigerant and other cold lines.

It might also be necessary to provide oil traps in the suction line. Air conditioner compressors are lubricated with a bath of oil mixed with the refrigerant. While most of it remains in the compressor case, some does become entrained with the discharged gas and enters the system. The oil flows harmlessly through the system and eventually returns to the compressor. Because it enters the compressor slowly (without a large liquid slug), it won't damage the compressor.

It might be necessary to place oil traps in the suction line to help the oil return to the compressor. If the line runs uphill to the compressor, oil can become trapped at the evaporator. The vapor easily flows uphill, but it might not move fast enough to carry the oil with it. Putting traps in the line forces the oil to accumulate and block the line. The suction pressure of the compressor then forces the oil up and carries it to the compressor.

Most manufacturers call for oil traps in the suction line every 10 ft when running uphill. Consult the installation manual for your system for specific guidelines.

Modular split

Modular split systems are similar to central split systems. Both have a single compressor and condenser usually located in an outdoor case. A central split system has a single evaporator coil, but modular units often have multiple evaporators.

The multiple evaporator coils can be installed in various locations in the building, providing cooling throughout the space. These evaporators are often packaged with an integral fan, return louver, and supply diffusers. These self-contained units can be installed on a wall or ceiling. Models are available to fit above a drop ceiling, to surface mount to a plaster ceiling, to fit partially into a wall cavity, or to screw onto a wall surface, giving the installer a great deal of flexibility.

Tubing runs from each evaporator to the common compressor. There, all of the evaporators share the compressor and condenser. Each evaporator's refrigerant line has a solenoid valve that allows refrigerant to flow only when the evaporator requires cooling. Any evaporator's thermostat can start the single compressor and condenser fan.

These systems are often used in applications where central split units are not practical. They are an excellent choice for buildings without a central air handler (for example, a house with hydronic heat). Install the evaporators in strategic locations throughout the space for uniform cooling.

Modular split systems are also available with a single evaporator, fan, and compressor installed at the space to be cooled. A remote condenser coil and fan are placed wherever it is convenient. These systems keep the working components indoors, out of the weather, for easier service. Most of these units are installed above a drop ceiling, with the air inlet and outlet louvers exposed below and the compressor exposed above. A set of refrigerant tubes and a power line for the fan run out to the condenser from the main compressor unit.

If the cavity above the ceiling receives sufficient ventilation, the condenser can be installed there as well. This arrangement is sometimes used to provide extra cooling in a small area (like a computer equipment room). It works well where the space above the ceiling is a plenum return for a larger system. The larger system must be able to handle the additional heat load the modular unit creates.

Chillers

Water chillers are useful where a central cooling system must service a wide area. As noted earlier, water is an excellent carrier of heat energy, and chillers use this to great advantage. Cool water flows to air handlers throughout a building, providing a central system with a single cooling unit. This can simplify design, maintenance, and repairs.

Chillers can use any type of compressor used with direct expansion air conditioners, but centrifugal and screw types are the most common. They are available in a very wide range of capacities, from less than 100 tons to several thousand tons.

Most chillers include a water-cooled condenser that discharges heat to a cooling tower or other heat rejection unit. The evaporator and condenser heat exchangers form a pair of shells (with the evaporator generally larger), and the compressor is mounted above both. All interconnecting piping and fittings are installed and sealed at the factory, and the unit is delivered with a full refrigerant charge inside.

It is essential that there be a constant flow of water through the chiller's evaporator whenever the machine is running. The refrigerant's evaporation temperature might be below 32°F, and stationary water in the shell will freeze. The ice damages the chiller, rupturing the refrigerant tubes and allowing water to get inside.

Prevent this damage with a flow switch interlocked into the chiller's operating controls. The switch must prove that water is flowing before the compressor can run. Some systems also electrically connect the chiller pump's motor starter to the flow switch as a backup.

Many principles for hydronic heat piping apply to chilled water systems. Requirements for system expansion, air elimination, and pumping were described in Chapter 5. Piping layouts for chilled water systems are generally similar to the hydronic heating systems described in Chapter 7.

Almost all chilled water systems use air handlers to cool the space's air. Water coils remove sensible and latent heat from the air, reducing its temperature and humidity.

Large installations often divide the total cooling load between multiple chillers, with each having an adjustable unloader to set the maximum amount of cooling it can deliver. These multiple chillers and their limiters allow large machines to handle a small fraction of the design load without frequent starting and stopping.

Variable-speed pump drives are also used to save on electric costs when loads are below the design maximum. Whenever the system does not require the full amount of cooling, less pump flow will deliver an adequate amount of cooling water. Too much water delivery for the load only reduces its temperature change in the chiller and in the cooling coils. Reducing the flow at times of low load restores the temperature change to the proper level and can save significant amounts of energy.

Some multiple chillers are installed on secondary piping loops. They are arranged exactly like multiple secondary piped boilers. Each has its own pump and main loop bypass line. This allows each unit to work independently of the other, to feed chilled water to the main system loop delivering water to the building. The individual air handlers might be on secondary loops of their own, allowing simple control and independent operation of each.

Ice storage. Using ice to provide partial cooling capacity is becoming more common as energy costs increase. While ice itself does not save energy, operating costs might be reduced with ice storage systems. First, consider an ice storage system's operation (refer to Figure 8.2).

Ice is a means of storing heat energy, or more accurately, a means of storing an energy loss. Every pound of ice requires 144 Btu of cooling to freeze, and it absorbs the same amount of heat when it melts. Therefore, ice storage systems make sense where there is an incentive to store cooling power for later use. There are two reasons that this might be the case.

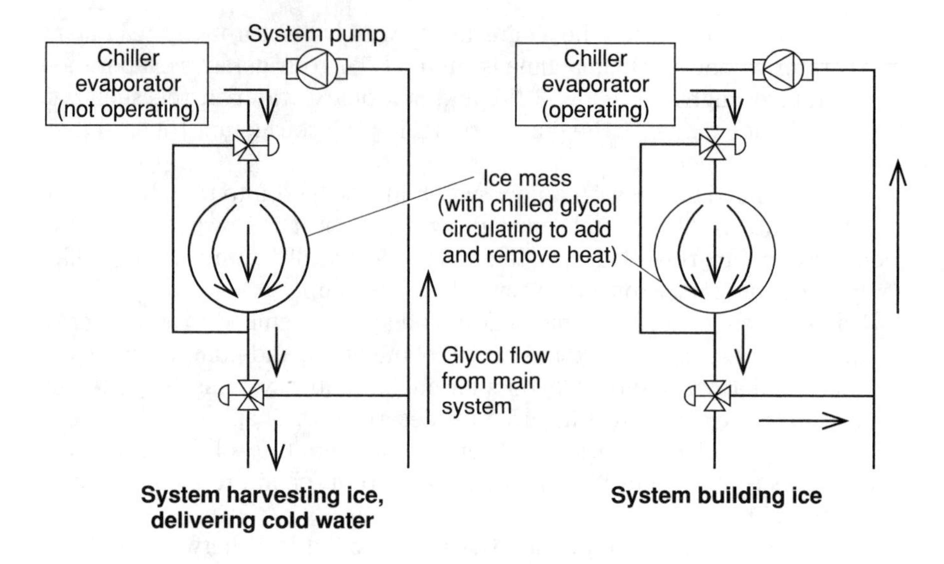

Figure 8.2. Ice storage system. Ice storage vessel contains bulk water or water-filled plastic cells.

First is economic. As noted earlier, many utilities penalize customers for their peak electrical demand used in a given month. Unfortunately, most building's peak cooling loads coincide with peak electrical demands. If the chillers do not have to run at full capacity (or at all) during the day, the total peak load drops and operating costs are reduced. Letting the chiller make ice at night (when the building is not occupied) reduces the total daytime load. Also, some electrical utilities cut the cost of power used at night. As an incentive to help reduce their peak loads, the cost of power at night might be half that of the day rate. This provides even more economic benefit to an ice storage system.

The other reason that ice storage might be helpful is to make up for a capacity limitation. An ice system might be an economical way to serve a short-lived, high cooling load. Installing extra cooling equipment to handle the load might be too expensive or impossible due to facility limitations. The "too small" system can build a bank of ice at night large enough to supplement its output during the day. Nighttime operation provides the extra daytime cooling needed without adding extra cooling capacity.

Ice storage systems use a chiller that can operate below normal temperatures. It must routinely produce temperatures in the 25°F range, and some chillers cannot handle these low temperatures. Low head pressures in the system must be accommodated with low-temperature chiller outputs.

The system is usually simple, but the proper operation of the controls is essential. Antifreeze circulates through the refrigerating system. It substitutes for the hydronic loop water usually used, and delivers cooling to the building and the ice mass. During the day, when the system is not making ice (and might be using it), the chiller's antifreeze goes to the cooling loads like a typical system. The ice bank is in the chilled water circuit. Melting ice (and the possible output of the chiller) provides sufficient cooling to the building to meet peak heat loads and avoid high peak utility charges. As the ice bank melts, 144 Btu of heat are absorbed for each pound of liquid water produced.

There are a variety of systems for handling the ice. Some enclose pure water in hundreds of individual spheres or bricks, with enough air inside to allow for expansion as it freezes. The spheres or bricks are placed in a large tank. Antifreeze circulates around them inside the tank. It either freezes or thaws them, depending on the part of the cycle that is running.

Other systems use bundles of fine tubes that carry the antifreeze inside drums of plain water. The tubes alternately freeze and thaw the surrounding water during the cycle. Still others use bulk ice production, where antifreeze-carrying pipes are immersed in large tanks of water. During the freeze cycle the pipes become embedded in ice. A ring of cold water forms around the pipes' surfaces as the ice thaws. This helps carry heat to the remaining ice.

No system is "best." There are advantages to each type. All of them operate similarly, and the working cycles of each are basically the same.

The chiller is set for low-temperature operation when it makes ice. Valves divert the output of the chiller to the ice bank system. The subfreezing antifreeze solidifies the ice bank, and it is ready to deliver its cooling.

When it is necessary to reverse the system and thaw the ice, the valves switch position. Antifreeze flows through the chiller and the ice bank. It then goes to the building's air handlers to cool the space. Heat from the building warms the antifreeze and it is ready for additional cooling.

The chiller might precool the antifreeze before it goes to the ice. This allows the ice to last longer. The chiller is also used when the ice is meant to provide supplemental cooling for the chiller.

It is easy to estimate the amount of cooling a mass of ice can provide. As noted earlier, 1 lb of ice absorbs 144 Btu when it melts. Therefore, to provide 1 ton of cooling (12,000 Btu/hr) requires

$$\frac{12{,}000 \text{ Btu/hr}}{144 \text{ Btu/lb}} = 83.3 \text{ lb of ice melted per hour}$$

If, for example, we wished to provide 50% of a 150-ton cooling load for a 4-hr peak, the amount of ice needed would be

$$83.3 \text{ lb/ton/hr} \times 0.5 \times 150 \text{ ton} \times 4 \text{ hr} = 24{,}990 \text{ lb}$$

This is approximately 3000 gal of water, a substantial amount. Ice storage systems consume a lot of space and are usually placed outdoors or underground. There is no harm done in placing an ice bank outside. Freezing winter weather can't harm it.

Servicing ice storage systems requires a thorough knowledge of the control and piping system. Automatic timers are usually used to switch the system from ice making to ice melting (or ice *harvesting*) modes. Valves must reliably close and open to allow the proper flow of coolant through the various parts of the system.

Systems that do not freeze properly might have several problems. The chiller might not be cooling the antifreeze mixture sufficiently. Confirm that it is producing no more than 25°F antifreeze when it is building ice.

A possible problem to watch for is contamination of the ice water with antifreeze. Systems that use coolant-filled tubing or pipes immersed in a bath of water might leak. The antifreeze destroys the ice-building capability in a tank if it leaks into the water. Use a commercial coolant freeze-point tester to confirm that the water is clear of antifreeze.

Low-temperature air systems. Ice storage systems have encouraged the development of low-temperature air conditioning systems. The very cold water available from a melting ice bank, or the chiller's antifreeze solution, makes them a natural for this use. However, not all low-temperature systems use ice storage. It is just a side benefit of using ice.

Low-temperature air conditioning uses air delivered to the space at temperatures just above freezing. Often the air is no more than 37°F to 40°F, compared with the comparatively warmer 50°F of standard systems. Using low-temperature air provides several advantages and a few possible problems.

Advantages of low-temperature air delivery include low operating and system installation costs, excellent humidity control, less noise, and smaller ducts. Disadvantages include the potential for bad drafts and the need to run a very cold antifreeze solution through the chiller. If an ice storage system is already being used, however, this last point is not a problem.

Note that low-temperature air systems do not usually use direct expansion air conditioners. The very low air temperatures require that the refrigerant's evaporation temperature be below freezing. Ice would quickly form on the coil's surface, stopping all air flow and any air conditioning. Chilled water (as melted ice) or antifreeze can, however, be safely used at temperatures very near 32°F.

Energy savings from delivering very cold air is derived from the lesser volume of air needed to remove the required heat. The colder the air delivered to a space, the less volume that is needed. Consider the equation for air driven heat transfer:

$$\text{cfm} = \frac{\text{Btu/hr}}{(1.08 \times T_d)}$$

This can be modified to show that the same amount of heat can be removed with two different temperature differences.

$$\text{cfm}_{\text{low temp}} = \left(\frac{T_{d,\text{ high}}}{T_{d,\text{ low}}}\right) \times \text{cfm}_{\text{high temp}}$$

A standard air conditioning system delivers 55°F air into a 75°F room. Therefore, its temperature difference is 20°F. A low-temperature system uses 40°F air, so it would have a temperature difference of 35°F as it enters the same 75°F room. Using these figures, the previous equation becomes

$$\text{cfm}_{\text{low temp}} = \left(\frac{20°\text{F}}{35°\text{F}}\right) \times \text{cfm}_{\text{high temp}} = 0.57 \times \text{cfm}_{\text{high temp}}$$

This means that the system needs only 57% of the air volume of a standard system to give the same cooling with a low-temperature system. Fan energy savings are substantial with this volume reduction. The fan law for power use states that horsepower varies with the cube of the fan speed (and, therefore, the delivered air volume). Horsepower needed to deliver the example's 57% volume is

$$\text{hp}_{\text{low temp}} = 0.57^3 = 0.19 \text{ hp}$$

The fan requires about ⅕ the power to deliver the same cooling, for a savings of 80%. With large systems the financial savings are substantial.

The risk of chilling drafts from a low-temperature system must be considered. Special ceiling-mounted air delivery diffusers are available for cold air systems. These pick up and mix room air with the delivered air to help increase the discharge temperature close to the outlet. Air temperatures within 1 to 2 ft of these diffusers are generally only about 10°F below the ambient room temperature. The installer must follow the usual good practices for air delivery.

Keep diffusers away from areas where people will be working, and allow the air to diffuse into the space gradually. Avoid allowing the air to blow directly against a wall or other object that might turn it toward occupied areas.

Cooling Towers

All air conditioners must have a way to eliminate the heat absorbed from a cooled space. Dry coolers were discussed earlier, and a review of cooling towers, fluid coolers, and evaporative condensers is included in this section. These units allow greater operating efficiencies than dry coolers, and their use should be considered where operating costs are a consideration. The subject of water treatment for these units is covered in Chapter 5.

First, a few definitions are in order. The term *cooling tower* is often used to describe almost any unit that uses evaporating water to dissipate heat. That is not, however, a complete definition of cooling towers.

Direct contact towers cool only the water flowing through them. No fluid other than the water being cooled flows in a direct contact cooling tower. Heated water enters the tower near the top, sprays or cascades over a set of baffles, and leaves from the bottom sump. It leaves the tower and runs to the equipment to be cooled and is pumped back to the tower for the cycle to repeat.

Evaporative condensers and fluid coolers are *indirect contact* cooling towers. Water is not the working fluid that is ultimately cooled. It cools the primary fluid that flows through a set of pipes or tubes inside the unit. The fluid-carrying tubes and fins are cooled by water cascading over them. Water evaporates into the air, removing heat from the fluid flowing inside the tubes.

Evaporative condensers are specifically designed to cool air conditioning systems. Hot, gaseous refrigerant from the compressor enters the unit at the top and is cooled by the cascading water. As it cools the refrigerant changes to a liquid. An evaporative condenser's distribution system is specifically designed to handle refrigerant.

Fluid coolers, on the other hand, do not handle condensing refrigerant. Simple tubing carries the fluid to be cooled while water cascades over the tubing.

In both evaporative condensers and fluid coolers the water remains in the tower. It is repeatedly pumped from the lower sump to the tower's top. Water treatment prevents it from becoming fouled with minerals or biological contaminants.

If properly sized, towers can cool water to very close to the surrounding air's wet bulb temperature. In most climates this is often well below the dry bulb figure, and this lower temperature can improve operating efficiency.

The tower's cooling effect is associated with both the convection of heat from the hot water contacting the air and the loss of water by evaporation. The ambient air's humidity (dew point) and the tower's efficiency determine the amount of evaporation that will occur.

Some water evaporates as it runs through the tower and over the baffles, fins, and tubes. More than 1000 Btu of heat are rejected with every pound of water evaporated. This is the heat of the water's evaporation, and its loss reduces the temperature of the remaining water to close to the ambient air's wet bulb temperature.

Direct contact coolers

Water running through a direct contact cooling tower is cooled from a high temperature to a lower one. This reduction is the *range* of the tower. Note that this difference is equal to the increase in the water's temperature as it runs through the chiller's condenser. The range gets smaller if the flow rate increases. Tower efficiency or capability does not affect the water's range.

The water is cooled to within about 10°F of the ambient air's wet bulb temperature. The difference between the water's exit temperature and the ambient air's wet bulb temperature is the tower's *approach*. Unlike the range, the efficiency of the tower influences its approach. Most towers are rated for an approach of approximately 10°F. A much larger tower is required to provide a lower approach than this, but a 10°F approach is usually sufficient.

Most cities in the United States have designed wet bulb temperatures well below 80°F. If a 10°F approach is used, the leaving water temperature is less than 90°F. This is acceptable for most HVAC chillers or heat pumps. It would not be practical to try to cool the leaving water any further than this.

Cooling towers are rated by their water flow rate, its outlet and inlet temperature, and the entering air's wet bulb (or both wet and dry bulb) temperature. Many HVAC towers are rated by their capacity in tons, allowing 3 gpm and 15,000 Btu/hr for each ton (to include the thermal load of the compressor). The water is expected to be cooled from 95°F to 85°F, and an ambient wet bulb temperature of 78°F is assumed. Note that this standard includes a 7°F approach. This might be more generous than most systems require.

Compare these ratings and their assumptions to the expected working conditions for any specific application. The condenser discharge temperature, water flow, and design wet bulb temperature must be confirmed for the equipment selected and the climate the system is installed in. The manufacturer can help you select a tower for nonstandard conditions.

Many towers use sprayed or cascading water running over the tower's *fill*. The fill increases the water's surface area. It either forces the water to splash from one layer to the next, or it makes the water form a thin film over a set of curved surfaces. The thin film of water provides a large surface area and does not drift as much as splashed water. Air runs through the tower, either by natural convection or with a fan, causing some of the water to evaporate.

Film-type fill is usually more efficient than splash-type fill. The water has more surface area exposed to the air flow and it takes longer for it to drop to the bottom of the tower. A compact film-type tower has the same heat rejection ability as a bulkier splash-type tower.

Cooling towers can be further defined by the type of air flow and the relative directions of the air and water flows. Natural draft towers do not use a fan for air circulation. Wind, the buoyancy of warmed, humidified air, or the capturing of air with a forced water spray cause sufficient air to flow through the tower. These units are usually used for very low-cost HVAC applications where the heat load is small.

Most cooling towers use a mechanical draft. A centrifugal or axial fan (depending on the layout and external pressure restrictions) forces air through the tower. Fans installed on the air inlet are *forced draft* designs, while fans used at the tower's outlet are *induced draft* designs (see Figure 8.3).

Both types work well, but forced draft towers usually have a lower profile than induced draft units. However, induced draft towers are less prone to air recirculation. The high exit speed of the air blows it well away from the tower, so there is less chance of it returning to the inlet.

Towers with air and water flows running in opposite directions are *counterflow* designs. Usually, counterflow systems use water falling from top to bottom and air flowing from bottom to top.

The other main type of tower, the *crossflow* tower, uses air and water streams running perpendicular to each other. Water runs primarily vertically, while the air runs horizontally, or vice versa. Figure 8.3 shows both crossflow and counterflow towers.

All combinations of induced draft, forced draft, counterflow, and crossflow are available. While almost all of them have the hot water enter at the top, the air entry and exit locations can vary. Single or multiple air entry locations are available.

Towers used in cool climates can cause large vapor plumes. Visible clouds of water vapor spew out of these units in cool, damp weather. The warm,

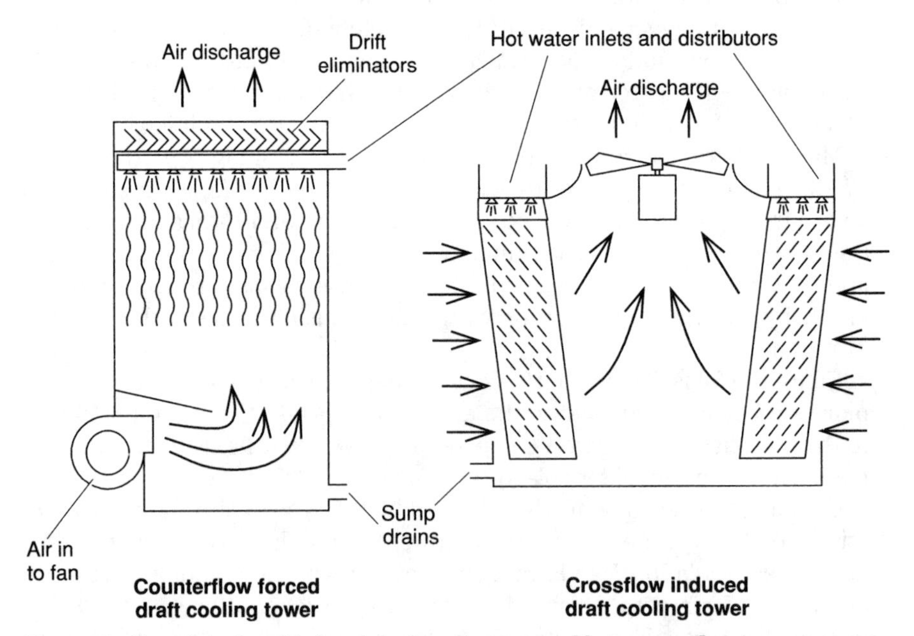

Figure 8.3. Forced draft and induced draft cooling towers. Most counterflow towers use centrifugal fans to provide the required static pressure.

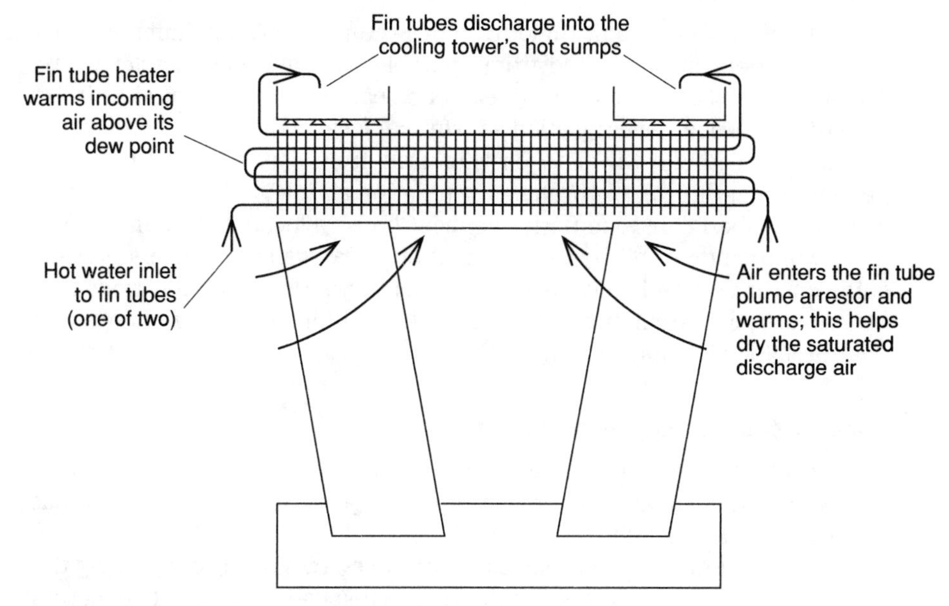

Figure 8.4. Induced draft tower plume arrestor.

saturated discharge air cools on contact with the ambient air, and the water vapor condenses into a cloud. This might be objectionable for a variety of reasons, and it can be controlled. Induced draft towers can be fitted with plume arresters (see Figure 8.4).

Fin tube heat exchangers are installed near the top of the tower's inlet. Warm water to be cooled by the tower runs through the heat exchangers before it enters the tower. The heat exchangers heat the ambient air drawn into the upper part of the tower, above the area where water flows. This warmed air is physically above the tower's wet section and does not pick up any of the tower's water. The heat exchanger water heats the incoming air well above its dew point. The main part of the tower operates as usual, and the air warmed by the fin tubes mixes with the saturated tower discharge. The extra sensible heat from the warmed air heats the entire mass of discharge air, discouraging plume development.

The discharge of liquid water from a cooling tower is its *drift*. Baffles (*drift eliminators*) installed on the tower's outlet catch most of the drift and collect it into large droplets. These fall off into the tower's sump. If excessive water discharges from a tower, the first thing that should be checked is the drift eliminators.

While tower drift is usually not dangerous, it might be contaminated with molds and bacteria. *Legionella* bacteria survive in a water system that remains below 130°F. It is important to minimize drift and make sure a good tower location is selected. Building codes often specify the minimum distances towers can be located from operable windows and air inlet vents. The intention is to prevent the spread of disease via contaminated drift. Be sure the proposed tower location satisfies all local codes.

Besides locating a tower to prevent drift from coming in contact with people (either directly or through a ventilation air inlet), the tower must be installed properly for good performance. Choose a location that allows free air flow into and out of the tower. Locations near walls or barriers prevent sufficient air flow and limit the tower's operation. Also, avoid locations that encourage air recirculation. These include locations under overhangs and near probable downdrafts.

Capacity controls and freeze protection

Some cooling towers must be used during times of low load when outdoor conditions are cold or below freezing. This type of operation requires that the tower's heat rejection be limited.

Mechanical draft towers are usually controlled by slowing or stopping the fan motor. Many towers are available with dual-speed motors, or two motors sheaved to a common shaft, to allow the fan to run at either full or half speed. The half-speed setting usually provides about half the tower capacity at $\frac{1}{8}$ the power consumption. Shutting the motor off completely limits most towers to about 10% of their full capacity.

Towers used in freezing weather need external shutter dampers around the outside to limit the entry of ambient air. The warm water won't freeze if the outside air is sufficiently restricted. Shutter dampers operate with the ambient air temperature and the sump water temperature controls. Any time the air is below freezing, or the sump water is sufficiently cooled, the shutters close.

Some shutters modulate between fully open and closed or assume either full open or a preset minimum opening. The type of control system determines the shutter's action. Simpler controls usually allow two positions, while more complex systems provide a fine degree of control.

Other systems used for outdoor freezing applications use an indoor sump and an on-off pump control. These are good for systems that handle intermittent loads. When heat rejection is needed, the pump starts the system water flowing. The warm water supply pipe has a drain valve into the sump that is kept closed whenever the pump is running.

Tower operation is normal, with warm water delivered to the top of the tower and cooled water draining to the indoor sump. There is no outdoor pan under the tower to hold water there. When the tower is temporarily shut down, the pump is stopped and the supply pipe drain opens to the indoor sump. All of the remaining water in the supply line drains into the sump where it is protected from freezing. The tower is restarted by closing the supply pipe drain and turning on the pump.

Some towers include sump heaters to prevent freezing. These are useful in marginally freezing weather, but should not be relied upon to prevent freezing during extremely cold temperatures. They can only keep the sump warm and do not affect water running through the tower. If the water isn't shielded from the cold air, a freeze is likely. These heaters can handle the occasional cold day or night. It is usually impractical to completely drain cooling towers during this time, and heaters can allow them to continue running.

Evaporative condensers

The most common indirect contact cooling tower used in HVAC systems is the evaporative condenser. Much like a direct contact cooling tower, the heat of water evaporation improves the efficiency of heat rejection. Evaporative condensers provide the most efficient condensing system of all because they use only one heat transfer step. Condensing refrigerant directly loses heat to the surrounding water and air without the need for a separate water-cooled condenser feeding a direct contact tower.

Hot refrigerant vapor feeds the highest point of the coil and flows down as it is cooled. Water sprays or cascades over the tubes, removing heat from the refrigerant flowing within. As the vapor cools it condenses, and liquid refrigerant emerges at the bottom of the coil.

Many evaporative condensers include a liquid refrigerant receiver at their outlet. This tank collects the liquid for delivery to the thermal expansion valve. Refrigerant receivers help control the volume in the system as exterior temperatures vary. They provide a place for the liquid to collect and prevent it from flooding the expansion valve under severe conditions.

Another reason that a receiver tank might be necessary is to provide a liquid seal for the outgoing liquid line. The receiver's discharge is below the level of the liquid, keeping a constant liquid supply available. Also, the receiver is a likely place for air and other noncondensable gases to collect. A purge valve installed on the tank allows for their removal.

Some evaporative condensers are set up to provide liquid subcooling. An extra set of coils cools the condensed refrigerant below its saturation temperature. A line runs from the receiver tank's outlet to a coil located in the lower section of the evaporative condenser. The coil is exposed to the falling water and incoming air. Air at this location is the coldest anywhere in the tower, so refrigerant in the subcooling coil receives the most cooling. The refrigerant is usually cooled about 10°F below its saturation temperature, helping to keep it from vaporizing on its way to the thermal expansion valve.

Free cooling

Cooling towers can be set up to provide free cooling for chilled water systems. Like air economizers, outdoor air cools the building. Cooling towers can cool water to within 10°F of the ambient air's wet bulb temperature. This means that there are many times when the water is cold enough to use directly for comfort cooling. Any time that the wet bulb temperature is below approximately 45°F the opportunity exists for free cooling from the tower. There are several ways to accomplish this.

The simplest is to use refrigerant cycling in the chiller. Sometimes called *vapor migration* systems, a set of valves on the chiller allows refrigerant to pass directly between the evaporator and the condenser. The compressor and thermal expansion valve are not used.

Cold water from the tower passes through the condenser section as usual. Refrigerant vapor comes in contact with the cold tubes, condenses, and turns into a liquid. As the vapor condenses on the tubes it loses heat. Condensation is a warming process, and the refrigerant delivers heat to the water as it changes from a vapor to a liquid. If the chiller's condenser is mounted above the evaporator (as most are), the liquid refrigerant flows by gravity into the evaporator case.

At the evaporator, the comparatively warm tubes (carrying the building's cooling water) vaporize the arriving liquid refrigerant. Evaporation (vaporizing) is a cooling process. The warm water flowing through the coils delivers heat to the refrigerant and causes it to change from a liquid to a vapor. This cools the building water flowing through the tubes. The refrigerant vapor rises back to the condenser section and the cycle repeats.

Note that the chiller is not running. Because heat is flowing from the hotter evaporator to the colder condenser, the process doesn't need energy from the compressor. As the refrigerant repeatedly condenses and boils, it absorbs heat from the evaporator tubes and delivers it to the condenser.

Another system to deliver free cooling uses a heat exchanger installed between the cooling tower and building's chilled water loop. A set of valves allows the cooling tower to cool either the chiller's condenser or the heat exchanger. Whenever free cooling is wanted the chiller is shut down and the valves are set to send the tower's water to the heat exchanger.

It is important to use a very high-efficiency heat exchanger with this type of system. The less temperature difference between the water arriving from the tower and the water going to the building, the better. Plate and frame exchangers are popular for this service because of their large surface areas.

Both the vapor migration and heat exchanger systems keep the cooling tower water physically separate from the building's cooling water. Some systems, however, allow the tower water to flow directly through the building's chilled water piping. The major concern with this arrangement is possible contamination of the building's piping system.

Many of these interconnected free cooling systems use water filters to catch particles before they enter the building system. A *side stream* filter might be used to continuously clean some of the sump water. These do not handle the full tower flow, but only part of it. They should run long enough to ensure that most of the dirt finds its way to the filter.

Other systems use a full flow filter downstream of the tower to clean the water before it enters the building's chilled water loop. These are called *strainer cycle* systems because of the high degree of filtration required. Strainer cycle systems are prone to clogging and fouling. Open cooling towers can pick up large quantities of dust, insects, seeds, and even small animals. These can flow into the system and quickly clog the filter. The result is a complete lack of cooling for the building. Often the chiller cannot run if the filter is opened for cleaning because of the interconnected piping.

Refrigerant Regulations

In the past decade concern has grown over the possible damage that some refrigerants might cause to ozone in the earth's stratosphere. In addition, refrigerants are thought to play a role in global climate changes that might be taking place. These concerns have led to worldwide agreements and legislation in the United States that affects the way HVAC service technicians do their work. The regulations fall into two main categories: those that govern the manufacturing, recycling, and handling of certain classes of refrigerants, and those regulating the qualifications of service technicians.

Refrigerants fall into several categories, depending on their chemistry. For the purposes of regulations, there are three major classifications: chlorofluorocarbons (CFCs), hydrochlorofluorocarbons (HCFCs), and hydrofluorocarbons (HFCs). All CFCs and HCFCs contain chlorine. This subjects them to legislation.

Chlorine in the stratosphere causes ozone to break down. Under special conditions found above the Antarctic a few months of each year, chlorine splits the three-atom ozone molecule apart. Ozone concentration drops significantly because there is no, or very little, sunlight available at these times to restore it. Both CFC- and HCFC-type refrigerants can supply chlorine to drive this depletion. Other refrigerants have no chlorine and do not harm the atmosphere in this way.

Refrigerant manufacturing and handling

All CFC and HCFC refrigerants have an *ozone depletion potential* (ODP) number assigned by the U.S. Environmental Protection Agency (EPA). This rates the relative amounts of chlorine that each refrigerant can deliver into the upper atmosphere. The higher the rating, the more potential harm they can cause.

All commercially used CFC refrigerants have been found by the EPA to have ODPs of 0.2 or higher. The EPA has classified all of these as Class 1 ozone-depleting substances. HCFCs are Class 2 depleters because their

ODPs range from 0.01 to 0.05. See Table 8.1 for a list of common CFC and HCFC refrigerants and their ODPs.

All ozone depleting substances are regulated by the U.S. Clean Air Act of 1990 and an international treaty (the Montreal Protocol). This specifies the proper use and handling of all ozone depleting substances. Production of these materials will eventually be banned by all signatories to the Montreal Protocol. Class 1 materials are being banned more quickly than Class 2 substances. Nonchlorine containing HFCs are not subject to any restrictions.

Manufacturing of all CFC refrigerants ceases on January 1, 1996. Over the last several years the amount of CFC production has been more and more restricted, concluding with the outright ban in 1996. CFCs will become increasingly difficult to find and their prices will escalate as inventories decline.

HCFCs are also subject to a manufacturing ban. Production levels will be frozen in 2010 and their production will be banned in 2020. This will cause these materials' prices to increase as the cutoff date approaches. While the impact on the HCFC ban will be significant, it probably won't be as serious as the ban on CFCs.

Many HCFC systems now operating will have reached the end of their economic lives by the year 2020 and be ready for replacement anyway. The refrigerant ban will make it easier to justify replacing obsolete equipment.

Technicians that work on systems containing CFC or HCFC refrigerants must follow specific rules for handling these materials. Section 608 of the Clean Air Act of 1990, and additional regulations passed in 1993, govern refrigerant venting, recycling, certification, leaks, sales, disposal, and record keeping, including the following:

TABLE 8.1 Refrigerant Ozone Depletion Potentials

Refrigerant	ODP
CFC-11	1.0
CFC-12	1.0
CFC-113	0.9
CFC-114	0.7
CFC-115	0.4
CFC-500	0.7
CFC-502	0.2
HCFC-22	0.05
HCFC-142b	0.01
HCFC-22/142b	0.03
HCFC-123	0.02

- No person shall knowingly vent any ozone depleting refrigerant into the atmosphere while maintaining, servicing, repairing, or disposing of refrigerants or equipment. Four types of releases are permitted:
  ~Minor amounts of refrigerant that escape while making good faith efforts to follow recycling and disposal rules;
  ~Refrigerants released during normal system operation (not including maintenance, repair, or service);
  ~Small amounts of mixtures of nitrogen and HCFC-22 used for leak testing (not including other releases of HCFC-22 and nitrogen mixtures);
  ~Refrigerant releases associated with disconnecting hoses used for recycling and test equipment.
- Technicians must evacuate all equipment to specific vacuum levels before opening them for service. There are exceptions to the vacuum requirements for small HCFC-22 equipment or equipment using CFC-13. CFC-11 and 12 must be evacuated from equipment, whatever the system's capacity.
- Evacuation requirements are waived for equipment that leaks so badly that it cannot attain the required vacuum levels. In these cases it is expected that the system will be evacuated after the leaks are repaired.
- A refrigerant recovery system must be used to evacuate the equipment being repaired. The refrigerant has to be returned to the original equipment or to another system owned by the same person or business. Specific rules apply to the recovery equipment used, its need for registration, and requirements for minimum recovery amounts.
- It is not permissible to transfer refrigerant to another person unless it has been cleaned to the American Refrigeration Institute (ARI) 700 standard of purity. It must be tested and certified by a licensed agency that it meets these requirements before its sale is permitted.
- Refrigerant to be recycled to other parties must be formally reclaimed. There are specific procedures and requirements for the reclaiming company regarding record keeping, refrigerant loss rates, and certification.
- Comfort cooling systems containing more than 50 lb of refrigerant must leak no more than 15% of their charge per year. The system's owner must maintain records showing the amount of refrigerant added to these systems. Technicians adding refrigerant to these systems must provide an itemized invoice to the owner showing the amount added. The owner is responsible for keeping these records and confirming that they do not suggest excessive leakage.
- Refrigerant must be removed from equipment before its disposal. The last person in the disposal chain (landfill owner or scrap dealer) is responsible for seeing that refrigerants have been removed before the equipment is scrapped.

Service technicians are profoundly affected by these rules. They are responsible for capturing and reusing as much of the refrigerants as possible, and will be fined if they do not. While refrigerants are technically not hazardous wastes, the message is clear: Minimize the release of ozone depleting refrigerants to the atmosphere. Service technicians have primary responsibility for the safe handling of these materials.

The increasing cost of refrigerants provides another incentive to conserve them. The reduced or eliminated manufacturing levels, along with taxes imposed to discourage their use, have made CFC refrigerants extremely expensive. It is in the technician's best financial interest to allow as little to escape as possible.

Technician licensing

Refrigerant handling regulations are not the only thing included in the Clean Air Act of 1990 that affects technicians. All persons that service refrigeration systems containing CFCs or HCFCs must be licensed to show that they are trained in the law and know how to prevent refrigerant losses.

Certification is obtained by taking and passing an EPA approved exam given by an approved certifying organization. Certification has been mandatory since November 14, 1994. There are four types of technician certifications available:

- Type 1 certification must be obtained for servicing small appliances.
- Type 2 certification is required for servicing or disposing of high or very high-pressure appliances (except motor vehicle equipment and small appliances).
- Type 3 certification is required for disposing of or servicing low-pressure equipment.
- Universal certification is required for technicians that service all types of equipment.

The EPA has classified CFC and HCFC refrigerants into three classes depending on their operating pressure. Included in these are

Low pressure	CFC-11 and HCFC-123
High pressure	CFC-12, CFC-500, CFC-502, CFC-114 and HCFC-22
Very high pressure	CFC-13 and CFC-503

Proof of certification is required to purchase any CFC or HCFC refrigerant from a dealer. Sale of refrigerants to noncertified individuals is prohibited.

Rules covering refrigerant recycling require that technicians own and use recovery equipment. Technicians must certify to the EPA that they own (or rent) the equipment and that they are following proper procedures for its use. Even the number of service trucks the technician owns must be registered with the EPA.

Most recovery equipment must be tested and certified by an EPA approved agency to show that it meets the requirements of ARI 740 for system evacuation. Recovery equipment manufactured before November 15, 1993, does not need to be tested, but it must evacuate systems to the required levels. All new recovery equipment must be tested and certified. Recovery equipment manufacturers provide proof of certification with the equipment for the technician's records. Table 8.2 lists specific recovery requirements for ozone depleting refrigerants. Note that more stringent requirements apply for new equipment compared to older units.

Equipment used for removing refrigerant before the refrigerant-containing equipment is scrapped does not need certification or registration. It must, however, evacuate systems to the same performance standards required by registered units. This means that new self-built units can be used only for removing refrigerant before disposing of worn-out equipment. Homemade units cannot be used for evacuating equipment that is meant to be repaired and restored to service.

CFC alternatives

The increasing cost and growing scarcity of CFC refrigerants has made substitute refrigerants popular. While the same problems will face HCFC systems after 2010, the immediate need is for alternatives to the popular CFCs. Refrigerants CFC-11 and CFC-12 are commonly used for comfort cooling systems. Refrigerant CFC-11 is used in many centrifugal chillers and CFC-12 is used in some reciprocating direct expansion systems (including commercial refrigerators). CFCs 113 and 114 are less commonly used, but also run some centrifugal compressors. The refrigerant of choice to replace these CFCs is HCFC-123. Unfortunately, this is not a "drop-in" replacement.

TABLE 8.2 Refrigerant Recovery and Evacuation Requirements

	Required evacuation (in. mercury) with equipment manufactured:	
Type of appliance	Prior to 11/15/93	After 11/15/93
HCFC-22 appliance or component containing less than 200 lb of refrigerant	0	0
HCFC-22 appliance or component containing 200 lb or more of refrigerant	4	10
CFC-12, 500, 502, or 114 appliance or component containing less than 200 lb of refrigerant	4	10
CFC-12, 500, 502, or 114 appliance or component containing more than 200 lb of refrigerant	4	15
CFC-13, 503 very high-pressure appliance	0	0
CFC-11 or HCFC-123 low-pressure appliance	25	29.05

HCFC-123 does not perform as well as CFC-11. A loss of about 10% capacity and 5% efficiency usually occurs when it is substituted. Power consumption increases, while the amount of cooling drops. This can make an already marginal system inadequate to serve its requirements. Beyond the efficiency and performance problems, HCFC-123 has other limitations. Compatibility with rubber and plastic seals and gaskets and toxicity concerns are also potential problems.

Many elastomeric materials are attacked by HCFC-123. Open drive chillers (where the motor is not hermetically sealed inside the refrigerant-containing case) that use CFCs often require disassembly and replacement of seals and gaskets when substituting HCFC-123. Inspect the bearings and other parts while the machine is opened. The drive train should also be replaced. Either a new impeller or gear set can restore much of the efficiency loss when the machine is converted to HCFC-123.

Converting hermetic chillers (where the drive motor is enclosed in the refrigerant case) for HCFC-123 use can be impossible. It will attack the motor's electrical insulation, causing it to break down and burn out. The only solution is to remove the motor and replace it with a unit that uses compatible wire, or have it rewound. Either adds to the cost of the conversion. Conversion can be so difficult that it might be more practical to install a new chiller that does not use CFCs.

Toxicity concerns over HCFC-123 are also an issue. Because of potential health risks, the allowable concentrations for HCFC-123 exposure are 100 ppm (parts per million) maximum, with a 10-ppm level for an entire work day. This limitation imposes strict requirements on system venting, leak detection, and chiller room ventilation.

Another possible alternative (although less popular) for CFC-11 replacement is HFC-245a. It operates slightly less efficiently than CFC-11, causing energy consumption to increase about 5%. The biggest concern with this nonozone depleting compound is its flammability. At sufficient concentrations it can ignite and feed a fire.

Other alternatives to CFC-11 will no doubt be developed in the next several years. Unfortunately, it does not appear that a "magic mix" will ever be available for simple replacement of CFC-11, CFC-113, or CFC-114.

Refrigerant CFC-12 is used in some centrifugal chillers and reciprocating or rotary compressors. Common applications include building cooling systems, refrigerators, and automotive air conditioners. The most compatible replacement for CFC-12 is HFC-134a. This material contains no chlorine and, therefore, has an ODP of zero. It has little effect on system efficiency or capacity. Problems do exist, however, because of its incompatibility with mineral-based lubricating oils.

Most CFC-12 systems use mineral oils. These are insoluble in HFC-134a and fall out of solution. The internal surfaces of the evaporator and condenser tubes become coated with liquid oil, preventing proper heat transfer. The thermal expansion valve might also function improperly if it has to pass a slug of oil.

Ester-based oils are compatible with HFC-134a and most equipment now using CFC-12. If existing mineral oil in the system can be removed and replaced with ester oil, HFC-134a can be safely substituted. This oil replacement is possible with semihermetic compressors. Unfortunately, there is no way to remove all of the mineral oil contained in hermetic compressors.

A possible universal replacement for CFC-12 is a mixture of HCFC-22, HFC-152a, and HFC-124. Various mixes of these materials are being tested to find the best combination for various systems. They are compatible with mineral oils already present in CFC-12 reciprocating systems.

Research on new refrigerants is going on at an accelerated pace. DuPont, the leading manufacturer of refrigerants, has introduced a line of compounds under the "Suva" name. These are replacing their widely known "Freon" products as manufacturing resources are diverted to HCFC and HFC production. DuPont and other manufacturers will continue to introduce new refrigerants to the marketplace that do not harm the ozone layer.

Installation requirements

Concern over the possible health effects of refrigerants has made some localities regulate the installation of building cooling and refrigeration equipment. The intention is to limit the risks associated with having moderate to high quantities of refrigerant in one area.

The American Society of Heating, Refrigeration, and Air Conditioning Engineers (ASHRAE) publishes two standards concerned with protecting building occupants from potential refrigerant related health problems. Standards 15 and 34 work together to define the "best practices" for HVAC equipment installation requirements. While not all communities' codes reference these, it is in the installation technician's best interest to be familiar with them. You can get copies of the actual standards from ASHRAE, but the basic requirements are outlined here.

Standard 34 defines refrigerant's flammability and toxicity classifications with numbers and letters. Numbers range from one to three, with their ratings being

1	Nonflammable
2	Low flammability
3	Highly flammable

Refrigerant's toxicity ratings are given with the letters A and B.

A	No evidence of toxicity
B	Evidence of toxicity at concentrations below 400 ppm

Refrigerants are given a number and letter rating to show what their flammability and toxicity properties are. For example, HCFC-22 is rated A1, meaning it is nonflammable and is not toxic.

The type of building an air conditioning system is installed in also influences ASHRAE requirements. For comfort cooling systems, the following classifications exist:

Institutional	Occupants cannot leave without assistance. Examples include hospitals, nursing homes, and jails.
Industrial	Occupancy is restricted to those with authorized access. Examples are manufacturing facilities, processing, and storage buildings.
All others	These include places of public assembly, residential, and commercial properties.

Standard 34 also defines a cooling system's likelihood for leaks into the occupied space. Those with direct expansion coils in the ventilation air stream are considered *high probability* systems, because a leak in the coil is likely to be carried into the space. Chillers and any system where the refrigerant coils are not directly contained in the air being cooled are considered *low probability* systems. A refrigerant leak in one of these systems is not likely to be inducted into the space.

Considering all of this, the ASHRAE standards define the following requirements:

- The amount of refrigerant the largest system can have without requiring a separate equipment room. This is given as the number of pounds of refrigerant per 1000 ft^3 of occupied space.
- Equipment room layout to prevent refrigerant vapor from accumulating and posing a hazard to service technicians.
- The equipment room's ventilation system. This includes both normal ventilation and purge levels.
- Relief piping for refrigerant-containing systems.
- Required safety monitoring equipment.

Maximum occupied area amounts. Table 8.3 gives the maximum amount of various refrigerants allowed in a building before an equipment room is needed. Note that in most cases where central cooling systems are used an equipment room will be required. Buildings served with multiple, small PTACs (or modular split systems) probably do not need an equipment room. The trigger is the amount of refrigerant in the largest single system compared to the total occupied building volume. The total amount of refrigerant in all systems is not considered.

Equipment room layouts and requirements are detailed in section 11.13 of Standard 15. This gives requirements for equipment headroom, doors, safety monitoring equipment, ventilation, and prohibitions on open flames

TABLE 8.3 Refrigerant Safety Classifications and Occupied Space Limits

		Maximum quantity in occupied areas: lb/1000 ft^3			
		Institutional occupancy, leak probability		Other occupancies, leak probability	
Refrigerant	Safety class	High	Low	High	Low
CFC-11	A1	0.8	1.6	1.6	1.6
CFC-12	A1	6	12	12	12
HCFC-22	A1	0.7	9.4	9.4	9.4
CFC-113	A1	0.95	1.9	1.9	1.9
HCFC-123	B1	N/A	0.004	N/A	0.004
HFC-134a	A1	8	16	16	16
HFC-152a	A2	N/A	1.2, max = 550 lb	N/A	1.2

Note: N/A—Not allowed for comfort cooling systems.

and personnel access. Specific layouts are not prescribed, but the need for open air flow with no chance of air stagnation is described. The intention is to have the ventilation system move air in all parts of the room with little chance of a pocket of refrigerant remaining inside.

Equipment room ventilation. Equipment room ventilation is handled at two levels: one is for normal operation and the other covers purge (vapor removal) conditions. Normal ventilation is used when the building is occupied and is the larger of

- 0.5 cfm per square foot of the equipment room's area or 20 cfm per person normally in the equipment room
- The air volume needed to limit the equipment room's temperature increase to 18°F over the ventilation (outside) air's temperature. This is determined with the formula

$$\text{cfm} = \frac{\text{Btu}}{19.4}$$

where Btu is the heat generated in the equipment room.

Purge ventilation volumes are found with the formula

$$\text{cfm} = 100 \times \sqrt{\text{Mass}}$$

where mass is the number of pounds of refrigerant in the largest system enclosed in the equipment room.

These two airflow requirements can be achieved with multiple fans. While the normal ventilation system might need to be heated in the winter, the purge system does not require this. It is only used if a refrigerant leak occurs and the room must be cleared of vapor.

Like any other ventilation system, a method of supplying and removing the air must be provided. The air exhausted from the equipment room, in normal and purge modes, must not discharge into adjacent enclosed spaces. It must go directly outside.

Refrigerant exhaust systems must pick up from both high and low locations. Most refrigerants are at least three times heavier than air, and ventilation outlets located only near the ceiling (for heat pickup) do not effectively remove refrigerant.

Pressure relief piping. Standard 15 calls out specific requirements for vent and pressure relief fittings used with chillers and other refrigerant-containing equipment. All of these should be piped outside. The vent pipe must terminate at least 15 ft above the surrounding ground and more than 20 ft from any window, exit, or building ventilation inlet.

Purge unit vents and any safety relief plugs must be connected to the vent piping. The vent pipe must include a drip leg just prior to going outside that can hold at least 1 gal of liquid refrigerant. A capped service valve must be provided on the drip leg to drain any accumulated refrigerant. This must be done at least every 6 months, and appropriate refrigerant handling procedures must be followed with the drained liquid.

Safety monitoring. An automatic safety monitor must be used in the equipment room. Rooms containing type A1 refrigerant require oxygen deprivation sensors, while all others require refrigerant concentration sensors. Oxygen deprivation sensors must trigger an alarm and start the room's purge ventilation system if the oxygen level drops below 19.5% volume. Refrigerant detectors must be set to the individual material's TLV limit.

Sensors must be installed to continuously detect oxygen loss or refrigerant presence. The requirements include

- Continuous operation without switches that can inadvertently turn them off.
- Permanent mounting.
- Stability over the range of temperatures, humidities, and pressures at which the monitor must operate. In addition, they must have zero calibration drift or be self-calibrating.
- Have a long service life and require little maintenance.
- Provide external contacts to trigger the alarm and start the room's purge ventilation system.
- Sensitivity to oxygen loss or refrigerants only (to avoid nuisance trips).

Sensors should be installed as a life safety device. Therefore, they must be located to detect problems as quickly as possible. They should be installed where airflow is likely to expose them to leaking refrigerants. Monitors with multiple sensors can be used to accommodate large or oddly shaped rooms. At a minimum, the sensor should trigger an alarm audible and visible both inside and outside the equipment room, as well as start the purge ventilation system. It can also be integrated into a central building automation system to alert emergency personnel in the event of a leak.

Chapter

9

Water Source Heat Pumps

Water source heat pump (WSHP) systems are unique in the heating and air conditioning industry. These units, like air-to-air heat pumps, are reversible air conditioners that can either cool or heat a space. Unlike air-to-air heat pumps, however, water source units can pass heat between each other. The same units are as useful in the core of a building (that almost always demands cooling) as they are at the building's perimeter where both heating and cooling might be needed.

System Concepts

Water source heat pump systems are able to move heat within a building from areas where it is not needed to areas where it is. Unlike standard heating and cooling systems, there are many times when the building is kept comfortable without adding or removing energy (except the "overhead" for system operation). No other system can do this.

A complete WSHP system (as shown in Figure 9.1) requires the following basic components:

Two or more WSHP units. These are relatively small (up to 5-ton) heat pumps that exchange heat to or from a flow of air and a flow of water. While heating, energy is absorbed from the water flow and discharged to the air. The water's temperature drops and the air is heated. While cooling the space, the heat pump absorbs energy from the air and discharges it into the water. The water's temperature increases. Each building heating and cooling zone has its own WSHP.

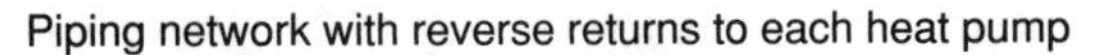

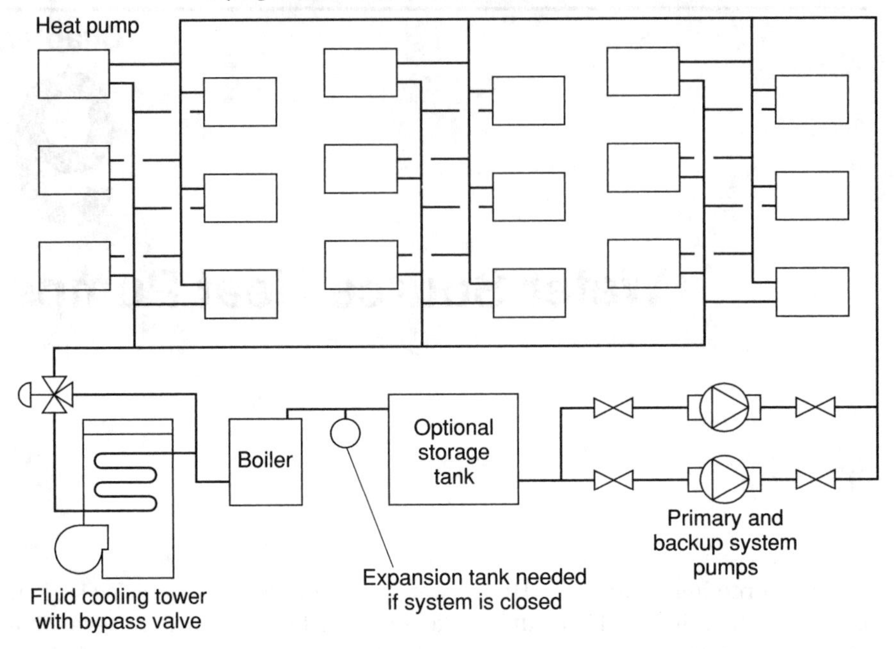

Figure 9.1. Water source heat pump system schematic.

An auxiliary heat source. This can be a fossil fuel or electric boiler, or a set of duct-mounted electric heaters. These are used only when the building requires a net addition of heat, as in very cold weather.

A way of rejecting heat. Most systems use an indirect contact cooling tower (fluid cooler) to eliminate heat when the building needs to be cooled. This is only used in hot weather or during times of high internal heat loads that cannot be used elsewhere in the building.

A piping system (the *loop*) connecting all of the other parts together. This includes a circulating pump, expansion tank, filters, and air eliminators like those used for hydronic heating or chilled water systems. The piping loop is the "highway" for heat exchange in the building. As the water flows through the heat pumps, energy is added to it or taken out as needed.

Heat pumps should be installed to provide individual zone-controlled heating and cooling. Zones might be as small as a single conference room or an individual office, or as large as a set of interior rooms. Each space within a zone should have similar heating and cooling demands so that all require the same type of conditioning simultaneously.

The heat pump heats or cools the air and discharges it into the zone's space. A thermostat monitors space temperature and switches the heat pump between its heating, standby, and cooling modes. No manual changeover is necessary. The transition from one mode to another is completely automatic and controlled by the thermostat.

A heat pump takes heat from the space and discharges it to the circulating water loop. Meanwhile, heat energy required for heat pump warming comes from the same water loop. Sharing the loop allows each heat pump to freely exchange the building's energy. The net effect of both operations is heat transfer from where it is not wanted (the zones being cooled) to where it is needed (the zones being heated). No external energy is supplied, excepting that required to run the heat pump's compressors and the loop's water pump.

Water source heat pumps are useful in buildings that experience wide swings in heating and cooling loads. During the spring and fall it is common for some areas to need heat in the morning and air conditioning in the afternoon. It's not uncommon for a warm spell in the winter to cause a cooling requirement long after the central air conditioning system has been shut down for the season. Water source heat pumps can handle either situation automatically, with no intervention from the building operation's staff.

Throughout the day, as the sun loading moves into and out of various parts of the building, the system automatically delivers heating or cooling as the zone thermostats request it. Heat removed from the "too warm" parts of the building passes through the piping loop to those parts that need heat. This is especially useful where computer rooms or other heat-generating sources always needs cooling. In cold weather the waste heat from these areas is automatically "recycled" to help heat other areas of the building.

During the summer and winter buildings usually need general cooling or heating. It might be that not enough waste heat is present in the winter to fully heat the building's exterior zones. Similarly, in the summer there is no need for the heat circulating in the system's water. For these situations, a central boiler (or duct-mounted electric heating coils) and a cooling tower provide the means of adding heat to or removing it from the building.

Heat transferred from one part of a building to another with these heat pumps is not really "free." Energy is consumed by the heat pumps' refrigeration systems to get heat into and out of the circulating water. These costs might be significant.

If a single heat pump has a coefficient of performance (COP) of 4.0, then it transfers four times as much energy as it consumes. If, for example, two heat pumps in a system have COPs of 4.0, the general heat balance is as follows: The first heat pump removes, say, 12,000 Btu from the "too warm" space. It puts this heat, along with its own 3000 Btu of working energy, into the water loop. The total heat added to the loop is 15,000 Btu.

The other heat pump picks up the 15,000 Btu from the water loop and delivers it to the "too cold" space. It adds an additional 3750 Btu from its compressor to this heat for a total delivery of 18,750 Btu. Overall, 12,000 Btu are removed from the warm space and 18,750 Btu are delivered to the cold space.

Transferring our original 12,000 Btu cost 6750 Btu of energy for the delivery. Looked at another way, the 18,750 Btu of "free" heat really cost 6750 Btu. This is the overhead penalty imposed by the heat pumps. A small amount of additional energy cost (not considered here) is also associated with running the circulating pump, but it is minor compared to the heat pumps' use.

This means that heat pumps might not always be the most economical option. If the electricity costs to provide the overhead are higher than the costs of the fossil fuels required to heat the space, the fossil fuel alternative might be better.

For the previous example, the 6750 Btu of electricity might be more expensive than 18,750 Btu of heat supplied by a fossil fuel. Note that the required cooling might be satisfied by supplying outdoor economizer air. Check the complete system and the design goals regarding flexibility, installation difficulty, temperature zoning, etc., before settling on a choice.

Heat Pump Operation

Water source heat pumps are designed for comfort heating and cooling operations on their air sides. The water side drives heat into (and removes it from) a flow of moderate-temperature water. A standard combination condenser and evaporator coil is used for the air delivery system. The small, integral high static pressure fan drives air through the coil and the supply duct system. A shell and tube heat exchanger is usually used on the water side.

Operation changes from heating to cooling mode by switching a refrigerant reversing valve. In the heating mode, the compressor's discharge is directed to the air coil. This heats air flowing through it and condenses the refrigerant. It flows back to the reversing valve, on to an expansion valve, and then to the water heat exchanger. The water delivers heat to the cold refrigerant and evaporates it. The refrigerant returns to the compressor, ready to repeat the cycle. During this process, the air flowing through the condenser coil heats to approximately 115°F. The water is cooled as heat energy transfers from it to the air.

Changing the reversing valve's position switches the roles of the air coil and water heat exchanger. The air coil becomes the evaporator and cools the air flowing through it. The water heat exchanger is now the condenser, and heat flows to it.

Choosing individual heat pumps for an application requires estimating both the heating and cooling loads for the individual zones. Select units that satisfy the higher requirement. All of the individual heating and cooling loads should be added to arrive at the net total building heating and cooling requirements. The capacity of the building's heating and heat rejection equipment can be reduced from the total estimated need. See the section on "System capacity estimating" later in this chapter.

One potential limitation of WSHPs is their relatively low air output temperature. Zones with very high heating loads (due to large glazed areas or high levels of cold outside air infiltration) might require intense heat delivery. The moderate air temperature output of the heat pump might not always suffice.

Some WSHPs use multiple-stage space thermostats. These turn on additional heat in areas with severe loads. The thermostat switches the heat pump into its heating mode when it needs heat. If the temperature continues to fall below its second stage setpoint (usually set about 2°F below the dial setting) it energizes the auxiliary heat. This only runs when the heat pump cannot deliver enough heat to the zone, and is the first to be shut off when the thermostat is satisfied.

Water in the system's circulating loop remains between approximately 65°F and 95°F at all times. The piping does not need insulation (if it runs through heated areas) because its temperature is close to the space temperature. No condensation will occur on it, either, no matter whether the system is supplying heating, cooling, or a mixture of each. The moderate loop water temperature prevents condensation.

Condensation often does occur, however, at the heat pump's evaporator coil when it is in the air conditioning mode. Install a trapped drain at the condensate outlet to dispose of the water. Because any unit can operate in the cooling mode at any time, it is important that the condensate drains be kept clear. A clogged drain causes the pan to overflow and water to leak out of the unit.

The operating range of the loop temperature provides the system's diversity. Providing a wide temperature range in the system permits individual heat pumps to share energy. Increased heating loads cause more heat to be drawn out of the loop, reducing the water's temperature. Higher cooling requirements cause more heat to be added to the loop, increasing the water's temperature.

Wide loop temperature ranges also reduce the operation of auxiliary heating or cooling systems. Short-term heating or cooling loads that cause a small to moderate change in the loop's temperature won't start the auxiliary heating or cooling systems. The loop's temperature swings enough to carry the momentary load and can return to normal afterward.

Some WSHP systems do not use a boiler in the hydronic loop. Electric heating coils in the heat pump's supply air ducts provide building heat. When a zone's thermostat requests heat, an automatic control operates either the duct's electric heater or the heat pump's compressor and reversing valve. Selection depends on the loop's temperature. If the loop temperature at the heat pump is above 65°F, the compressor runs and the system draws heat out of the piping system. Water temperatures below 65°F prevent the compressor from running, and operate the duct's electric heaters instead.

Piping Layouts

Water source heat pump systems require several auxiliary components besides the piping and the heat pumps themselves. These include a water pumping system, boiler, cooling tower, and controls. The piping system links all the parts together. They are usually constant volume systems with water pumped to each heat pump in proportion to its capacity. Typical flow rates range from 2.4 to 3.0 gpm per 1000 Btu/hr of total heat pump capacity.

For example, design a system for 2.5 gpm per 1000 Btu/hr. It uses eighteen 1.5-ton heat pumps and nine 2.5-ton heat pumps. Proper loop flow is

$$\text{Pump gpm} = \frac{\text{Total Btu}}{(1000\ \text{Btu/hr} \times 2.5\ \text{gpm})}$$

$$= \frac{((18 \times 1.5\ \text{ton}) + (9 \times 2.5\ \text{ton}) \times 12{,}000\ \text{Btu/ton})}{2500}$$

$$= \frac{(27\ \text{ton} + 22.5\ \text{ton}) \times 12{,}000\ \text{Btu/ton}}{2500}$$

$$= \frac{49.5\ \text{ton} \times 12{,}000\ \text{Btu/ton}}{2500}$$

$$= 238\ \text{gpm}$$

From the heat pumps, the water goes to either the boiler or cooling tower, depending on the system control's operation. Finally, the water returns to the pump.

See Figure 9.1 for a typical schematic of a heat pump system. The heat pumps are connected in a reverse return arrangement to avoid the need for balancing valves. Balancing valves are used, however, to achieve proper flow through the cooling tower and boiler. With the heat pumps connected in a reverse return arrangement, these are the only balancing valves needed in the system.

This arrangement works for closed piping systems. This is the most common arrangement used with WSHPs. Closed systems prevent the entry of oxygen, corrosives, and particulate or biological contaminants. The only open space in the system might be inside a nondiaphragm expansion tank. Diaphragm-type tanks can even prevent this contact with air. Air control and removal are the same for WSHP systems as described for hydronic heating designs.

The loop water circulating pumps should be chosen with care. Water must always circulate through the system, so it is worthwhile to select very high-efficiency units. Because the pumps run constantly, an investment in high-efficiency motors and pumps is cost effective.

Also, install a backup loop water pump. Water flow in the system must be maintained under all conditions. The backup pump should automatically take over if the main pump fails. Murphy's Law dictates that a pump failure happens when it is least convenient, most likely during very cold weather when no one is around. An automatic backup pump prevents building damage.

Note the optional storage tank. Storing water can provide a way to save heat during one part of the day for use during another. This is useful in buildings where operation at all hours is necessary.

A properly designed storage tank holds about 6 to 8 hours' worth of circulating system water. The water is, in effect, delayed for that time as it completes a single trip through the loop. Water heated by the system in the afternoon stays in the tank until the evening. After being delayed in the tank it returns to the piping system where its energy is available for use by the heat pumps.

This immensely increases the system's thermal inertia, providing a way of using heat generated from building lighting, equipment, and people during the day to help heat the building at night. The boiler or duct heaters can remain off longer by using this waste heat during peak heating periods.

You can estimate storage tank capacity with the formula

$$\text{Capacity (gal)} = \text{Pump gpm} \times \text{Hours} \times 60$$

If we wish to store a 6-hr supply of water in our previous example's system, the proper storage tank volume would be

$$\text{Capacity} = 238 \text{ gpm} \times 6 \text{ hr} \times 60 = 85{,}680 \text{ gal}$$

This is a very large tank for a moderately sized system.

Some heat pump systems use small or moderately sized (up to 1000 gal) storage tanks. While these might prevent fast changes in loop water temperature (for example, when a boiler starts), they do not store water for any meaningful time. No energy transfer from one time to another is available when these small tanks are used.

There is no significant benefit from using a storage tank for cooling a building in hot weather. Saving cool system loop water generated during nighttime operation only delays the cooling tower's start the next day. Cooling towers do not consume much energy, so very little energy is saved with a storage tank when the system is predominantly cooling the building.

If an open storage tank is not used, an expansion tank for the water loop is necessary. The small range of water temperatures permits the use of a relatively small expansion tank. Total temperature swing for the system is less than 40°F, allowing a small tank to keep operating pressures within normal limits. See Chapter 5 for information on how to estimate expansion tank requirements.

Pipe expansion due to temperature changes is also small. Plastic pipe can be used because the low temperature change reduces the amount of thermal expansion. PVC pipe, which has a high coefficient of thermal expansion, works well.

Closed piping systems do not use direct contact cooling towers. Instead, an evaporative fluid cooler is used. System water runs inside a coil that is cooled by the tower's water cascading over it (similar to evaporative condensers). The tower's water never contacts the system water in the coil.

Some WSHP installations use open water systems. The degree of "openness" can vary, from using an open water storage tank with an indirect contact cooling tower to a completely open design. These use a direct contact cooling tower, where system water is cooled in the tower and the same water flows through the piping system.

Open systems do not require an expansion tank. Open systems allow the water to expand and contract as needed, and the height of the water in the system establishes the pressure. The water's temperature has very little bearing on its pressure.

Open systems require much more water treatment than closed systems. Those with open storage tanks and indirect contact fluid coolers can absorb oxygen from the air in the tank. While no contaminants can enter the system from the indirect tower, the oxygen can cause corrosion throughout the system if it is not removed. Standard corrosion inhibitors and oxygen scavengers are required to prevent damage.

Evaporation might also be a problem with open systems. Dissolved mineral cycles of concentration must be monitored to prevent scale formation. Use phosphate scale preventatives and install a bleed or flush control in systems that experience significant evaporation.

The most critical water treatment systems are needed for open systems using direct contact cooling towers. A large amount of oxygen uptake and evaporation takes place in these systems, as well as the induction of dirt and biological contaminants. Precise water treatment and filtration prevents the piping system from clogging and fouling.

Open systems usually have high operating costs. The water treatment costs and ongoing maintenance work needed to keep them functioning makes them expensive. These ongoing costs more than make up for any money saved by using a direct contact tower instead of a closed fluid cooler.

All indirect contact tower fluid coolers need some water treatment. Water circulating in the sump requires standard biological treatment and scale and corrosion prevention. Water circulating in a closed system, however, does not require ongoing treatment. There is no risk of sending contaminants to the heat pump system.

Many WSHP loops use an antifreeze solution to eliminate the risk of damage in the winter. Most of these systems circulate water through the outdoor cooling tower, both summer and winter. Others use antifreeze because some piping must traverse unheated areas that could freeze. Finally, systems that run the tower only when cooling is necessary use antifreeze to protect the loop water "stranded" in the tower's piping.

Antifreeze reduces the loop water's heat transfer capacity and increases its pumping requirements. Consult with the heat pump manufacturer about adjusting the unit's capacities when antifreeze is used in the loop. Sometimes larger units are required to provide the same amount of heating and cooling.

Many systems use three-way valves at the cooling tower to prevent circulation through it during cold weather. This prevents heat losses, but leaves stagnant water in the tower's piping exposed to freezing temperatures (and requiring antifreeze). Other systems maintain tower flow. Heat losses are limited by closing exterior dampers, shutting off the cooling water cascading over the loop water pipes, and turning off the tower's fan.

A good compromise for freeze protecting the cooling tower (and limiting loop heat losses) is to use a shrouded (dampered) fluid cooler tower with a manual sump drain. Water always flows in the tower and the dampers keep cold air out during freezing weather. Sump water is manually drained when the building no longer needs cooling and is refilled at the start of the cooling season.

Towers without shrouds are usable if the loop water is removed from the outdoor piping at the end of each cooling season. Include manual drains and fill points for season startup and shutdown. Bleed all of the air trapped in the piping when refilling the tower. Fresh water contains oxygen, so add anticorrosion chemicals whenever large amounts of water are added.

Simple variations in the piping arrangements can provide even more flexibility for the system. A heat pump serving more than one room (both within the same zone) is controlled by a single thermostat. Install the thermostat in the area with the greatest heating or cooling need. All areas fed from that heat pump receive heating or cooling simultaneously, whenever the space with the thermostat requires it. This can cause over- or underheating (or cooling) in areas that do not have as large a load.

Water source heat pumps can have their air delivered into several semi-independent subzones. Each subzone has its own thermostat to prevent over- or underheating (or cooling). Subzones are created by installing a water coil in the duct feeding the subzone area. Water from the pump loop circulates through the coil whenever the subzone's thermostat is too cold or warm (see Figure 9.2).

The building loop water in the coil tempers the air discharged by the heat pump. It can't heat or cool the space because the coil only brings the discharge air to approximately room temperature. This happens whether the unit is in the heating or cooling mode. Loop water temperatures range from

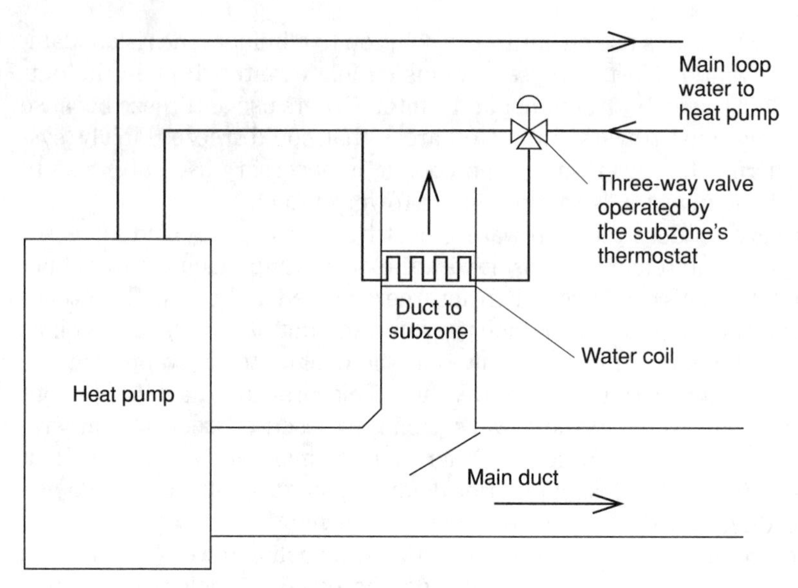

Figure 9.2. Subzone coil system.

65°F (when most of the heat pumps are in the heating mode) to 95°F (in a predominantly cooling mode). This moderate water temperature tempers the 50°F or 115°F air discharged from the heat pump.

Note that the water coil won't provide meaningful heat when the heat pump is in the cooling mode, nor can it cool when the unit is heating. It does, however, decrease the difference between the air leaving the subzone's supply diffusers and the ambient temperature in the area. The subzone air is delivered at approximately room temperature, so it no longer heats or cools the space.

Some closed WSHP loops are integrated with the building sprinkler network. Sprinkler heads are attached to one side of the hydronic loop (usually the supply). A connection between the loop and the sprinkler water supply admits water to the system if the loop's pressure falls (presumably because a sprinkler head has opened). Standard sprinkler head spacing must be maintained and the system has to be hydraulically designed as a sprinkler system.

An advantage to this system is lower costs, because a single piping system provides fire protection, heating, and cooling. Systems using shared piping must be designed per NFPA 13 to deliver the required sprinkler water flow and coverage. It is beyond the scope of this book to describe the details of integrated systems, but the NFPA can provide information on acceptable arrangements.

Before pursuing an integrated system installation, confirm with local building officials that it is acceptable. They might first need assurances that it meets all sprinkler code requirements, and that these functions will not be compromised by the heating and cooling functions.

Boilers—Special Considerations

Fossil fueled boilers used with WSHP systems must be chosen and installed with care. The circulating water is usually much cooler than that found in typical hydronic heating systems, and that has important implications for the boiler.

Most controls don't energize the boiler until the loop's temperature drops to about 70°F. The boiler remains on only until the water temperature increases to 80°F. The average water temperature for the boiler, then, is approximately 75°F.

Water this cold circulating through a boiler's heat exchanger might cause condensation to take place on its surface. Combustion products contain water vapor, and this can condense on the heat exchanger if its surface temperature is below 212°F. Cold water in the system can cool the heat exchanger's temperature enough to permit condensation.

Normally, hydronic heating boilers run until the water temperature reaches at least 130°F. This is usually high enough to heat the flame side of the heat exchanger beyond the flue gas' dew point. The low loop water temperature in a WSHP system's loop encourages condensation on the boiler's heat exchanger.

Some boilers can handle low-temperature feed water. Swimming pool heaters are a typical example. Their heat exchangers are made of corrosion resistant materials, or the heaters are designed to always run hot enough to prevent condensation. It is best to verify with the dealer or manufacturer that the unit being installed into a WSHP system reliably operates with low-temperature water.

Fortunately, boilers that cannot handle low-temperature water can be made to work acceptably in heat pump loops. This might be necessary when installing a heat pump system with an existing boiler. Install the boiler in a secondary loop, connected as shown in Figure 9.3. The boiler has its own secondary pump and aquastat, and the loop uses a three-way valve to obtain heat.

If the secondary loop's balancing valves are set correctly, the boiler's inlet water temperature never drops below a predetermined point. When the system is not calling for heat, the boiler's secondary loop recirculates within itself maintaining its aquastat temperature. When the system calls for heat, it opens the three-way valve and hot water discharges to the main loop.

By properly adjusting the balancing valves, just enough main loop water enters the system and mixes with the boiler's loop to provide the correct boiler inlet temperature. Table 9.1 provides the proper proportions of flow for the water flowing through the recirculation and system lines shown in Figure 9.3.

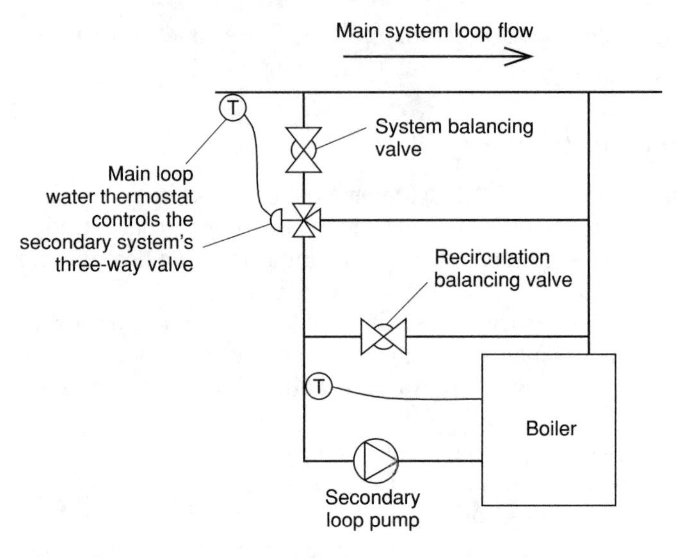

Figure 9.3. Boiler secondary loop connections. If balanced properly the boiler always operates at higher temperatures than the main loop.

TABLE 9.1 Heat Pump Boiler Recirculating and System Water Factors

Assumptions: 20°F rise through the boiler
65°F minimum main system loop temperature

Minimum boiler temperature (°F)	Maximum boiler temperature (°F)	Recirculating factor	System factor
130	150	0.765	0.235
140	160	0.789	0.211
150	170	0.810	0.190
160	180	0.826	0.174
170	190	0.840	0.160
180	200	0.852	0.148

Note: See Figure 9.3 for corresponding schematic.

Estimate the capacities of the boiler's pump and the proper settings for the valves by following this example: Assume a large building has a WSHP boiler requirement of 450,000 Btu/hr. A boiler unsuited for handling low-temperature water is available. Find the proper water flows for a secondary system as shown in Figure 9.3. The heat pump loop's water receives heat from the boiler when its temperature is 65°F. The boiler's inlet temperature is to be at least 130°F, and its outlet is to be 150°F. The temperature rise through the boiler is, therefore, 20°F. Select a pump flow to provide the 20°F rise through the boiler at its full flow. The proper pump flow is

$$\text{gpm} = \frac{\text{Btu/hr}}{(498 \times T_d)}$$

$$= \frac{450{,}000 \text{ Btu/hr}}{(498 \times 20°\text{F})}$$

$$= 45 \text{ gpm}$$

Some of the loop's 65°F water enters the secondary loop and mixes with the boiler's 150°F output water. This cools it to 130°F. The boiler heats it back up to 150°F, and some of this water goes into the main loop. The total amount of water in each loop is constant because the same amount of water enters and leaves each. The heated water entering the WSHP loop provides heat to the system. The proportions of boiler water and main loop water that should flow through the balancing valves must be correct to provide the intended temperatures.

For this example, we select a minimum boiler temperature of 130°F and a maximum boiler temperature of 150°F on Table 9.1. The table gives, for these temperatures, a recirculating water factor of 0.765 and a system water factor of 0.235. These factors, when multiplied by the boiler's secondary pump flow rate, give the gallons per minute that should flow through the recirculating and system lines when it receives heat. For our example, the proper flow rates for the lines are

$$\text{Recirculation gpm} = 45 \text{ gpm} \times 0.765 = 34.4 \text{ gpm}$$

$$\text{System gpm} = 45 \text{ gpm} \times 0.235 = 10.6 \text{ gpm}$$

Balancing the flows to these amounts ensures that the boiler always receives water no colder than 130°F. If the balancing valves are set incorrectly, the secondary loop temperature might not be correct, and thus the main loop might not receive sufficient heat. It might also allow the boiler loop's temperature to fall too low to prevent condensation.

This application is best served with automatic balancing valves. It's a little tricky to set both valves to the proper flow setting without a flowmeter. Automatic valves make the process very simple. Purchase and install valves that are factory set to the required flow rates. No other work needs to be done to get proper flows.

If automatic valves aren't available, try to use a flowmeter. They make the balancing process very simple and eliminate the need for a lot of the "back and forth" tuning that might be required without one.

If a flowmeter isn't available, the balancing valves can be set by monitoring the temperatures of the flow when the three-way valve is delivering water into the system. First, set the recirculation valve to 50% open and run the boiler and secondary pump. Keep the three-way valve closed (so no main loop water enters the secondary loop). Allow the water to heat to the aquastat's setpoint and give it time to stabilize the piping and the boiler's heat exchanger temperatures. Monitor the temperature at the boiler's inlet and outlet.

Next, set the system balancing valve approximately 50% open and the three-way valve to deliver hot water into the main loop. The main loop's pump must be running to prevent the hot water from immediately recirculating from the secondary loop's outlet back to its inlet. The boiler should start immediately when the cooler water hits the aquastat.

While the boiler is running and cool main loop water is entering the secondary loop, adjust the system balancing valve to give a 20°F temperature rise through the boiler. Leave the recirculation valve alone. This might require some fine tuning, as the temperature change might go above and below its final point as the valve setting is changed.

Once the temperature increase through the boiler reaches the correct 20°F value, the balancing valves are set correctly. Note the water temperature at the boiler inlet. It is likely to be higher than the planned 130°F minimum temperature. Table 9.1 assumes a loop water temperature of 65°F, and any difference between this and the actual temperature causes the measured reading to be different.

Multiple boilers are also very adaptable to WSHP loops. They can be connected in their own secondary piping loops as described for hydronic heating systems in Chapter 7. The loop's temperature controller can operate the multiple heat sources sequentially to help maintain the proper water temperature.

As is true with multiple-boiler hydronic systems, using more than one auxiliary heater on a heat pump loop provides several benefits. Additional security is provided by having some backup capacity in the heating system. If the total load is divided between two or three boilers, even if one unit fails, the system will still be able to function under most load conditions.

The second and third boiler in the system do not need three-way valves. They can be piped in the system with secondary pumps connected as shown in Figure 9.4. The boiler and secondary pump start when their heat is required.

Duct Heated Systems

Some heat pump systems do not use boilers to add heat to the loop water. Instead, electric duct heaters installed in the heat pump air supply ducts provide building heat. These installations save the costs of a boiler, while still allowing heat sharing.

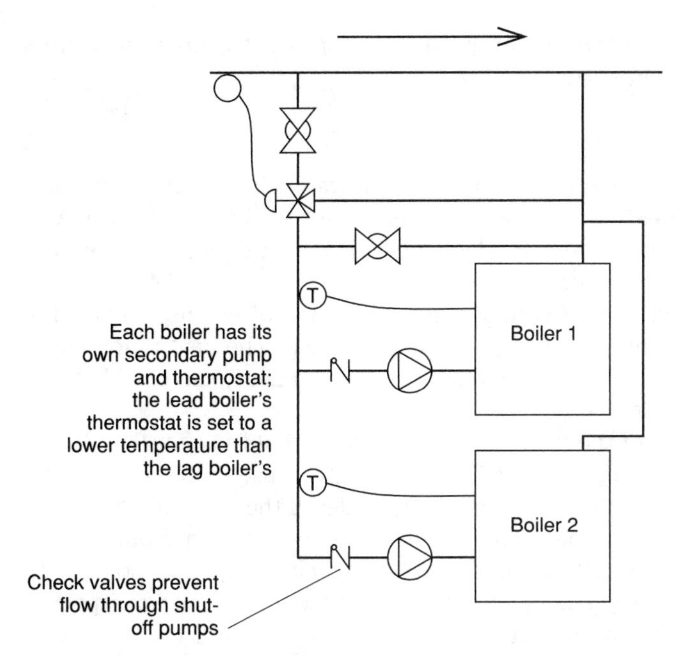

Figure 9.4. Multiple-boiler secondary loop connections. Only one set of balancing valves is required.

Air is heated by either the heat pumps or the electric resistance heaters. If the loop's water temperature is too cold to run the heat pumps in their heating mode, the zone's thermostat energizes the duct heaters. The changeover from running the heat pump to operating the electric duct heaters is automatic.

Note that if some of the heat pumps are cooling their zones, the loop's water temperature will rise. When it becomes warm enough to allow the system to use refrigeration-derived heat, it automatically does so.

Duct heaters are installed downstream of the heat pump to deliver heat to the entire zone. Most systems do not use multiple heaters to split a single zone into several subzones. Instead, loop water coils are installed to temper air delivered to "too warm" subzones. The benefit of using a water coil to cool the air is that the loop's water is heated while the air is cooled. This helps warm the water enough to run the more efficient heat pump's refrigerant cycle heat. As usual, no heat is wasted with a WSHP system.

Special Controls

Special controls help ensure that all of the system's individual heat pumps work together. The water flow and temperature must allow the units to share heat energy from warmer areas to cooler ones. Also, auxiliary heaters and cooling towers must run when needed.

Water flow

Most WSHP systems require constant system water flow. If the loop water is interrupted, no heating or cooling is available for the building. In the winter, this can cause the building and its contents to freeze, leading to extensive water damage from burst pipes.

If water flow fails in the summer, the total lack of cooling causes the building's interior temperature and humidity to soar. The heat pumps have to run a long time at full capacity to get the environment comfortable again after this type of heat soak.

Automatic startup of the loop's backup pump prevents a water flow failure if the primary pump breaks down. The best controls use a differential pressure switch installed in the pump's inlet and outlet. If the pressure difference falls to zero, flow has stopped. The switch starts the secondary pump automatically, restoring water flow. An alarm light or signal should also be activated to alert maintenance personnel that the primary pump needs repair.

Systems that continue running with no water flow until the automatic safety controls shut the heat pumps down. Units cooling the space shut off on high head pressure when the stagnant water in the heat exchanger is heated to over 150°F. Systems using PVC piping can be damaged with water this hot. Plastic pipe loses a great deal of strength at these temperatures and will sag. Extreme cases can cause the pipe to break away from its fittings, possibly causing a flood.

If the system circulating pump must be turned off for maintenance, all of the individual heat pumps should be shut down first. This prevents system damage from overheating, and keep the units from having to shut down on their safety circuits.

Some WSHP systems shut off the circulating pumps when the building is unoccupied. These use electrical interlocks wired to each heat pump. The shutdown control breaks a circuit to the heat pumps, disabling them while the loop's pump is turned off. These interlocks prevent the heat pumps from operating with no water flow.

The benefits of turning off the circulating pump for energy savings are limited. If high-efficiency pumps and motors are used, more efficiency is achieved with night setback zone thermostats than with a complete system shutdown. The thermostats prevent the temperatures from falling or rising too far from the occupied setting. Freeze damage or excess energy use during the "catch up" period is avoided. Tests have shown that unoccupied temperatures should not deviate more than 8°F from the occupied setting. Any greater difference than this uses more energy due to the long running period when the temperature must readjust to the occupied setting.

Loop temperature

A control system is necessary to keep the temperature of the circulating loop water within a specific range. More energy sharing between heat pumps can occur with a wide temperature range. Auxiliary heaters and cooling towers are kept shut off for longer periods, allowing short-term heating or cooling loads to be absorbed by the rest of the system.

On the other hand, allowing the loop temperature to span too wide a range reduces the efficiency of the heat pumps. More energy is required to draw heat from a cold water source or push heat into an already hot loop. Keeping the loop's temperature within specific bounds allows each unit to work efficiently whether it is heating or cooling its zone.

Most loops maintain a water temperature between approximately 65°F and 95°F. Different heat pump manufacturers might have other requirements, but these are generally acceptable if specific information is not available. System controls operate the boiler and cooling tower in stages as required to maintain the water temperature within these limits.

Supplemental heat is usually added to the loop water when its temperature drops to 70°F. If properly sized, the boiler won't allow the water temperature to drop below 65°F while it is operating. Electric heaters can be placed in the piping loop with no considerations made for condensation or other problems associated with low-temperature water. Install fossil fuel boilers with a secondary loop, as described previously, if the boiler should not handle cool water.

Boilers without secondary loops operate from a set of loop water aquastats. The aquastats should start the boiler when the temperature drops to 70°F and shut it off after it rises to 75°F. Install the temperature probes upstream of the boiler to prevent excessive starting and stopping (short cycling).

Secondary boiler controls are a little more complex. The boiler and its pump can be controlled by an outdoor temperature sensor and an aquastat in the secondary loop. They start and run whenever the outdoor temperature drops below a specific point. This allows the boiler to provide heat when the building is likely to need it. In most applications this corresponds to an outdoor temperature of approximately 55°F.

Secondary loop boiler controls have, therefore, two sources. Outdoor temperature starts the secondary pump and allows the boiler to fire. Secondary loop temperature determines when the boiler fires. If the secondary loop temperature drops below the aquastat's setpoint (usually at least 130°F), the boiler fires. The pump, however, runs continuously as long as the outdoor temperature is low enough. Little energy is consumed by the boiler when the loop is not drawing water from it, so its "continuous" operation does not significantly increase operating costs. The boiler's secondary loop piping should be well insulated to limit heat losses while it is in this "standby" mode. The three-way control valve used to deliver heat from the boiler's secondary loop to the main loop should be opened when the loop's water temperature drops below 70°F, and should be closed when it rises above 75°F.

Multiple boilers can be sequenced with an automatic control. If a single boiler cannot adequately maintain the loop water temperature, the next boiler starts. If the water temperature drops while one boiler is on, the loop controller starts the second (and third) boiler.

Cooling tower operation is also controlled by the loop's water temperature. Most systems begin heat rejection when the water temperature increases to approximately 90°F, and shut it off when it drops to 85°F. This is done in stages, using different parts of the cooling tower in sequence to increase the heat loss.

First, the tower's exterior dampers (if any) are opened. If the loop water stabilizes between 85°F and 88°F, no other action occurs. When the loop's temperature increases above 88°F, the tower's sump water pump and spray are started. These wet the loop water coils to allow evaporation to start. The pump runs whenever the loop water is above 88°F. More cooling is provided if the loop's temperature increases beyond 92°F. The cooling tower's air circulating fan runs, increasing the heat rejection. This should keep the loop's water temperature at or below 95°F.

As the loop water cools from 95°F, the previous actions are shut down in reverse order. First the tower fan stops, then the water spray is shut down. Finally the tower's dampers close. With the water below 85°F, no cooling is provided.

Towers used in the winter can be sequenced to provide cooling even if the sump water has been drained. In cold weather sufficient cooling is usually available without the sump spray from the fan blowing cold ambient air over the loop water coil. As long as the loop has freeze protection (antifreeze, sufficient insulation, constant flow, or tower dampers), these systems can work in cold weather with no problems. Building cooling can start automatically, even during a warm spell in the middle of winter.

The automatic temperature controls allow the loop's water temperature to rise from 75°F to 85°F without starting the auxiliary heater. Its temperature can fall from 85°F to 70°F without running the cooling tower. This 10°F to 15°F range is usually sufficient to allow the heat pumps to share heat throughout the building, automatically distributing energy from warmer areas to cooler ones.

Systems using electric duct heaters must prevent the heat pumps from operating as heaters if the loop water temperature falls too low. An aqua-stat locks all of the heat pumps out of the space heating mode any time the loop temperature drops below 60°F. A low-voltage control wire from the main control system to each heat pump opens a circuit connected to the compressor and reversing valve relays. Although the units cannot operate as space heaters, they can air condition the space if cooling is necessary.

The electric duct heaters run whenever the units are locked out of the heat pump heating mode. A relay diverts the thermostat's heat request from the compressor and reversing valve to the duct heater's contactor. Instead of starting the heat pump, the fan and electric heaters run when the space needs heat.

All heat pumps use safety head pressure controls to shut them off in the event that the loop's water flow stops. These controls cause the unit to go into a lockout condition, where no further operation is possible until they are reset. Most heat pumps are reset by turning the main power supply off for a few seconds. Resetting the unit from the space thermostat is not usually possible.

The intention is to prevent building occupants from resetting the heat pumps. Lockouts show that the system has problems and probably needs technical service. If the building occupants can easily reset a locked-out heat pump, it might lock out again if the same underlying problem still exists. Repeating this process indefinitely can seriously damage the unit.

System Capacity Estimating

Unlike standard forced air or hydronic heating and cooling systems, the multiple units used in WSHP systems allow a great deal of flexibility with capacity estimates. As already mentioned, individual heat pumps should be selected to satisfy a zone's larger heating or cooling load. Auxiliary heating and cooling systems, however, should be selected based on a part of the total building load.

It is very unlikely that every heat pump in the system will simultaneously be heating or cooling. Some zones will be satisfied at any given time, with the heat pump not running, while others will be heating or cooling. Even in extreme weather the system is only partly operating in one mode. This permits the use of lower capacity boilers and heat rejection equipment.

Most WSHP systems require a total boiler and cooling tower capacity of approximately 85% of the building's total heating and cooling requirement. For example, a building with an estimated heating requirement of 670,000 Btu/hr would need a boiler capacity of

$$\text{Boiler Btu/hr} = 670{,}000 \text{ Btu/hr} \times 0.85 = 569{,}500 \text{ Btu/hr}$$

Installing a boiler of this capacity provides adequate heat for the building under almost any situation. Using a 670,000-Btu/hr boiler would work, but the unit is really too large for the load.

The cooling tower capacity can also be reduced by taking advantage of the diversity provided by multiple heat pumps. Again, it is very unusual for all of the units to be running simultaneously. The units that are off do not load the system, and the cooling tower's capacity can be reduced.

The proper size for cooling towers also equals 85% of the total building cooling load. Systems installed in very humid climates, where the summer design wet bulb temperature exceeds 78°F, should have a cooling tower sized to 90% of the total building load.

These lower boiler and cooling tower capacities permit the installation of smaller, more economical systems. This saves money on the system's purchase and installation costs. Do not, however, reduce the loop pump's capacity. Provide full water flow under all conditions. Reducing the flow does not provide effective energy sharing between the heat pump units. As noted earlier, energy savings are available if high-efficiency motors and water pumps are used for the loop system. Their constant operation provides very good opportunities for energy savings.

Chapter 10

Ventilation

Space ventilation includes two complementary things: the introduction of outdoor (presumably "fresh") air into an enclosed space and the removal of noxious, stale, or objectionable air, gases, or odors to the outside. Improved ventilation is critical to keeping an acceptable climate inside a building and to achieving and maintaining indoor air quality (IAQ). Correct ventilation also helps a building to maintain the proper balance of humidity. Proper humidity levels are essential to prevent damage to the building and its contents and provide the occupants with a comfortable, healthy environment.

Air Quality Requirements

Ventilation systems must provide the correct amounts of outside air for the building's occupants. A significant question is: What is the correct amount? There is no simple answer.

The issue of indoor air quality has recently become very contentious. Building occupants complain about inside air and blame poor air quality for many health problems. While a few complaints might really be due to other causes, there is no doubt that many buildings can, and do, make their occupants sick. Poor air quality can cause respiratory problems, dizziness, sleepiness, and more serious chronic health problems.

Building conditions that cause indoor air related illnesses are often described as *sick building syndrome*. There is no formal definition of the term, but it is usually used to describe buildings in which several people come down with air quality related illnesses. Poor ventilation system design and operation can make sick building syndrome worse. It might even be to blame for the entire problem.

Air quality in buildings with little outdoor air supply (where indoor air recirculates or stagnates) quickly deteriorates when the space is occupied. Carbon dioxide gas exhaled by people accumulates, leading to discomfort and drowsiness. Poor ventilation does not expel the carbon dioxide or dilute it, and its concentration rises throughout the day. Some air quality recovery usually takes place at night while the space is vacant.

Other ventilation related problems include the accumulation of gases and dust generated from sources within the space. Many building materials give off chemicals, especially when new. Formaldehyde is commonly "exhaled" by carpeting, upholstered furniture, and many wood fiber products. This gas is very irritating to the eyes, nose, and throat. People allergic to it can be stricken with severe reactions that might require hospitalization.

Some outside air supply systems can make a bad situation even worse. Air intakes located in parking garages, near busy streets, or where other contaminants can enter can pollute the air inside the building. Auto exhaust, cooling tower drift, air exhausted from toilet rooms, laundry dryers, cooking grilles, or industrial exhaust all can be drawn into a poorly located ventilation system intake. The original outbreak of Legionnaire's disease was caused by contaminated cooling tower drift entering a ventilation air inlet. People died after inhaling the bacteria carried in with the water droplets.

Dust and dirt can spread with poorly designed or serviced air handling equipment. Improper filtration allows irritating materials from one part of a building to be carried to other areas. Ventilation systems with plenum air returns might carry the air through very dusty areas (sometimes with old asbestos insulation or fire proofing) and spread the contamination to other spaces.

Required Ventilation Volumes

Building code officials have been inconsistent in their reactions to the concerns about indoor air quality. Concerns about energy consumption in the 1970s forced many building codes to severely restrict the amount of outside air used in buildings. Indoor air quality concerns have reversed that trend, with some communities now insisting on very large amounts of outside air for many applications. Other areas fall somewhere in between, with others deferring to the installer using "best practices." This is deliberately vague.

Check with local code officials on the amount of outside air that must be supplied for any new system being installed. Also, confirm the amounts of exhaust air necessary for toilet rooms, laundry areas, etc. Some communities even have ventilation requirements for large unoccupied storage rooms and closets.

Those that refer to *best practices* or similar language make the designer or installer decide the right amount. This is not a trivial estimate. If an incorrect amount is selected and the ventilation air isn't adequate, the designer and installer can be sued. The contractor is often blamed for injuries allegedly caused by poor air quality.

Building code officials probably can give some guidance on what makes up best practices. Often ASHRAE Standard 62-89 (or its latest edition) is used as a reasonable ventilation specification. This calls for providing 20 cfm of outside air for each building occupant. All occupied areas must be ventilated in proportion to the number of occupants.

Older versions of this standard called for much less outside air. Many systems installed in the energy conservation conscious 1970s provide only 5 cfm per occupant. Some were set up to give 10% outside air and 90% inside air for all operating conditions. Other requirements specify lower per occupant requirements (5 or 10 cfm per person). These amounts might be inadequate for systems installed today.

Ventilation supply volumes of 20 cfm per occupant are fairly high. Buildings using variable air volume (VAV) cooling systems often have total air supply requirements less than this. Some of these buildings must use dedicated ventilation systems that operate independently of the comfort heating and cooling systems.

Other requirements call for specific quantities of air per square foot of occupied space. Many kitchens, for example, must receive 2 cfm per square foot of floor space. Specific requirements for other spaces can be found by consulting code officials.

Requirements for repair or retrofitting of ventilation systems in older buildings vary from one community to another. Some code officials allow older facilities to operate at the older ventilation requirements, while a few insist that they be modified to modern standards. Others fall in between, allowing older systems to remain as long as they continue to operate. If they need modification or substantial repair, however, they must be upgraded to the new standards. The decision depends on the local codes for the area.

Building Pressurization

Outside air introduced into a building and inside air exhausted to the outside are both forms of ventilation. Properly using and balancing each type of air is essential. The net difference between fresh air induction and exhaust contributes to the net pressurization of the space (compared to the outside).

As discussed in Chapter 1, stack effect and wind influence a building's internal pressure. Stack effect, caused by buoyant warm air rising to the top of a building surrounded by cold air, increases the pressure at the top of the building. The bottom of the building has less pressure than the top as the warm air inside rises away from it. The building expels warm air at its top (because of the higher pressure there) while "inhaling" cold outside air at its lower pressure bottom.

Wind affects a building's internal pressure. Wind blowing on one side of a building drives outside air in. The wind's velocity pressure causes infiltration on that side. Air leaves the other side of the building where the outside pressure is lower. While the total building pressure is usually not influenced by the wind, local areas in the building experience either infiltration or exfiltration. Flow direction depends on the direction and strength of the wind.

A properly operating ventilation system should be set up to slightly pressurize the interior space. Most systems should provide a positive pressure of about 0.02 in. water compared to the outside. This causes air inside the building to always try to leak to the outside.

This is not possible under all conditions. Very high winds will probably drive some air into the building on its windward side. Similarly, extreme cold weather might cause some air intake in the lowest areas of tall buildings. The goal, however, of a properly operating ventilation system is to provide positive building pressure under most conditions.

There are many benefits to a slight positive pressure. The reduction in infiltration eliminates uncontrolled drafts and dust intake. The HVAC system fully controls and handles all of the fresh air brought into the building and properly conditions and filters it before it enters the space. Humidity control is easier too. Less infiltration usually means that the air conditioning system has lower latent heat loads during the cooling season.

Positive pressure is attained by properly balancing the exhaust and intake air supplies to the building. There should be more air brought into the space than exhausted. A ratio of about 90% exhaust air volume compared to the intake air volume is usually correct. The total fresh air intake must exceed the capacity of all exhaust systems installed in the building by 10% to properly pressurize the space. For every 20 cfm of outdoor air supplied, exhaust 18 cfm (or allow that much to escape) from the space. Buildings with large exhaust volumes require even more outside air.

Select a building's ventilation supply air volume as the larger of the legally required amount and the amount needed to properly balance the exhaust system. If the legally required amount is higher, provide a way for most of the outside air brought into the building to escape. This must escape without relying on exfiltration or the building's exhaust system. Use special air relief dampers or fans to get rid of the excess air. Again, the exhaust volume should equal 90% of the legally required outside air supply.

If the building's exhaust volume is higher than the legal ventilation requirements, no special air relief systems are required. Provide 10% more air than the exhaust system expels to satisfy the occupants and give the correct pressure balance.

To obtain proper building pressurization, the total exhaust must be less than the fresh air intake. First, add all of the building exhaust volumes together. Hotels, restaurants, laundries, and medical centers often have high total exhaust volumes that require large amounts of air from the supply ventilation system.

For example, if the exhaust system's volume totals 7500 cfm, the minimum total supply air should be 10% higher, or

$$7500 \text{ cfm} \times 1.1 = 8250 \text{ cfm}$$

If this exceeds the total legally required ventilation air supply, then this is the proper supply volume. If, however, the building's occupancy load requires a higher outside air supply, provide a way for this air to leave the building. For example, if the building had a normal occupancy of 515 persons, and the local code called for 20 cfm per person, the requirement would be

$$515 \text{ persons} \times 20 \text{ cfm/person} = 10{,}300 \text{ cfm}$$

The exhaust system expels 7500 cfm of this, leaving the remainder (10,300 cfm – 7500 cfm = 2800 cfm) to pressurize the space. Provide a way for this air to leave the building.

Air relief systems are usually very simple. They include backdraft dampers installed in the building's walls or return air system, positioned to allow air to escape from the building to the outside. These *spill air* systems allow excess air to discharge based on the building's pressure.

If the backdraft dampers are chosen properly, they can operate as a simple automatic pressure control. The higher the interior pressure (compared to the outside), the further the dampers open. All backdraft dampers have minimum pressure requirements where they begin opening. If they are selected to open at a reasonable pressure (between 0.01 and 0.03 in. water), the total building pressure sets their opening. Higher building pressures make them open farther and discharge more air.

The relief air system must be large enough to handle all of the volume difference between the supply and exhaust systems. For the previous example this is 2800 cfm.

Many buildings receive all of their cooling air from the outside when operating in an economizer mode. Spill air dampers must handle this large volume of air. If they cannot, a powered exhaust might be required.

Powered exhaust systems usually use propeller fans installed behind outside louvers to force indoor air out. These are effective where large volumes of air must be expelled from a small area.

Large buildings might require more sophisticated air pressure relief systems. Stack effects can cause backdraft louvers installed low in the building to remain closed despite the need to expel air. Also, changing air conditioning requirements and outdoor temperatures might allow some air handlers to operate in economizer mode while others are not.

These changing effects make it impossible to predict specific building exhaust requirements. The exhaust system must operate in response to the conditions existing at any given time to properly set the building pressure. Differential pressure sensors are usually used at several locations to detect the differences between indoor and outdoor pressure. All of the sensors' outputs are added by the control system, and one or more exhaust fans start when pressure relief is necessary. Multiple fans can be staged with a series of pressure sensors. As pressure in the building rises, the fans run to keep the pressure from becoming too high.

Exhaust fans are usually installed directly behind backdraft dampers at outdoor louvers. If the fans do not operate (due to failure), some indoor air can relieve itself by passing through the fan housing and opening the backdraft dampers.

Ventilation Systems

A variety of ventilation systems are used in buildings, depending on the individual requirements and facility details. Usually the ventilation system and the comfort cooling and forced air heating systems share the same fans, ducts, and registers.

Depending on the volume of outside air required, a simple outside air louver can be installed on a constant volume HVAC system air handler. Outside air mixes with return air before the fan's inlet. Each flow's volume is set by dampers in the airstreams. The mixed air enters the air handler's heating and cooling sections for distribution into the building. Relief air discharges out one or more vents on the outside wall. These systems are usually adequate for fairly small jobs. There are, however, several limitations to these simple systems.

Ventilation air is required throughout the year, including when temperatures are well below freezing. Extremely cold outside air must be tempered before it is introduced into the space. This can consume additional energy and freeze unprotected steam, hot, or cold water coils and pipes. Nevertheless, the system should always supply the required volume of ventilation air.

Simple outside air intake systems are usually suitable for year-round operation in areas with warm or generally mild climates. The cold weather that is possible in many areas of the country requires HVAC technicians to take one or more special precautions. These include

- Use either direct expansion air conditioning systems, or drain all chilled water coils before the heating season begins. While this prevents a quick start-up of air conditioning if the weather becomes warm, it is the only positive way to prevent a water coil freeze.

- Use antifreeze in all hydronic systems with outside air inlets. Cold air can freeze both chilled water and hot water coils, so use it in both systems.
- Confine air heating systems to furnaces only (avoid using any water or steam coils).
- Use freeze-proof steam coils only. Standard steam coils are easily frozen.
- Use face and bypass dampers with hot water coils. Never reduce the flow of hot water to reduce heat output (unless antifreeze is used). Always maintain full hot water flow through the coil. Control the air handler's output temperature by regulating the amount of air flowing through the coil.
- Install a preheating coil in the outside airstream. This warms the air to at least 50°F, protecting downstream chilled water or modulated hot water and steam coils. Preheating coils usually operate based on the temperature of the incoming air. Most controls start their water flow whenever the outside temperature is below 40°F.
- Use a freeze protection thermostat in the airstream just ahead of the system's water coils. This shuts off the air handler's fan if the entering air temperature drops below approximately 40°F (or below the temperature that the preheating coil begins operating). Freeze protection thermostats give an extra level of protection against coil freezing.
- Install a heat recovery heat exchanger in the incoming airstream. It preheats the ventilation air with energy taken from the exhaust airstream.

Variable air volume and water source heat pump systems have their own special ventilation system requirements. Variable air volume units control space temperature by modulating the amount of air delivered to a zone, and heat pumps generally recirculate room air within a single zone.

Variable air volume systems must control the amount of cooling air supplied to the space. As the space thermostat senses the need for cooling, the VAV system increases the amount of airflow. Higher cooling loads (such as crowded conference rooms) automatically receive more air. Under typical operating conditions, however, most systems do not deliver 20 cfm per person. Heat and temperature balance is usually achieved with much lower air volumes.

Water source heat pumps cannot usually change air temperatures over a wide range. If air well below room temperature is delivered to them for heating, the discharged air temperature will not be high enough to warm the space. All of the heat energy will be used just warming the extra-cold air up to room temperature. Space heating cannot occur unless the discharged air is warmer than the space's temperature.

Most VAV and WSHP systems need separate, dedicated, independent ventilation systems. Design them only to supply outside air to the building and maintain the proper internal pressure.

Dedicated outside air supply systems need their own heating and cooling sources. During cold weather, the air temperature should be brought to approximately 65°F before it goes into the space. In hot, humid weather it might be cooled to decrease the load on the comfort cooling system. Chapter 3 describes the heating and cooling loads imposed on a building due to outside air ventilation or infiltration. These loads might be easier to handle with separate heating and cooling systems.

Much of the ventilation needs during mild weather (between 45°F and 65°F) can be handled by a standard HVAC system's air economizer. These provide free cooling and plenty of fresh air.

Air economizers

Economizer operation refers to the use of outside air to provide comfort cooling entirely with outside air. Technically, this is *air side economizer* operation, as opposed to receiving free cooling with a cooling tower to cool a chilled water system. Air side economizers have the added advantages of less energy use (no pumps or tower fan) and the supply of large amounts of ventilation air.

To operate effectively, an economizer must cool all of the internal and solar space loads that the building has. The cool outside weather eliminates any load associated with outside air infiltration or heating. All of the internal and solar loads are added to get the total cooling load that the economizer must handle. The amount of air that must be delivered is figured with the formula

$$\text{cfm} = \frac{\text{Btu/hr}}{(1.08 \times T_d)}$$

The temperature difference used in the equation is the difference between the outdoor air's temperature and the desired indoor temperature. The more outside air delivered into the building, the higher its temperature can be. Cooling takes place if a high enough volume of moderately cool air is delivered to the space. Similarly, lower air volumes require lower outside temperatures to provide a given amount of cooling.

Cooling capacity and its relationship to outdoor air temperature is given in Table 10.1. The table shows the amount of air required at various outdoor temperatures to provide 1 ton of cooling capacity (12,000 Btu/hr) into a 75°F space.

An air conditioning design rule of thumb (400 cfm of air delivery per ton of air conditioning load) allows full cooling with outdoor temperatures of 45°F to 50°F. Many fan systems provide this volume of air for each estimated ton of cooling load. It does not mean, however, that an air handler delivering this volume can only use economizer air with these cool outside temperatures.

TABLE 10.1 Volume of Outside Air Required to Provide 12,000 Btu/Hr of Sensible Cooling

Assumes an indoor temperature of 75°F

Outside (°F)	Required volume (cfm)
65	1110
60	741
55	556
50	444
45	370

The building is not near its full cooling load during economizer operation. No hot, outside air infiltration occurs, and no conductive flows (other than solar heating) can take place. The only cooling loads on the building are from internal heating and possible solar gains. The cool outside air also helps keep the building's surfaces cooler than they are on a full cooling day. All of these load reductions means that a much higher outside air temperature can be used for economizer cooling than is required for main system operation.

In practice it is common for air conditioning systems designed with the 400 cfm/ton rule to give satisfactory economizer operation up to outside temperatures of 60°F. Buildings that have been designed to handle very high air infiltration loads (which do not exist during economizer operation) can receive adequate cooling with outdoor temperatures of 65°F.

Remember that the air brought into a building with any ventilation system must be permitted to escape. Provide air relief dampers or fans sized to remove 90% of the air brought into the building via the ventilation system. As noted earlier, existing building exhaust systems (toilet exhausts, kitchen hoods, laundry vents, etc.) remove some of the air. If these are not sufficient, provide more relief air capacity to prevent overpressurizing the building.

Relief air outlets can be used in the return duct system. This is a convenient location for them, as the pressure in the duct is usually close to that in the building space itself. However, the outside air inlet also connects to the air handler's fan inlet. This is usually directly connected to the return duct (and, therefore, the relief air outlets). It is not possible to have an outlet and inlet connected from the same place (the air handler's return duct) to the same place (the outside). Air either comes into the building from the outside, or discharges from the building out. It can't do both at the same time.

To accommodate both relief air and fresh air inlets in the return duct system, install a pressure dropping damper between them. This causes a pressure loss in the duct from the space's return grilles to the fan inlet. The total pressure of the air on the fan's side of the damper is lower than the pressure on the return air grille's side. The lower pressure on the fan side allows it to draw outside air into the building. Higher pressure on the return air grille side permits excess air in the building to discharge outside. Figure 10.1 shows a suggested damper arrangement in a return air system.

In effect, the damper between the relief air outlet and fresh air inlet splits the return system into two parts. The section closest to the return air grille is close to the pressure in the space. The section past the damper (and closer to the fan inlet) runs at lower pressure.

Both the damper between the return air sections and the damper in the fresh air inlet must be properly balanced to ensure correct operation. If both dampers are fully opened at the same time, the system will be poorly balanced. A large amount of outside air is allowed to enter the space, but the relief dampers can't open and permit it to escape. The building's pressure increases to the point where exfiltration balances the air brought inside. Outside doors will be hard to close and drop-in ceiling tiles might be blown into the space above because of the excessive building pressure.

Rooftop units

Roof-mounted HVAC units are commonly used in commercial forced air heating and air conditioning systems. While they function like other hori-

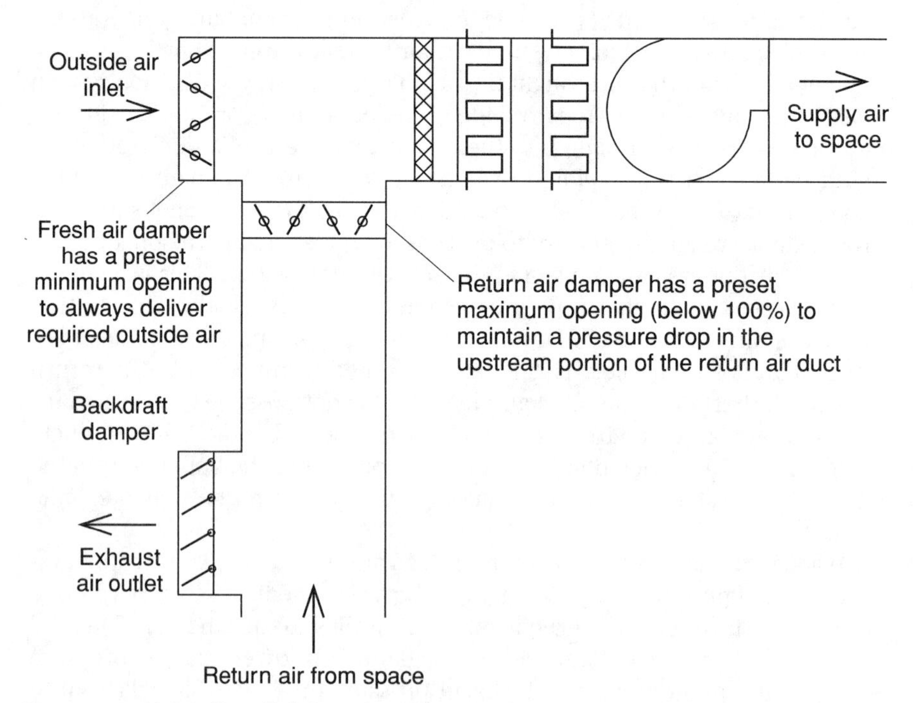

Figure 10.1. Air handler relief air, and outside air and return air inlet dampers.

zontal furnaces, a few accessories are available for them that can provide significant ventilation benefits.

Rooftop forced air heating and air conditioning systems can provide low-cost building pressurization control and ventilation, along with comfort heating and cooling. Rooftop units are fully integrated systems that can take care of most building ventilation requirements in one package. Using rooftop units to provide the proper amounts of ventilation air is usually much simpler than connecting air mixing boxes and louvers to standard furnaces and air handlers.

Low-cost options for most rooftop units include economizer air inlets and spill air relief outlets. These perform the complementary actions needed to always deliver fresh air to the occupants, provide free cooling during mild weather, and allow excess air to leave the interior space. They include complete control packages to operate the economizer whenever cooling is needed and outside air temperatures allow it. The outside air inlet can be set to a minimum opening at other times to maintain the required ventilation air intake.

Heat exchangers

Ventilation systems often benefit from the use of air-to-air heat exchangers. They provide a way to recover much of the heat energy in the interior air before it is lost through the exhaust system. This energy can be added to or subtracted from the incoming fresh air to preheat (or precool) it before it enters the space. The energy savings can be substantial.

Heat exchangers use at least two air paths. One carries the building's exhaust air to the outside and the other handles the incoming fresh air. The two airstreams never mix or touch each other, but the exchanger provides a way for each of them to share energy with the other.

Heat energy only flows from high-temperature objects or areas to lower temperature ones. This can help save energy when ventilating a building. The exhaust airstream is often cooler than the outside air in the summer and warmer in the winter. Heat flow through the heat exchanger transfers energy to or from the incoming outside airstream. In the winter the outgoing warm exhaust air transfers some of its heat to the incoming outside air. The reverse happens in the summer (cool exhaust air absorbs heat from the outside air supply). In both cases the outside air supply temperature is brought closer to the indoor temperature and the heating or cooling load is reduced.

There are four basic types of heat exchangers used for HVAC systems: plate types, coil runaround systems, heat pipes, and heat wheels. Their operations are fundamentally different, but they all save energy. One significant difference is that heat wheels can transfer both sensible and latent heat.

Plate type. Most HVAC heat exchangers use many parallel aluminum or steel plates arranged to allow air to flow between them. These function exactly like plate and frame water heat exchangers. One airstream flows through the spaces between plates one and two, three and four, five and six, and so on. The other airstream flows through the spaces between plates two and three, four and five, six and seven, and so on. Heat energy flows through plates two, three, four, five, and six (see Figure 10.2).

Although heat does flow, there is no contact or air contamination between the airstreams. They are kept entirely separate. Most plate heat exchangers have the airstreams flow in opposite directions past each other. This counterflow arrangement allows each stream to reach a final temperature close to the other stream's starting temperature and improves their efficiency.

These units can transfer a substantial amount of sensible heat, with some capabilities for latent heat transfer. If the temperature of the warmer airstream drops below its dew point, water condenses on the heat exchanger's plates. This causes the air to lose latent heat. This changes into additional sensible heat (with a temperature increase) in the air flowing in the colder airstream.

If condensation of moisture is expected, install a drain (with a trap) for the heat exchanger. Most units include drain fittings to dispose of water collected during operation.

Frost control with plate heat exchangers is usually done with an air bypass around the unit's outside air supply section. If a thermostat detects the leaving exhaust airstream is below approximately 35°F, the air bypass is opened. Cold outside air does not flow through the exchanger, but bypasses it instead. Meanwhile, warm exhaust air continues to flow through its sec-

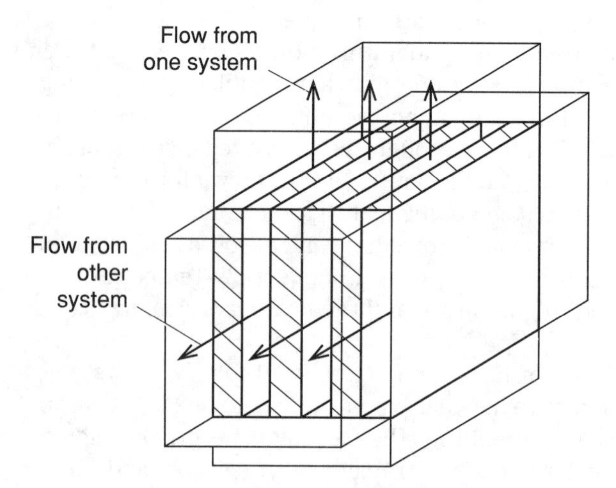

Figure 10.2. Plate-type air-to-air heat exchanger.

tion (where the ice formed) and eventually melts the ice. When all of the ice melts, the exhaust temperature increases and the air bypass is closed. Other frost prevention systems use outdoor air preheating or electric defrosting heaters installed inside the unit.

Plate heat exchangers are efficient, but have some limitations. These include:

Both the supply and return airstreams must be located close to each other. Both airflows run through the same unit and have to be ducted to its location. This can make retrofit installations of plate heat exchangers difficult in existing buildings with widely separated exhaust and supply ventilation systems.

They can be difficult to clean. Exhaust and outside supply air can carry a significant amount of dust and dirt. The airstreams must be filtered to prevent dust and dirt accumulation within the heat exchanger. Once dirt gets inside it is almost impossible to remove because most heat exchangers cannot be opened for cleaning. They are sealed to prevent the possibility of air cross-contamination.

There is no simple way to "turn off" the heat transfer. Since air flows through both sides of the exchanger, heat moves from the hotter to the colder temperature stream. Air must bypass around one or both exchanger sections (possibly with a defrost control bypass) to prevent heat transfer, making additional ducts and dampers necessary. This can be important if the incoming ventilation airstream is to be used for economizer cooling. The warm exhaust air heats the incoming air, destroying much of its ability to cool the space.

Coil runaround. Another type of primarily sensible heat exchanger is the coil runaround system. These use at least two water coils installed in the ventilation exhaust and supply airstreams. A closed-loop water (or antifreeze) circuit connects the coils and fluid runs through the system from one coil to the other. Heat is carried along with the circulating fluid. Figure 10.3 illustrates a coil runaround system.

Heat moves from the warmer airstream to the cool fluid circulating through the coil. The air's temperature drops and the fluid's increases. The fluid circulates to the coil in the colder airstream, where it releases the heat it picked up at the hotter coil. This airstream warms up from the heat released by the fluid and the fluid cools down from the heat loss. The cycle repeats as the fluid returns to the warmer coil and absorbs more heat.

Coil runaround systems are inherently less efficient than plate-type systems. Heat must transfer twice: once from the warmer airstream to the fluid and again from the fluid to the cooler airstream. This double transfer reduces the total efficiency. More heat remains in the warmer airstream, and less goes into the cooler one.

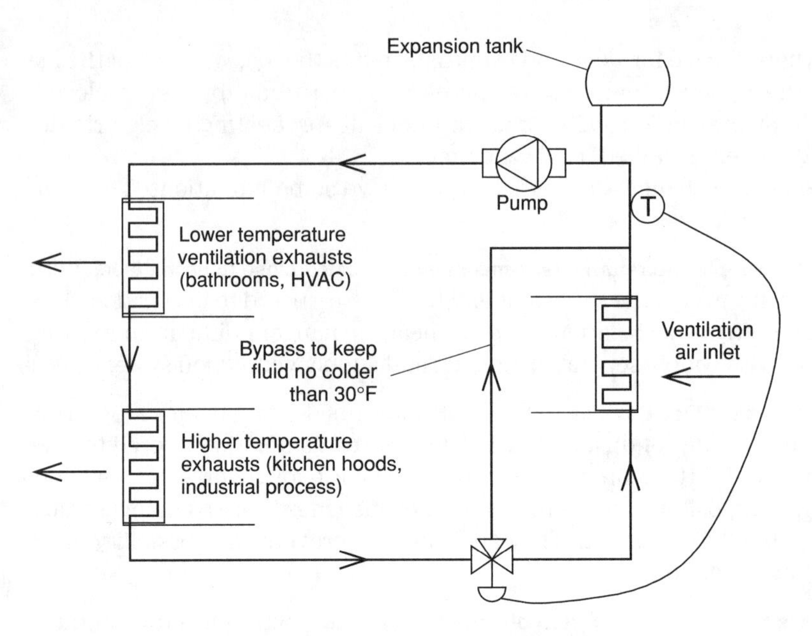

Figure 10.3. Coil runaround heat exchanger system. Multiple coils can be used on both the exhaust and ventilation air systems.

There are several advantages, however. They are very flexible regarding the physical layout of the ventilation supply and exhaust systems. Multiple coils can be installed in several locations, with heat picked up or delivered to each. The coils' sizes can be chosen to match the anticipated heat availability at each location. Larger coils can be installed where more heat flow is available. This allows the balancing of heat flows with different airflows and temperature differences.

Most of the coils used for runaround heat exchangers have more rows of tubes than units selected for standard hydronic heating or cooling systems. Deeper coils allow them to operate more efficiently over narrower temperature ranges. Heat can be effectively transferred from air that is only a few degrees different from the circulating fluid.

Heat transfer can also be turned off. Simply shutting off the circulating pump stops all heat flow. The pump can be connected to automatic controls to shut off whenever economizer operation is wanted, preventing heat transfer.

Defrost control is also simple with coil runaround systems. Fluid flow through the coil is bypassed with a three-way valve whenever the air temperature of the exhaust coil's air output drops near freezing. The valve modulates to regulate the fluid flow, always keeping the outgoing air temperature just above freezing.

Heat pipes. Heat pipe exchangers are entirely passive devices. As with plate units, heat can transfer through them in either direction to warm or cool incoming air. Their operation depends on the internal vaporization and condensation of a working fluid inside.

Heat pipes are usually built as a fairly long, cylindrical tube, sometimes covered with fins. Tube diameters range from about ⅝ in. to more than 1 in. Most HVAC heat pipes are installed horizontally, with some systems using automatic tilt controls to regulate the heat flow.

Inside a heat pipe is a low-pressure working fluid and an absorbent wick running the length of the tube. The fluid is a type of refrigerant that vaporizes and condenses at the temperatures used for comfort HVAC systems. The wick runs from end to end along the heat pipe's inside diameter (see Figure 10.4).

Heat pipes transfer a large amount of heat from one end to the other by alternately vaporizing and condensing the fluid. The fluid is vaporized (boiled) at the hot end of the pipe and the gas flows to the cool end. There it condenses (liquefies) and soaks the wick. The liquid travels through the wick and returns to the hot end where the process repeats.

Heat is absorbed by the vaporizing liquid at the hot end and is released when it condenses at the pipe's cold end. Because the vaporization and condensation processes are taking place at the same pressure, both ends of the tube always try to reach the same temperature.

Heat transfer improves by returning the condensed fluid back to the hot end as fast as possible. Most systems do this by tilting the exchanger slightly so the hot end is lower than the cold end. Liquid fluid runs inside by gravity from the cold, condensation end to the hot, vaporization end. Conversely, heat transfer can be almost stopped by tilting the heat pipes so

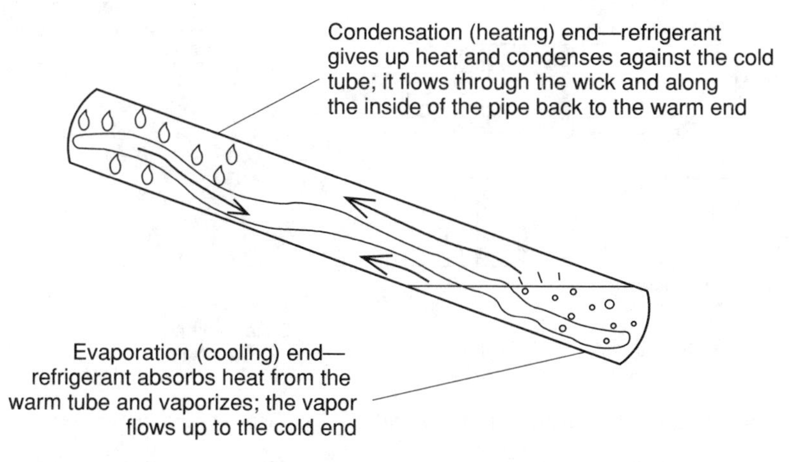

Figure 10.4. Internal components of a heat pipe. The wick assists return of the liquid refrigerant in installations with little slope.

that the cold end is lower than the hot end. This helps confine liquid fluid to the cold end, with only a small amount moving to the hot end through the wick.

Systems used in HVAC applications use either horizontally installed heat pipes for two-way heat transfer, or allow a small amount of tilt to improve heat transfer in one direction. An automatic control (pneumatic or electric) tilts the internal heat pipe assembly as required to adjust the amount and direction of heat transfer. If, for example, frost begins to accumulate on the exhaust side of the heat pipes, the assembly is tilted toward the supply air (cold) end. Heat transfer stops and the exhaust tubes defrost.

Like plate-type heat exchangers, heat pipes must be installed in a location where all of the supply and exhaust air is available. Heat pipes are usually less than 8 ft long, so at most only 4 ft is installed in each airstream. This means that large systems need tall, narrow ducts at the supply and exhaust points.

Heat wheels. Heat wheels are large, circular assemblies porous to the flow of air. Shaped like a thick phonograph record, the disk rotates around an axle in its hub. Both the supply and exhaust airstreams flow through the disk from one face to the other. The disk is made of wide, thin metal bands wound around the hub. The bands are spaced about ⅛ in. apart to allow air flow past and through them. Figure 10.5 illustrates a heat wheel exchanger. Wheels range in size from 4 ft to more than 15 ft in diameter and from 6 in. to more than 20 in. thick. Larger sizes provide greater capacities.

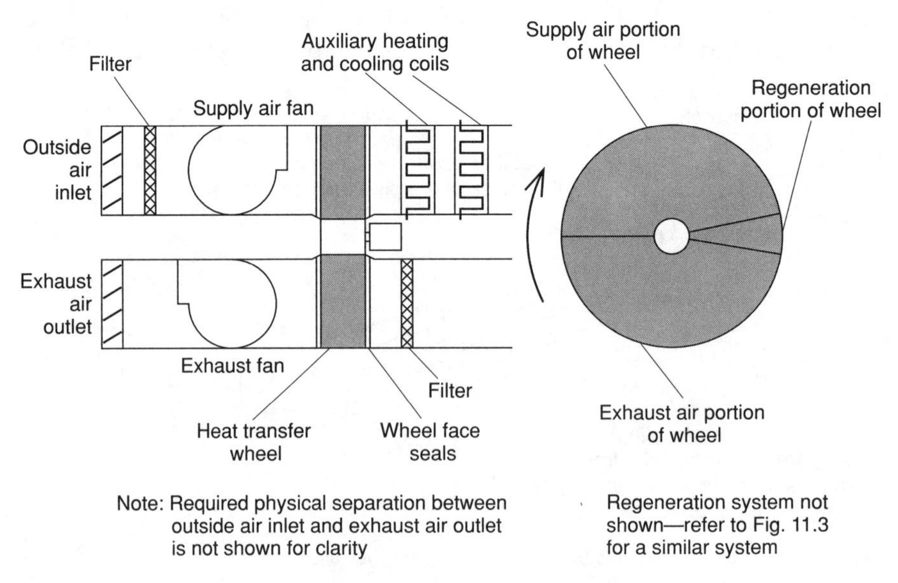

Figure 10.5. Typical rotary heat wheel installation.

Both the supply (fresh air) and exhaust airstreams run through the disk. One wedge-shaped area of the disk handles the supply air and another handles exhaust. The disk rotates continuously, with each part passing alternately through both airstreams. It is heated and cooled by the airflows, and its capacity can be regulated by its rotation speed. If no heat transfer is needed, the rotation is stopped.

As the warm airstream passes through the relatively cool disk, it gives up heat. The air leaves the disk cooler than it started, and the disk's temperature is increased where the air went through it. The disk then rotates out of the warm airstream and into the cooler flow of air. Here, the warm disk gives up the heat it absorbed into the cooler airflow. The air is heated and the disk is cooled. It continues to rotate, leaving the cooler airstream and entering the warmer one. This process repeats as long as both of the streams flow and disk rotates.

This heat transfer allows sensible heat to move from the warmer airflow to the cooler one. Some wheels can transfer both sensible and latent heat (these are called *total heat* wheels). This is done by trapping moisture from one airstream onto the wheel's surface and releasing it into the other airstream. These wheels are coated with a desiccant material that traps water vapor. As moist air passes over it, some of the water vapor's molecules are caught and adsorbed by the coating. Two things happen: the latent heat of the leaving airstream is reduced and its sensible heat is increased. This causes the air and the wheel to attain a higher temperature than if no water vapor removal had occurred.

The temperature increase depends on the amount of moisture removed from the air. The more latent heat removed, the more sensible heat is created. This temperature increase is offset when the wheel turns to the next area of the exchanger.

When the heated, moistened area of the wheel moves into the path of the drier airstream, most of the moisture adsorbed by the wheel is released as vapor. Like evaporation, this is a cooling process. The transfer of water from one airstream to the other causes heat to transfer, as well. The drier airstream picks up latent heat and loses sensible heat on its trip through the wheel. Therefore, its temperature becomes lower than it would have been if no water vapor transfer had occurred.

The temperature drop (cooling) is almost as much as the temperature increase (heating) in the "humid air" section of the exchanger. The net effect is that some sensible heat transfers from the dry airstream into the humid one. Also, some heat is picked up by the wheel in the final section as it passes through the regeneration area.

Total heat wheels must be regenerated to allow them to readsorb more water. Regeneration means drying the desiccant enough to bring its vapor pressure lower than that in the humid airstream. Most systems' dry airstreams aren't dry enough to do this without a separate regeneration area.

To do this the wheel rotates out of the second airstream and enters the regeneration zone. Here, heated air (at 150° to 200°F) passes through it to complete the drying process. Usually only a small amount of hot regeneration air is needed to completely dry the wheel and ready it for the moist airstream again.

This moisture transfer effect is very useful in ventilation systems running in warm, moist climates. Outside ventilation supply air is drawn into a fan and blown through the wheel and into the building. This fresh airstream makes up the moist airflow supplied to the heat exchanger. As the air passes through the wheel much of its water vapor is left on the wheel's surface.

Building exhaust air is drawn over the moistened section of the wheel to remove the water and discharge it outside. The exhaust air dries the wheel's surface and helps cool it for more effective regeneration. A small flow of hot regeneration air dries the wheel further before it again rotates into the outside ventilation airstream. Some systems can use waste heat from the air conditioning system's condenser to regenerate the wheel at almost no cost.

Energy savings during the cooling season can be substantial with a total heat wheel exchanger. Many cooling systems have their greatest ventilation loads associated with removing the incoming air's moisture. Total heat wheel exchangers can remove much of this moisture with little energy cost.

Both the supply and exhaust airstreams pass through the wheel and can leak at the duct-to-wheel interfaces. The ducts might have a soft seal that lightly scrapes the wheel's face to reduce leakage, but these don't really provide a tight closure. If the exhaust air is contaminated (for example, a bathroom exhaust) it is important that no crossover from the exhaust to supply occurs. Proper application of the supply and exhaust fans prevents contamination. No contamination can occur if the fresh air zone of the wheel is at a higher pressure than the exhaust air zone.

Install both the supply and exhaust fans toward the wheel's "outside" face. The outside ventilation supply fan should draw its air from the outside and through a filter. From the fan the air should be blown through the wheel and into the building. The outside ventilation supply air must pass through the fan before it reaches the wheel. Exhaust air should be drawn out of the building, through a filter, and then through the wheel to reach its fan's inlet. The exhaust air must pass through the wheel before it enters the fan inlet. The fan can then immediately discharge the exhaust air outside.

Arranged in this way, both fans allow leakage only from the supply air to the exhaust stream. Normal operation, or a failure of either fan, doesn't allow exhaust contamination of the fresh airstream. All leaks run from the fresh airstream to the exhaust air outlet.

Both airstreams should be kept clean to prevent clogging of the wheel's deep, narrow passages. Use filters on both the supply and exhaust airstreams. Heat wheels are difficult to clean once they are contaminated, sometimes requiring removal and washing to clear dirt trapped far inside.

Energy estimates. While total heat wheels can transfer more energy, even the savings from a sensible-only exchanger can be significant in cold weather. Consider an example:

A building in winter has an interior temperature of 70°F, while the outside temperature is at its design point of 10°F. Ventilation exhaust and supply airflow through a heat exchanger installed to preheat the incoming air. Both the exhaust air and outside supply airflow at 2000 cfm. Find the amount of energy needed to heat the outside air with and without a heat exchanger. Assume the efficiency of the heat exchanger is 60%.

The energy needed to heat 2000 cfm of air from 10°F to 70°F is

$$\text{Btu/hr} = 1.08 \times \text{cfm} \times T_d$$
$$= 1.08 \times 2000\ \text{cfm} \times (70°\text{F} - 10°\text{F}) = 129{,}600\ \text{Btu/hr}$$

This is a substantial heating requirement. If the heat exchanger could reduce this loss by 60%, the energy required from the heating system to warm the incoming air would be

$$\text{Btu/hr} = 129{,}600\ \text{Btu/hr} \times (1 - 0.6) = 51{,}840\ \text{Btu/hr}$$

Note that the heat loss with the exchanger is found in two steps. First, find the heat loss without a heat exchanger. Next, multiply that result by the exchanger's efficiency subtracted from one.

The amount of energy savings is the difference between the heat requirements with and without the heat exchanger.

$$\text{Savings} = 129{,}600\ \text{Btu/hr} - 51{,}840\ \text{Btu/hr} = 77{,}760\ \text{Btu/hr}$$

This is a very large energy savings for the "cost" of a slight increase in static air pressure caused by the heat exchanger. One year's energy savings can often pay for the purchase cost of a unit in extreme climates.

Heat exchanger energy savings are the result of the temperature change of the incoming outside air. The added heat energy (in the winter) warms the incoming air (and cools the exhaust air). From the previous example, find the temperatures of the airstreams leaving the heat exchanger. We know that each airstream is flowing at 2000 cfm, their original temperatures, and the amount of heat gained (and lost). The equation for finding the final temperature is

$$\text{Final temperature} = \text{Start temperature} + \left(\frac{\text{Btu/hr}}{(1.08 \times \text{cfm})}\right)$$

For the outside air coming in, the final temperature is

$$\text{Final temperature} = 10°\text{F} + \left(\frac{51{,}840\ \text{Btu/hr}}{(1.08 \times 2000\ \text{cfm})}\right) = 34°\text{F}$$

Exhaust air leaving the building loses heat, so its Btu per hour flow is negative. This flow's final temperature is

$$\text{Final temperature} = 70°\text{F} + \left(\frac{-51{,}840\ \text{Btu/hr}}{(1.08 \times 2000\ \text{cfm})}\right) = 46°\text{F}$$

While the incoming air still needs to be preheated before it is introduced to the space, much less heat is needed.

One concern with heat exchanger operation in very cold weather is frost accumulation. If the indoor exhaust air is cooled below its dew point, water vapor condenses out as liquid. Further cooling below 32°F, or contact between the liquid water and cold heat exchanger surfaces, allows ice to form. This can cause serious operating problems.

Frost in the heat exchanger reduces the air flow. All exhaust flow stops when the heat exchanger's internal passages are choked with ice. Proper heat exchanger control prevents frost buildup.

Some energy savings are also available with a heat exchanger in the summer. Sensible-only heat exchangers usually do not remove enough heat from the incoming airstream to justify their costs in warm climates. Most cooling loads associated with ventilation are caused by latent heat. Humidity brought in with outside air can heavily load a cooling system. Consider the following example and the benefits provided by using a sensible heat exchanger.

The same building considered previously is now being maintained at 75°F and 50% relative humidity in the summer. The 2000 cfm of outside air is at 95°F and 75% humidity. First, figure out the sensible and latent cooling loads from the outside air without a heat exchanger. The sensible heat load is

$$\text{Btu/hr} = 1.08 \times 2000\ \text{cfm} \times (95°\text{F} - 78°\text{F}) = 36{,}720\ \text{Btu/hr}$$

The total (sensible and latent) heat flow is

$$\text{Total Btu/hr} = 4.5 \times \text{cfm} \times \text{Enthalpy difference}$$

For the example, the enthalpies can be found from Table 1.3 and are

Inside air	28.0 Btu/lb
Outside air	52.8 Btu/lb

The total heat flow is

$$\begin{aligned}\text{Total Btu/hr} &= 4.5 \times 2000\ \text{cfm} \times (52.8\ \text{Btu/lb} - 28\ \text{Btu/lb})\\ &= 223{,}200\ \text{Btu/hr}\end{aligned}$$

Latent heat flow is

$$\begin{aligned}\text{Latent heat flow} &= \text{Total heat flow} - \text{Sensible heat flow}\\ &= 223{,}200\ \text{Btu/hr} - 36{,}720\ \text{Btu/hr} = 186{,}480\ \text{Btu/hr}\end{aligned}$$

If the incoming air is not cooled below its dew point (approximately 84°F), all of the latent heat remains in the air and loads the air conditioning system. If its temperature is reduced to the dew point, some water condenses in the exchanger. This represents latent heat removed from the incoming air. Also, a drain is needed on the heat exchanger to remove the condensation. The more that is removed in the heat exchanger, the less work the air conditioning system has to do.

Find the temperature of the incoming air after the 60% efficiency heat exchanger. Again, consider that the heat exchanger can only transfer sensible heat.

$$\begin{aligned}&\text{Btu/hr extracted from the incoming air}\\&= \text{Sensible heat flow} \times \text{Efficiency}\\&= 36{,}720 \text{ Btu/hr} \times 0.6 = 22{,}032 \text{ Btu/hr}\end{aligned}$$

The air's final temperature is

$$\text{Final temperature} = 95°\text{F} + \left(\frac{-22{,}032 \text{ Btu/hr}}{(1.08 \times 2000 \text{ cfm})} \right) = 84.8°\text{F}$$

This is very close to the air's dew point temperature. While some condensation might occur, the total amount would be very small. Almost all of the latent heat is delivered to the air conditioning system. Therefore, the cooling load with the heat exchanger is

$$\begin{aligned}\text{Total Btu/hr} &= (36{,}720 \text{ Btu/hr} - 22{,}032 \text{ Btu/hr}) + 186{,}480 \text{ Btu/hr}\\&= 201{,}168 \text{ Btu/hr}\end{aligned}$$

Note that the savings is small. Only 22,032 Btu/hr out of 223,200 Btu/hr is saved, so the real effectiveness was less than 10%. This can make economic justification of a sensible heat exchanger difficult in air conditioning climates.

Spot Make-Up Air

Specialized air supplies are often installed to supply air to process or combustion locations. These prevent indoor air from being consumed or exhausted, allowing the building to keep its total positive pressure and cut losses of conditioned air. Most HVAC systems use spot make-up air to save energy. Common applications for make-up air supplies include

- Kitchen range hoods
- Combustion air for boilers and furnaces
- Air for laundry dryer exhausts

- Paint booth ventilation (and other industrial exhausts)
- Any area needing supply air to offset an adjacent space's ventilation exhaust

Most spot air supplies deliver just the amount of air required to offset the exhaust, or slightly less. The relationship between the exhaust and make-up supply volume depends on the type of unit being serviced, but they rarely are used to pressurize the space.

Kitchen range hoods, for example, must not have hood-mounted make-up air systems that supply all of their exhaust requirements. Proper operation depends on them exhausting about 75% of their air from the kitchen (to remove steam and smoke from cooking appliances). Completely supplying all of the hood's air with its integral make-up source prevents it from working correctly. Additional make-up air can be supplied elsewhere in the kitchen to supply the majority of the hood's exhaust air. This won't interfere with proper exhaust system operation and smoke removal.

Some spot make-up systems can use entirely untempered air. This saves the cost of heating or cooling general ventilation air that would be used if the spot system was not available. Systems supplying combustion air and built-in range hood systems usually do not need tempering.

Untempered outside air should be kept away from people and from equipment that could be damaged or frozen by the cold air. Some buildings provide separate rooms for equipment with heavy exhaust requirements, allowing untempered make-up air to be confined to their immediate location. Commercial dryers, for example, are often installed so their chassis are in a separate space from the operator's area. The operators can be in an air conditioned or heated environment while the dryer receives untempered air at its rear inlet vents.

Supplying less than 100% of the spot exhaust requirement does not mean that the building should have a general negative pressure. The total of all spot make-up supply systems and the main HVAC system's ventilation supply should provide more air than is removed. The key is to select the proper locations for supply and exhaust points within the space.

Install spot make-up supply points as close as possible to the point of use. This prevents drafts and keeps the untempered air from migrating too far from its source. Ventilation supply and exhaust systems can confine objectionable odors and moisture close to their sources. By keeping the air pressure in specific areas lower than others, all leaks migrate from spaces outside the area to its interior.

Installing more exhaust points than ventilation supply points to an area keeps it at a lower pressure. Spaces immediately surrounding the area should receive extra ventilation supply air. This causes a general air migration from the surrounding areas in toward the space with the exhaust.

Areas that need or benefit from negative pressure (compared to their surrounding spaces) include

- Kitchens (to prevent odors from reaching diners)
- Bathrooms
- Manufacturing spaces next to office areas
- Indoor swimming pools
- Health care facilities' isolation areas

Similarly, areas that must be kept cleaner than others should be kept at a slight positive pressure compared to their surroundings. Additional air supplies to those areas, with exhausts located in surrounding spaces, makes air leak from the clean space to other areas. If the pressurizing supply air is properly filtered, very little dust will contaminate the clean space.

Providing areas with supply and exhaust ventilation to confine odors and dust depends on sufficient air velocity to prevent diffusion. If the ventilation system does not alter the areas' pressures enough to provide sufficient air velocity, little confinement occurs.

Diffusion is the process that causes odors, dust, and gases to flow away from their original release area. Stopping diffusion requires that the air's speed be faster than the diffusion speed. The speed of the airflow (in feet per minute) is determined by its cubic feet per minute divided by the number of square feet of area it flows through. Either increasing the volume or decreasing the area makes the airflow faster. Conversely, if the volume decreases or the area increases, the speed drops. An air speed of at least 50 to 100 fpm is necessary to prevent diffusion in most cases with little air "cross movement" or surrounding turbulence.

This effect causes what many people think of as ventilation failures. A system working perfectly does not appear to be able to clear the air because the velocities are too low. For example, a small bathroom fan might exhaust 50 cfm. If it runs with the bathroom door open, the air's velocity through the doorway (assuming a 6-ft, 6-in. by 2-ft, 6-in. door) is

$$\begin{aligned} \text{Velocity} &= \frac{\text{cfm}}{\text{Area}} \\ &= \frac{50\ \text{cfm}}{(6.5\ \text{ft} \times 2.5\ \text{ft})} \\ &= 3.08\ \text{fpm} \end{aligned}$$

This is too low to prevent diffusion of the room's air back into the space outside of the door.

If, however, the door is closed, all of the air will have to flow through a 1-in.-high undercut at the door's bottom. The air's speed is now

$$\text{Velocity} = \frac{50 \text{ cfm}}{(0.0833 \text{ ft} \times 2.5 \text{ ft})} = 240 \text{ fpm}$$

This is adequate to completely isolate the bathroom from the surrounding area.

If the fan is sized to provide 75 fpm with the door open, its capacity would have to be

$$\begin{aligned} \text{cfm} &= \text{Velocity} \times \text{Area} \\ &= 75 \text{ fpm} \times (6.5 \text{ ft} \times 2.5 \text{ ft}) = 1219 \text{ cfm} \end{aligned}$$

This would satisfy the requirement for a high enough air speed with the door open.

Isolating a space or zone requires the proper airflow in the right direction and possibly some physical barriers. Air speed, not just flow volume or pressure difference, provides isolation. Make sure the speed is at least 50 fpm to prevent diffusion in the wrong direction.

Kitchen and Dining Room Ventilation

The most common specialized ventilation requirement in commercial installations is kitchen and dining room ventilation. They should be considered together because they both contribute to occupant comfort and air quality, and should be designed and installed to complement each other's operation.

Air should always be caused to migrate from the dining room (and other surrounding spaces) into the kitchen. This confines cooking odors, smoke, and steam to the kitchen. By maintaining the kitchen at a lower pressure than the dining room, all airflow will be in the right direction. It is best to consider the requirements of each area separately and then see how they interact as a complete system.

Kitchen requirements

Commercial kitchens require substantial exhaust systems to remove air above cooking ranges and ovens, as well as steamy air over automatic dishwashers. These exhaust systems are often dictated by local codes that specify minimum air exhaust quantities at each piece of equipment. In addition, kitchen exhaust systems have specific fire safety requirements that directly influence the systems that are acceptable.

The exhaust hood and associated ducting must be installed per NFPA Standard 96. As a minimum, 16-gauge steel or 18-gauge stainless duct walls are needed. Velocities in the ducts must be at least 1500 fpm (and, as a general rule, below 2000 fpm). Cleanout openings (sealed when not being used) must be provided at direction changes and every 6 ft along horizontal runs. Fire protection for adjacent walls, roofs, and other building surfaces must be provided. Most kitchen exhaust fans must discharge upward to prevent flames or hot gases from impinging on the building's roof.

The exhaust system must not include fire dampers or fan shutoffs. It is the intent that the exhaust hood carry away any flames or hot gases produced by combustion. Any supply air or make-up systems, however, must be shut down in the event of a fire. A local fire suppression system must be installed at the hood, often a dry chemical or water fog type. It must be compatible with grease fires and activate the building's fire alarm system.

Range hoods (used over steam, grease, or very high-temperature equipment) must positively exhaust enough air to provide two effects:

- Remove the complete volume of contaminated air that the equipment below it produces.
- Develop enough air velocity at the equipment to confine the contaminated plume to the area immediately beneath the hood.

Range hoods should be installed so that their edges extend at least 6 in. beyond the boundaries of the equipment being served. Multiply the width by the depth of the hood's open bottom face to get its area in square feet, and multiply this area by one of the following factors:

- 50 fpm for steam or "normal" heat-producing equipment (soup kettles, dishwashers, and ovens)
- 85 fpm for grease-producing equipment (fryers and griddles)
- 150 fpm for charcoal broilers and other extremely high heat-producing equipment

For example, a hood must cover a 3-ft wide oven and fryer as well as a 24-in.-diameter soup kettle. Their installation totals 9 ft, 6 in. long by 3 ft deep. Find the size of the hood required and its cubic feet per minute. Assume that the back of the hood is located against a wall. The total size of the hood should be 10 ft, 6 in. wide and 3 ft, 6 in. deep to provide the needed 6-in. overlap along the front and sides. The air volume needed is

$$\text{Oven: } 3 \text{ ft wide} \times 3 \text{ ft deep} = 9 \text{ ft}^2 \times 50 \text{ fpm} = 450 \text{ cfm}$$

$$\text{Fryer: } 3 \text{ ft wide} \times 3 \text{ ft deep} = 9 \text{ ft}^2 \times 85 \text{ fpm} = 765 \text{ cfm}$$

$$\text{Kettle area} = \text{Diameter}^2 \times \frac{3.14}{4}$$

$$2 \text{ ft}^2 \times \frac{3.14}{4} = 1.57 \text{ ft}^2 \times 50 \text{ fpm} = 78.5 \text{ cfm}$$

$$\text{Total cfm} = 450 \text{ cfm} + 765 \text{ cfm} + 78.5 \text{ cfm} = 1293.5 \text{ cfm}$$

This exhaust volume is needed to remove the contaminated vapors alone. Additional airflow is needed to confine the vapor plumes, and this volume is found by multiplying the difference between the hood's open area and the total cooking equipment top area by 50. For our example, the total hood area is

$$10.5 \text{ ft} \times 3.5 \text{ ft} = 36.75 \text{ ft}^2$$

The cooking surface's areas are

$$(3 \text{ ft} \times 3 \text{ ft}) + (3 \text{ ft} \times 3 \text{ ft}) + 1.57 \text{ ft}^2$$
$$= 9 \text{ ft}^2 + 9 \text{ ft}^2 + 1.57 \text{ ft}^2$$
$$= 19.57 \text{ ft}^2$$

Their difference is

$$36.75 \text{ ft}^2 - 19.57 \text{ ft}^2 = 17.18 \text{ ft}^2$$

The required plume capture is

$$17.18 \text{ ft}^2 \times 50 \text{ fpm} = 859 \text{ cfm}$$

Therefore, the total exhaust requirement for the hood is

$$1293.5 + 859 \text{ cfm} = 2152.5 \text{ cfm}$$

Some local codes might require specific, higher exhaust rates. If so, any larger exhaust can be accommodated without excessive energy use by either using ventilation air supplied elsewhere or by supplying untempered outside air directly into the hood. This air should be directed upward toward the exhaust ports to prevent drafts and problems with capturing the contaminated exhaust.

Most kitchens must have their own space ventilation air, often 2 cfm per square foot. This air must be supplied to the general area and directed to help remove heat, smoke, and steam away from people's work areas. If our kitchen had a total floor area of 775 ft^2, the total supply ventilation air for the occupants would be

$$775 \text{ ft}^2 \times 2 \text{ fpm} = 1550 \text{ cfm}$$

All of this air will eventually be exhausted out of the hood. This leaves the kitchen with a net air loss of

$$2152.5 \text{ cfm exhaust} - 1550 \text{ cfm supply} = 602.5 \text{ cfm}$$

Some local codes have specific minimum airflows that must be exhausted out of a hood. If the example hood must flow at least 3000 cfm it would certainly be acceptable for the job it must do. It would, however, increase the kitchen's net air loss to

$$3000 \text{ cfm} - 1550 \text{ cfm} = 1450 \text{ cfm}$$

Before deciding on the best way to supply the make-up air (either 602.5 or 1450 cfm), it is best to consider the dining room's requirements. If all of this air is forced to infiltrate from the outside or from the dining room it will cause serious drafts, noise, and problems with doors slamming closed. If it is all supplied with a make-up hood there will be little pressure difference with the dining space to control odor diffusion.

Dining room requirements

Dining rooms must be ventilated like any other occupied space. Often, 10 to 20 cfm per person is required by codes. If the dining room adjacent to our example kitchen is designed to accommodate up to 100 persons (and 20 cfm per person is required), it should be ventilated with 2000 cfm of outside air. This ventilation air can be used to supply at least some of the kitchen's make-up air.

In order to keep the kitchen at a lower pressure than the dining room, the systems should always be set to have about 250 cfm flow from the dining room to the kitchen through their interconnecting doorways. This causes air to flow to the kitchen, but will not interfere with doors or cause excessive drafts or noise. Any more air than this should be delivered from the dining room to the kitchen via separate relief dampers or fans (equipped with required smoke or fire dampers).

Using our example exhaust hood (needing either 602.5 or 1450 cfm of air from the kitchen's space), the dining room has 2000 cfm available for it. Allowing 250 cfm to transfer through the doorways means that either 352.5 or 1200 cfm should be transferred with transfer grilles or fans. Assume that the kitchen requires only the 602.5-cfm supply (with 250 cfm arriving through the doorways and 350 cfm passing through a fire-protected grille), the dining room still has to exhaust

$$2000 \text{ cfm} - 602.5 \text{ cfm} = 1397.5 \text{ cfm of air}$$

Allowing for proper pressurization, the dining room needs an air exhaust of 1200 cfm. This leaves 10%, or 200 cfm, behind to minimize outside air infiltration. Bathroom exhausts might take care of much of this requirement if fire exit corridors are not used as plenums to deliver air to them. Figure 10.6 summarizes all of these airflows and shows how they balance.

In this case the kitchen's exhaust requirement was less than the dining room's supply. Some systems have the opposite situation: not enough dining room or kitchen ventilation air to accommodate the exhaust require-

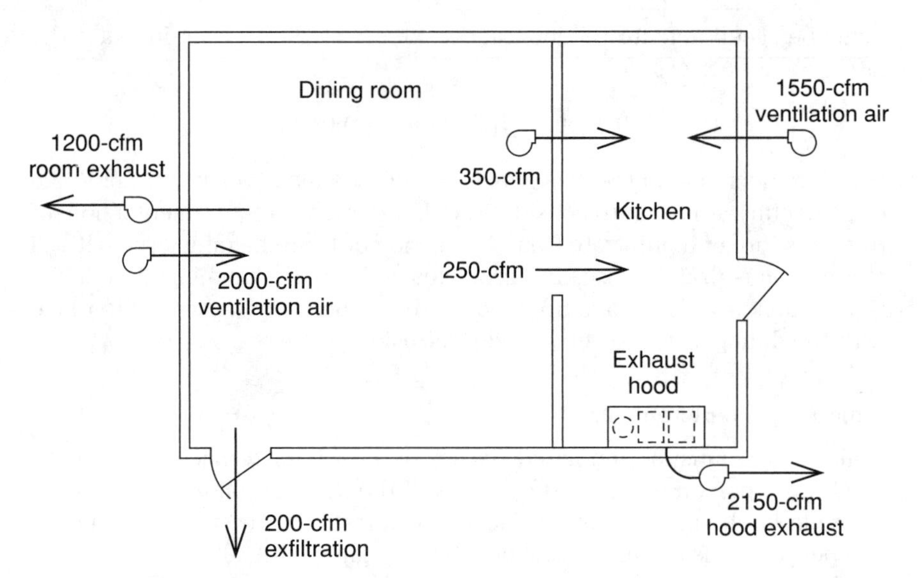

Figure 10.6. Example kitchen and dining room air balance.

ments. These cases are best handled with make-up hoods. Set the make-up volume equal to the difference between the total available from the kitchen and dining room's supplies and the total hood exhaust.

If, for example, a hood must exhaust 5000 cfm and the total ventilation air is only 2600 cfm, the net amount to be supplied by the hood's make-up system should be

$$5000 \text{ cfm} - 2600 \text{ cfm} = 2400 \text{ cfm}$$

Remember to never allow the make-up system to supply so much air that the hood's vapor or plume capture requirements are satisfied. It should only supply air to help prevent excessive negative space pressure.

Chapter

11

Humidity Control

Proper regulation of humidity is of increasing concern to building owners. Air that is too moist or dry can lead to serious problems with a building's function. Damage to the building and occupant illness can be caused by improper humidity control. One aspect of indoor air quality—the existence of bacteria, mold, and other microorganisms in the air—is often related to humidity regulation.

Dehumidification

Ironically, most humidity problems are not related to the HVAC system's operation. Humidity control problems are often caused by building construction deficiencies. The heating, ventilation, and air conditioning systems can, however, help control problems caused elsewhere.

Vapor barriers

Many facilities are built with improperly installed or nonexistent vapor barriers. These barriers are used to prevent water vapor, both inside and outside the building, from migrating through the outside walls. A vapor barrier prevents condensation at the warm-to-cold interface that exists inside the wall.

Water vapor migration is driven by the vapor pressure of the humidity in the air. It moves from areas with high vapor pressure to areas with lower pressure. Building total pressure and vapor pressure are both referenced to the outside air's pressure. It is possible for a building to have a higher total pressure than the outside air while it has a lower vapor pressure. Air leakage tends to move from inside to outside, but water vapor moves the other way. Therefore, ventilation systems that maintain a positive interior pressure might not keep humidity out.

Air's vapor pressure is determined by its total (or absolute) humidity level. The more water contained in 1 lb of air, the higher its absolute humidity and the higher the air's vapor pressure. This is different from relative humidity. Absolute humidity is measured in *grains* of moisture per pound of air. There are 6985 grains/lb, so absolute humidity is a measure of the ratio of the weight of water in the air compared to the air's weight. Relative humidity is the ratio of the amount of water vapor contained in air compared to the maximum amount it can hold at its temperature.

Absolute humidity does not vary with changes in sensible heat. Simply warming or cooling a parcel of air (as long as evaporation or condensation does not occur) changes only its temperature. The mass of moisture in the air does not change. In actual practice, however, air temperature can give some idea of its vapor pressure. Most of the time, other than in cool, very damp conditions or hot, dry ones, the air's vapor pressure changes directly with its temperature.

Cooler air cannot hold as much moisture as warmer air. Water vapor at cooler air temperatures is often at a lower pressure than at higher air temperatures. This is why a glass filled with a cold beverage "sweats" in a warm room. The water vapor in the warm air touching the glass' surface cools to the point of saturation. Vapor pressure just above the surface of the cold glass is lower than elsewhere in the room, so water vapor migrates there. It condenses out on the glass and drops off as a liquid. Given enough time, the cold surface will, because of its lower vapor pressure, reduce the humidity in the entire room.

This relationship of vapor pressure to temperature produced the often violated rule for vapor barrier installation: Always install the vapor barrier on the outside walls' predominantly warm side. In northern climates with cold winters this is the inside surface. Southern climates with long, humid summer weather should have the vapor barrier installed on the outside surface. Installing the barrier on the correct side prevents water vapor from migrating into the wall. The usually correct presumption is that the warmer side has the higher vapor pressure. Installing a vapor barrier on the wrong side of the wall traps moisture inside. Unable to leave, it causes damage.

Water vapor traveling through a wall from the warm to the cool side can condense in the internal insulation or on structural components. This liquid water can break down and corrode many building materials and provide a breeding ground for mold and bacteria.

It is very difficult to keep the internal areas of outside walls dry if improperly installed vapor barriers admit or trap moisture within. HVAC systems can only help to control the indoor environment. While they can indirectly reduce some internal moisture buildup, they are usually no substitute for proper construction.

An example of a building's poor construction, and the blame placed on the HVAC system can be found in a school recently renovated in Florida. Part of the renovation included installing a new energy-efficient cooling system with greater capacity to allow for future expansion. The original system used a central forced air system with hydronic reheat coils installed in ducts leading to each zone.

The air conditioner's central cooling compressor operated whenever any zone needed cooling. This was almost all the time. Each zone regulated its own temperature by operating its reheat coil. When a zone needed more cooling, water flow to the reheat coil was shut off and cold air was delivered to the space. Once the thermostat was satisfied, the reheat coil was turned back on. Delivered air temperature increased enough to prevent cooling the space. This system needed cooled air and hot water available whenever it had to run.

By requiring constant, simultaneous operation of the heating and cooling units for most of the year, the system used a lot of energy. Reheat systems are no longer permitted under Florida's energy conservation codes, so the new system could not include them. Also, significant operating savings were expected with the installation of a nonreheat system.

What was not known, however, was that the old system was doing an excellent job of dehumidifying the air. The constant operation of the air conditioning system always kept the humidity in the building low. There were never any obvious problems with outside wall moisture while the system was running.

After the new system was installed, inherent problems with the building's construction became obvious. The new, larger capacity air conditioning system did not need to run much of the time because it could quickly satisfy the zone thermostats' cooling requirements. The building had no mechanical cooling running for much of the day, and internal humidity levels rose. Without reheat coils, the boiler's warm-weather operation was not needed and it was shut down.

Shortly after the new system was put into operation, mold and mildew began to grow on the inside surfaces of the outer walls. It would often start under vinyl wall coverings that trapped moisture underneath. Gypsum board walls, ceilings, and other porous surfaces became constantly wet from humidity migrating in from the outside. When the warm, moist outside air cooled enough, the vapor condensed and left its water in the walls and ceiling.

This moisture migration had always been taking place, but the old cooling system had been removing it with constant air drying. Now that the system operated for only short periods of time, the building air's humidity increased too far to keep the perimeter surfaces dry.

Health problems and sickness increased, too. Local and state health officials were called in to investigate. Although the real problem of moisture migration was always present (and caused by the lack of an exterior vapor barrier), most persons and the press blamed the new HVAC system. Although it might have been installed "too large" for the job and was not running with capacity controls, the root problem was the building itself. This was not generally recognized, however, and the HVAC system's designer and contractor were blamed for installing a defective cooling setup.

There are many other incidents where migrating moisture is blamed on a malfunctioning HVAC system. When the contractor gets the system working properly and the problem remains, the owner often assumes he called the wrong contractor. Be aware of the causes of interior moisture, mold, and mildew problems. If the air conditioning and ventilation systems are working properly, explain to the building owner that corrective architectural work is called for.

Note that vinyl wall coverings are often installed onto previously painted walls during a building "face lift." This material is an excellent vapor barrier and can trap moisture inside walls. If this material is used in southern climates where hot, humid weather is common, it causes problems in buildings without an exterior vapor barrier.

Some general contractors and architects specify foil-backed gypsum board as an interior vapor barrier. This usually does not work. There is no way to properly seal each seam between the gypsum boards to maintain a continuous vapor barrier throughout the wall system. Many other errors are made with vapor barriers too. While the description of proper barrier techniques are beyond the scope of this work (or the scope of work for HVAC technicians), it must be understood. Mechanical systems are often blamed for building construction failures.

Humidification

Too little humidity in a building can cause problems too. These are usually more subtle, relating to personal comfort and correct operation of office equipment and paper handling. Most "too dry" problems occur in well-ventilated buildings when the heating system is running. As noted in Chapter 1, cold, humid outside air becomes very dry once it is heated to indoor comfort temperatures. Increased ventilation makes the indoor humidity lower. Conversely, buildings with very low fresh air ventilation rates can become too moist in the winter. Windows fog and condensation forms inside cold perimeter walls. These problems are the exception, however. Most heated buildings are too dry during cold weather.

The relative humidity of air decreases as it is warmed. Air at 10°F and 70% relative humidity dries to 7% relative humidity when it is heated to 70°F. This is extremely dry air. It draws moisture out of any available source, including people, paper, wood, and fabrics. This *parasitic humidification* increases the air's humidity while drying everything in the space that contains moisture. This drying of building contents can cause problems with static electricity (often critical around computers), paper handling (important with copiers and printers), and wood products. It also adds some moisture to the air, increasing its relative humidity.

Any increase in the air's humidity uses energy. Adding water vapor increases the amount of latent heat in the air. In order for the total energy in the air to remain constant, its sensible heat decreases the same amount as the latent heat increases. This causes its temperature to fall. The cooler, humidified air increases the load on the building's heating system. This increased load occurs from both parasitic humidification and any deliberate moisture additions. For example, many buildings without humidification systems often have relative humidities of 20% to 25% during very cold weather. The increase of the outside air's humidity from 7% to the higher level takes place entirely from parasitic moisture additions.

The amount of heating system energy needed to compensate for humidification is found from its enthalpy (total heat, Btu/lb) increase. For example, air at 10°F and 50% relative humidity has a total heat content of 3.1 Btu/lb. If it is heated without any humidification, its enthalpy rises to approximately 18 Btu/lb. Sensible heat energy added by the heating system is

$$18 \text{ Btu/lb} - 3.1 \text{ Btu/lb} = 14.9 \text{ Btu/lb}$$

If the building contents humidify this air to approximately 20% relative humidity, its total energy is 20.1 Btu/lb. The extra energy needed to compensate for the moisture addition's cooling also must be provided by the heating system. Therefore, the total energy added by the building's heating system and its moist contents is

$$20.1 \text{ Btu/lb} - 3.1 \text{ Btu/lb} = 17 \text{ Btu/lb}$$

The additional amount of energy needed to compensate for parasitic humidification is about 2.1 Btu/lb of air. This increases the total amount of energy supplied by the heating system by 15% over the energy delivered without any humidification.

Note that none of the moisture was added "intentionally." All of the moisture additions that occurred did so without deliberate humidification. Water given up from people and interior objects required the same amount of heating system energy to warm the space as water deliberately added with a humidifier. Extra moisture added with a humidifier increases the amount of energy needed, but not significantly. The additional energy needed with intentional humidification is the difference between that used by the parasitic humidification and that required for the desired humidity level.

Humidity levels below approximately 45% can increase static electricity problems. These can be critical in areas with computers or in electronic component fabrication areas. Static charges can build up to several thousand volts and quickly destroy delicate equipment and parts. Relative humidities lower than 30% can cause enough static electric charges to accumulate to annoy building occupants.

Ideal humidity levels for interior spaces vary with the individual application. Most office spaces should be humidified to 30% to 40% relative humidity. Computer rooms, paper handling areas, museums, and libraries can use humidities up to 50%. Values lower than these can increase personal discomfort, paper brittleness, and static electricity.

Continuing with our example of ventilating with 10°F, 70% relative humidity outside air: The total heat of the air after it is warmed and moistened can be found in Table 1.3. The values important for indoor humidification are

- 70°F, 30% relative humidity: 21.9 Btu/lb, 1.63 Btu/ft^3
- 70°F, 40% relative humidity: 23.5 Btu/lb, 1.74 Btu/ft^3
- 70°F, 50% relative humidity: 25.3 Btu/lb, 1.87 Btu/ft^3

These enthalpies are not much higher than the 20.1 Btu/lb the air contains when it is parasitically moistened to 20% relative humidity. In fact, the extra 1.8 Btu/lb of energy needed to humidify the air to 30% relative humidity represents only 9% more than providing parasitically humidified air. There are very significant improvements in occupant comfort, however. Instead of their skin and nasal passages having to supply moisture, they will be "moisture neutral" to the surrounding air. They won't be dried-out from the space's air.

The amount of humidity a space can have in the winter is limited by the outside temperature. Cold outside surfaces, particularly windows, allow water vapor to condense if the interior humidity is too high. Assuming the building has a correctly installed vapor barrier, window sweating is the largest problem caused by overhumidification. If the temperature of the inside surface of the windows becomes lower than the air's dew point, water forms on the glass' surface. When this occurs, the space has reached its maximum interior humidity.

Attempting to increase the humidity while the windows are fogging will fail. The glass forms a very efficient dehumidification system, with more water condensing as the interior air's vapor pressure increases. Condensation on the glass can cause problems. Wooden window frames and the surrounding walls rot and the water infiltrates into the building structure.

If condensation begins to form above ceilings and inside perimeter walls the damage can be severe. Window conditions should help decide the proper interior humidities at very low outside temperatures. Humidification must be reduced when windows begin to sweat.

With an inside humidity of 30%, double-glazed windows can usually remain free of condensation at outside temperatures down to less than –10°F. At this humidity level single-glazed windows usually don't sweat until the outside temperature is below 20°F. If the windows are heated, as happens with radiant heating systems, the humidity levels can be higher (or the outside temperatures lower) before condensation starts.

Estimating system capacity

Humidifiers are rated by the amount of water they can add to air in a given time. Most are rated in pounds per hour, gallons per hour, or gallons per day. Small residential units are often rated in pints per day. Enough water vapor must be added to a space to increase its humidity to the desired level.

Figuring out the required humidifier capacity is not difficult. First, the humidity of the outside air at design conditions should be estimated. Table 11.1 gives the amounts of water contained in air at various winter temperatures. Select the reading closest to the expected lowest temperature that will be encountered while humidifying the space.

As noted previously, interior humidity should be allowed to drop during extremely cold weather. If very cold weather is expected, don't use the heating system design temperature (often the record cold reading for the area) as the lowest outdoor humidification temperature. Instead, select the lowest temperature to be used before reducing the interior humidity.

TABLE 11.1 Water Content of Cold Air at 60% Relative Humidity

Temperature (°F)	Humidity ratio (lb of water/ lb of air)	Humidity ratio (lb of water/ ft^3 of air)
–30	0.000088	0.0000081
–25	0.000118	0.0000108
–20	0.000158	0.0000143
–15	0.000210	0.0000187
–10	0.000276	0.0000244
–5	0.000362	0.0000316
0	0.000472	0.0000408
5	0.000612	0.0000523
10	0.000789	0.0000667
15	0.001012	0.0000845
20	0.001292	0.0001068
25	0.001640	0.0001340
30	0.002073	0.0001675
35	0.002566	0.0002050
40	0.003008	0.0002378

Next, estimate the amount of outside air ventilation or infiltration to be humidified. Use either the design ventilation volume or the infiltration estimates used during the heating system capacity estimates (given in Chapter 3).

Finally, apply the design interior humidity needed and estimate the rate of required humidifier water addition. Use the following example to see how these calculations are done.

Assume a building is to be kept at 40% winter humidity and 70°F when the outdoor temperature drops to 0°F. Double-glazed windows can remain condensation free when the outside temperature is this cold. The building is 75 ft wide, 60 ft deep, and 25 ft high. It is expected to have 1.5 infiltration and ventilation air changes per hour during the heating season. Find the required humidifier capacity in pounds of water released per hour. Also, convert this capacity to gallons of water per day.

First, note from Table 11.1 that the humidity ratio of 0°F air is 0.0000408 lb water per cubic foot of air. This moisture is brought into the space with the outside air. Next, estimate the cubic feet per minute expected from air infiltration. The building's volume is

$$75 \text{ ft} \times 60 \text{ ft} \times 25 \text{ ft} = 112{,}500 \text{ ft}^3$$

The volume of air infiltration is found with the formula from Chapter 2:

$$\text{cfm} = \frac{\text{ACH} \times \text{Space volume}}{60}$$

For this example, the formula gives an outside air infiltration volume of

$$\text{cfm} = \frac{1.5 \text{ ACH} \times 112{,}500 \text{ ft}^3}{60} = 2813 \text{ cfm of outside air}$$

Table 1.3 shows the humidity ratios of heated air at common humidification targets. From this table, note that the inside air at 70°F and 40% relative humidity has a humidity ratio of 0.000470 lb water per cubic foot of air. This is the final amount of water that the interior air should have.

The total amount of water required to be added by the humidifier is found by subtracting the humidity ratio of the entering air from the air's target humidity ratio. This difference is then multiplied by the outside air infiltration volume.

$$\begin{aligned} \text{lb water/min} &= (0.000470 \text{ lb/ft}^3 - 0.0000408 \text{ lb/ft}^3) \times 2813 \text{ cfm} \\ &= 1.21 \text{ lb water/min} \end{aligned}$$

Multiplying this answer by 60 gives the required humidifier delivery in pounds per hour:

$$\text{lb/hr} = 60 \times 1.21 \text{ lb/min} = 72.6 \text{ lb/hr}$$

To convert the delivery to gallons per day, multiply the pounds per hour delivery by 24 (because there are 24 hours in 1 day) and divide by 8.3 lb/gal.

$$\text{gal/day} = 72.6\ \text{lb/hr} \times \frac{24\ \text{hr/day}}{8.3\ \text{lb/gal}} = 210\ \text{gal/day}$$

Dehumidifiers

Most comfort HVAC systems do not require special dehumidification systems. The air conditioning system, operating properly (with long on times) adequately controls interior relative humidity. Sometimes, however, additional moisture removal might be needed. This might be necessary in areas where very dry air is required for medical or manufacturing processes, or if the comfort cooling system is not matched to the application.

Air conditioning systems that don't control inside humidity leave the space feeling cool and damp. Water vapor from the outside infiltrates the space while the refrigeration compressor is off. Moisture enters with ventilation air and leakage around the building's perimeter. If the air conditioner does not run almost constantly, interior relative humidity levels rise.

This problem can be eliminated with a dehumidifier. It can run as required, removing moisture when the cooling system is off. Humidity levels decrease without having to overcool the space.

Operating costs won't significantly increase, either. The dehumidifier reduces the load on the cooling coils, allowing that system to run for even less time. Note that this combination still might not be entirely satisfactory for many occupants. The quick cool down that an oversized air conditioning system provides can give the sensation of a drafty environment. Even if air movement is very low, the rapid temperature drop might make people feel like cool air is blowing on them.

Cooling systems that are too small for their application can also benefit from the use of a dehumidifier. If the main system runs constantly to remove humidity and cool the air but cannot keep up, it can be given a boost with extra air drying. This is especially true with cooling systems using chilled water. These often operate with water temperatures around 40°F and leave the air with a 45°F dew point. Dehumidifiers often dry the air to a dew point of 30°F to 35°F.

With the dehumidifier removing large amounts of latent heat from the air, the load on the cooling system is reduced. All it must provide is sensible cooling with no dehumidification. This can often save over 50% of the load on the cooling system, making it adequate for the space.

Dehumidifiers cannot, however, compensate for a poorly constructed building. A badly or incorrectly installed vapor barrier can admit much more water than most dehumidification systems can remove. Drying the air in a building usually won't dry out moisture hidden within outside walls, ceilings, or floors. This water remains in place where it can damage the structure.

Dehumidifiers fall into two general types: refrigerant and desiccant systems. They operate in different ways.

Refrigerant systems

Refrigerant system dehumidifiers work by cooling the air below its dew point and reheating it back to the interior temperature. Water vapor is removed at the dehumidifier's cooling coils and disposed of. The cool, moist air is warmed back to its previous temperature (or slightly higher) and distributed into the space. See Figure 11.1 for a typical installation.

Most refrigerant dehumidifiers cool the air with a direct expansion coil and reheat it with the unit's condenser coil. Air passing through a refrigerant dehumidifier (with condenser reheat) ends up at a higher temperature than when it entered the unit. Energy consumed by the dehumidifier is added to the air as sensible heat. Nevertheless, the total amount of heat in the air after it runs through the unit is less than it started with. For every 10 Btu of latent heat removed by a refrigerant dehumidifier only about 3 Btu of sensible heat is added. This slight reheating can help to increase the feeling of dryness in the space. Because the relative humidity of air decreases as it is warmed, the slight heating from a refrigerant dehumidifier helps dry it.

Refrigerant dehumidifiers are often used with comfort air conditioning systems. The net reduction in the air's total heat after it passes through the dehumidifier reduces the load on the building's cooling system. Relieved of the need to remove latent heat from the air, it is used for sensible cooling (temperature reduction) only.

Desiccant systems

Moisture can be removed from air through chemical action. Some materials, such as solid silica gel or zeolite (called *water sieves*), have a great affinity for water vapor. They readily adsorb water vapor on contact, and release it when heated. The term *adsorb* technically refers to the fact that water vapor is deposited onto the material's surface. Something that *absorbs* water, on the other hand, soaks it into its interior.

Few comfort air conditioning systems use liquid desiccant dehumidifiers. These systems rely on the affinity of certain compounds and water vapor. Highly concentrated solutions of lithium bromide salt or glycol compounds absorb moisture out of the air. The solution is sprayed down through a tower while the air to be dried passes upward to mingle with the desiccant droplets. The air is dried and the desiccant picks up the humidity.

Wetted desiccant is picked up at the bottom of the tower and pumped to a regeneration heater. This heats the desiccant and drives out the accumulated water. The water vapor removed from the desiccant is expelled outside and the dried desiccant returns to the tower.

Most desiccant dehumidifiers used for HVAC systems are solid types. A bed of desiccant granules, or a structure impregnated with desiccant coating, is exposed to the airstream. Moisture from the air is adsorbed onto the desiccant and accumulates on its surface. After they have been wetted, these desiccant materials are heat regenerated to remove their accumulated moisture, allowing them to absorb more.

Some HVAC solid desiccant dryers use a bed of desiccant-coated beads (see Figure 11.2). Air is forced through the beads where it loses its moisture onto their surfaces. Beads closest to the incoming air are the first to adsorb water. Those deeper in the bed stay dry until the first ones become saturated.

Once most of the beads are wetted, the unit must be regenerated. The HVAC air flow through the unit is shut off or bypassed and high-temperature regeneration air is driven through the bed. The hot, moist air from the bed is discharged outside to remove the water from the building. After the desiccant has dried, the HVAC airflow is restored and air drying continues.

Because of the need to interrupt the air flow through the desiccant, most bead bed driers use two beds. While one is drying the ventilation air the other bed is regenerating. Dampers direct ventilation air and regenerating air alternately to each bed as required. An automatic timer switches the airflow from one bed to the other to maintain full-time operation.

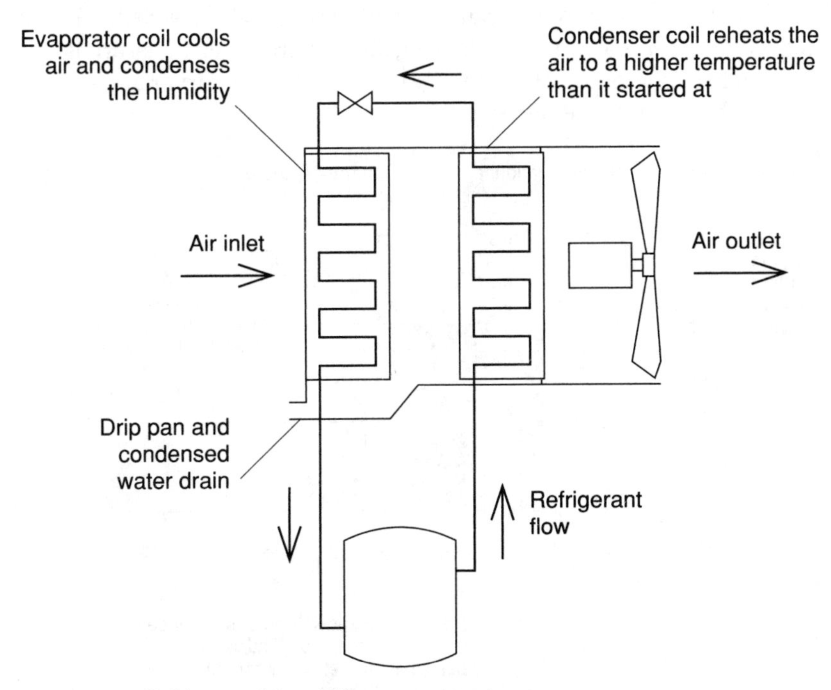

Figure 11.1. Refrigerant dehumidifier arrangement.

Most HVAC desiccant air driers use rotary wheels (Figure 11.3). These are built much like the rotary wheel heat exchangers described in Chapter 10. Desiccant compound is coated onto the front and back surfaces of a wide ribbon of aluminum or steel. This is rolled into a disk shape with sufficient clearance between each "layer" of the coated metal to allow air to fully contact the desiccant. The final shape of the assembly is a large, thick disk that air passes through from one face to the other.

Unlike heat exchangers, however, only one system airstream is used in desiccant dehumidifiers. Air enters one face of the wheel with high humidity and leaves the other face with lower humidity. During operation the wheel rotates on its axle to periodically expose the wet desiccant to a hot regeneration airstream.

Wheel rotation speed is adjusted for the HVAC airflow volume to allow optimum moisture removal. If air is allowed to pass through it too quickly, it won't dry to its design dew point. Generally, the slower the air velocity through the desiccant bed, the drier the air becomes.

Regeneration air must be supplied hot enough to reactivate the desiccant. Most systems use air at approximately 200°F to thoroughly dry all of the material. The hot, moist air leaving the regeneration section of the wheel is discharged outside.

System air leaving the desiccant dryer is hotter than when it enters. The latent heat of vaporization of the water is added to the air as sensible heat. This causes the temperature of the leaving air to increase as moisture is removed. Each Btu of latent heat (humidity) removed by the dryer appears as a Btu of sensible heat in the outlet air.

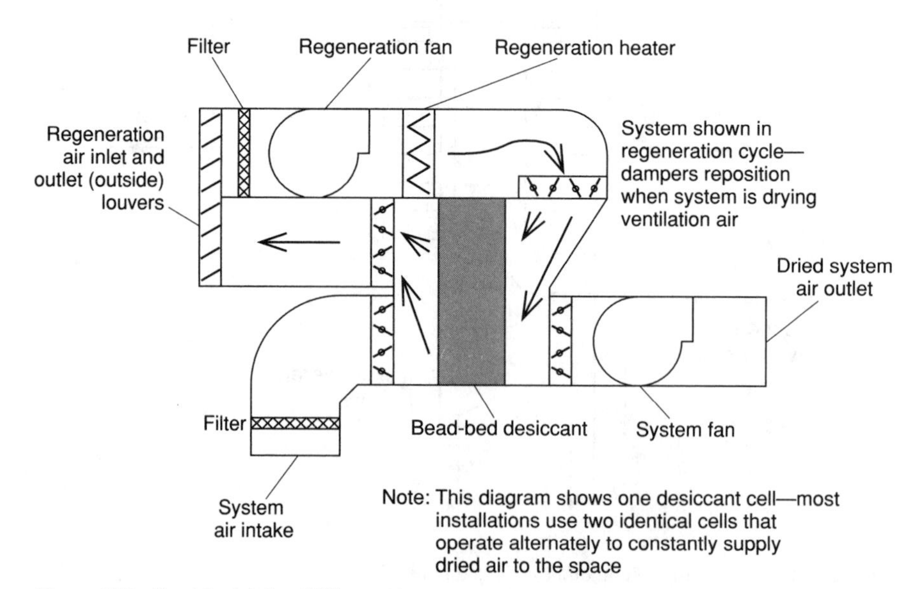

Figure 11.2. Bead-bed dehumidifier system.

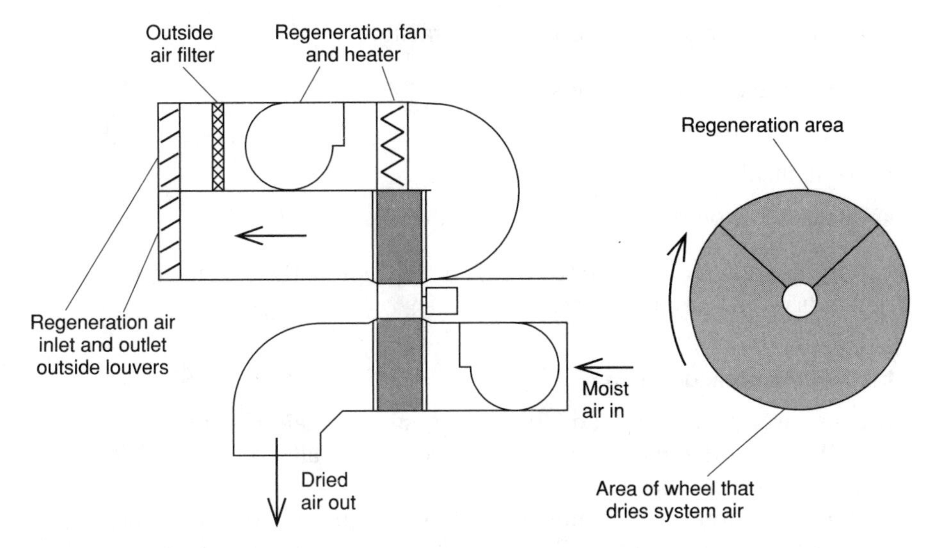

Figure 11.3. Rotary wheel air dryer.

Some additional heat is added to the outlet air from the reactivation heater's carryover and the action of the desiccant itself. The reactivation heater leaves the desiccant hotter than the incoming ventilation air. This hot desiccant sensibly heats the first process air that passes through it until its temperature falls.

Finally, the desiccant material itself also releases some heat from the chemical action of adsorption. However, most of the heating caused by desiccant air drying is caused by the release of latent heat.

Most desiccant air dryers used to supplement HVAC systems cool the discharge air with an air-to-air heat exchanger. Building ventilation exhaust (or outside air on its way to the regeneration heater) runs through one side of the heat exchanger. Simultaneously the warm, dehumidified air passes through the other side. This can significantly drop the final temperature of the air before it is supplied to the interior space.

Humidifiers

Several humidifier types are used for HVAC systems. Some include their own heat sources, while others rely on the building's heating system to warm the moistened outlet air. No one system is always better than another.

Systems that include their own heating units (or other integral heating system) include

- Direct process steam injectors
- Self-contained electric boilers
- Heated pans

Humidifiers that do not provide heat include

- Evaporative stationary pads
- Rotary drums or belts
- Mechanical atomizers
- Ultrasonic atomizers

All these add moisture to the air. The selection of the specific type to use depends on the circumstances.

Installation considerations

It is important that any humidification system remain clean during operation. Water contamination inside an HVAC system allows mold and bacteria to grow. This must be avoided.

Humidification systems must be installed to prevent the surrounding areas from becoming wetted. Close attention to installation details is necessary to prevent the unit from spoiling the interior environment with water.

Steam, boiling, and atomizing humidifiers installed in a duct system must be provided with plenty of free space downstream of the discharge. This is necessary to allow liquid water to evaporate before it can impinge on surfaces and possibly saturate them. This distance will vary with the type of humidifier and the temperature of the surrounding air.

Increasing air velocity, and presumably its turbulence, does not reduce the tendency for the humidifier to wet surrounding areas. Faster airflows carry liquid water farther downstream, increasing the chances for surface wetting. On the other hand, areas with stagnant air (often found just before or after most duct size transitions) can also encourage poor humidifier operation. Any liquid water droplets released into the flow fall out of the airstream and wet the surfaces they touch.

Placing these humidifiers too close to duct turns, coils, fan inlets, or the inlet side of high-efficiency filters can cause problems. They are best used in straight duct sections with at least 10 to 12 ft of undisturbed airflow past the discharge point. This gives any water an opportunity to evaporate before it hits a surface.

No internal duct liners should be used within 10 ft of a humidifier. This insulation can become saturated with water and carry it away from the unit. Besides allowing biological growth, it can wick water up to duct seams and let it leak out.

Also, it is best to install the humidifier into a hot air duct. Evaporation is much faster and more complete if the water is exposed to the hottest air in the system.

Another potential problem with some humidification systems is mineral and scale buildup. Any time water is evaporated, the concentration of remaining minerals and salts increases. Mineral deposits form in systems that are not regularly flushed with fresh water.

Lime scale forms an ideal growth medium for bacteria. Microscopic pores in the scale's surface give bacteria and their slime coating a place to adhere to. The small crevices in the scale provide a protected space for bacteria to reproduce. Scale buildup also clogs water systems and humidifier parts. Hard water quickly fouls any evaporative humidification system that does not regularly remove concentrated minerals.

Humidifiers can "fool" downstream temperature sensors. The vapor plume released from some humidifiers might be hotter or colder than the surrounding air. Some of these even change temperature as they travel and evaporate into the air. These air and water fog mixtures should be kept away from temperature sensors. Smoke detectors can also be tripped by a water vapor plume, and can be actuated by condensing humidity if they are cooler than the surrounding moistened air.

Allow at least 12 ft of space past a humidifier before reading the airstream's temperature or installing a smoke detector. Some systems with high air velocities might need more free space to allow the air's conditions to completely stabilize.

Steam

Facilities with clean steam available might be candidates for process steam injectors. Steam from a building heating system is injected directly into the air whenever more humidity is required. Most injectors are installed in supply air ducts. The air movement helps the condensed vapor droplets revaporize before they enter the building's space.

Steam humidifiers are designed to keep their discharge plumes dry. Process units use incoming steam to heat the vapor to be discharged. Because they reduce the steam's pressure to zero well before it is discharged, this heating removes any liquid water that might have formed. The discharge steam is slightly superheated as it enters the airstream.

Direct steam injectors can only be used when any chemical carryover that might be in the steam is certified as not harmful. Some boiler treatment chemicals, including scale controls and anticorrosives, can be irritating if they are inhaled or otherwise exposed to the skin and eyes. Direct process steam systems must never be used with boilers using such chemicals.

Some boiler treatment materials are certified by the U.S. Food and Drug Administration (FDA) for contact with food. These are supposed to be nontoxic, but their possible effects on building occupants from inhalation and eye tissue contact might be harmful. Because of this, it might be better to avoid process steam injection systems. Unless it is known that truly harmless chemicals are used (or no chemicals at all), the possibility for health problems exists.

Where suitable process steam is not available, small electric boilers are often used. Water is admitted to the boiling chamber where it is heated. Steam is released from the unit into the supply airstream. Most units used in commercial applications have automatic fresh water flushing systems to periodically remove standing water from the boiler chamber. Enough fresh water enters the unit to remove the mineral and salt deposits. If done often enough, automatic flushing can drastically reduce clogging.

Both process steam and self-contained boiler humidifiers operate at temperatures that kill bacteria and mold. As long as they are regularly used, no microorganisms can grow in the steam or boiler chambers. The discharged steam, however, can encourage these growths. The steam can condense inside the air handler, forming a cool, moist pool where bacteria and mold can thrive.

When the steam plume hits the air part of it condenses. A visible fog of fine water droplets forms and is carried into the air for at least 3 ft downstream. Once the water vaporizes, the fog disappears. As noted previously, the humidifier must be installed in a location where the liquid fog will not wet adjacent surfaces.

Heated evaporative pan

Some package air handlers use heated pans of water in the airstream to humidify the passing air. When humidity is required, an immersion heater in the water, a surface heater installed onto the pan's bottom, or a radiant electric heater above the water surface is energized. Water evaporates out of the pan and into the airstream.

Most pan humidifiers use automatic water level controls to open and close the unit's fill and drain valves. When the water sensing probe becomes dry, the fill valve opens. Once the level reaches the depth of the fill probe, the incoming flow is shut off. An overflow drain carries water away if the inlet water valve becomes stuck open.

Many units include automatic flush systems. After several fill and run cycles, both the drain and fill valves open. Fresh water runs through the pan, removing all of the concentrated minerals and other deposits.

Most pan humidifiers do not operate at temperatures high enough to discourage biological growth. Mold and bacteria can grow in the pan and on adjacent wet surfaces. These must be removed by regular, manual cleaning.

Unlike steam humidifiers, pan units do not discharge a plume of vapor under pressure into the airstream. This has two implications: a larger pan unit must be used to provide the same amount of moisture, and there is less chance for the unit to cause adjacent duct surfaces to become wet.

Stationary pad

Small-capacity humidifiers include stationary pad units. Air flows over an open, unheated pan of water holding several felt (or other absorbent material) pads. Water evaporates from these pads as the air flows over them.

These units are often used for small, residential applications. They are the simplest and least expensive humidifiers available. Because no energy is added to the water to encourage its evaporation, little humidification is provided for a given size. They also require the most work and maintenance.

Minerals in the water are left behind in the pad during the evaporation process. Available surface area for water evaporation is reduced by the minerals, and a biological growth media is provided. The pads must be regularly changed to maintain their humidification capacity and to limit mold and bacterial growth.

Most pad units include no automatic flushing system. Minerals can accumulate on any surface the water touches. Again, manual cleaning and scale removal is required to keep them clean and operating properly.

Scale preventive chemicals can be used to help stop the buildup of hard lime. These chemicals force the concentrated minerals to drop out of solution as a soft "mush," making removal easier. They do not, however, reduce the frequency of cleaning. The minerals and inevitable biological growth must still be removed frequently.

Simple evaporative pad humidifiers cannot be turned off and on. Baffles or shields can be used to divert air flow away from the pads, but these are rarely under automatic control.

Rotary drum or belt

One step above the pad humidifier is the rotary drum or belt humidifier. These use a moving, porous, wetted media pad in the airstream to transfer moisture to the air. The air flows through the pad and evaporates water carried along on its surface. These units provide much more effective humidification than stationary pads. By forcing the air to flow through the pad instead of past it, much more contact between the water and air is achieved. Many residential humidifiers use rotary pads or belts.

Most permanently installed units also include water flush and automatic fill systems. Many of these are simple overflow sump designs. A constant flow of water to the wetting pan is started whenever humidity is needed. This flow exceeds the water volume removed by evaporation, so most of it flows out the overflow drain. While much of the water is wasted, it does continuously carry away any minerals that might tend to accumulate. By carefully adjusting the water inlet flow, the amount of waste can be controlled. It should be open enough, however, to always provide a continuous overflow out of the drain.

Humidifiers without automatic overflow or flush systems must be cleaned as often as stationary pad types. The continuous evaporation of water causes minerals to accumulate. They should be cleaned regularly to prevent clogging and discourage biological growth.

Belt and drum humidifiers are started and stopped by turning the drum or belt motor on or off. With no fresh water being introduced into the airstream, no further humidification can occur.

Units installed on forced air furnaces are often used with a bypass system. A small amount of heated air is taken from the supply plenum and ducted to the humidifier. This evaporates the water as needed, and the moistened air is ducted to the return air plenum. Humidity is supplied to the entire system and the heated air helps it evaporate.

Atomizing systems

Some humidifiers use mechanically atomized water to add moisture to the air. An electric motor spins a centrifugal pickup in a pan of water, spraying the water into a fine screen. The impact of the water on the screen breaks up the water into a mist of fine droplets that are discharged into the space.

Some discharged water droplets might not be very small and can settle out to wet nearby surfaces. The size of the droplets is most influenced by the mesh of the screen used to break up the water flow. Finer meshes provide better atomization, but limit the humidity discharge.

Ultrasonic atomizers use high-frequency sound to agitate water to the point of atomization. A shallow pan of water is used with its bottom surface in direct contact with a piezoelectric crystal or speaker. Ultrasonic electric signals cause the crystal or speaker to vibrate and this energy is coupled into the water. Part of the water at the top surface is shaken away from the liquid and driven into the airstream.

Ultrasonic humidifiers atomize the water more finely than mechanical ones. Their smaller droplets evaporate more quickly in the airstream and can cause less adjacent surface wetting.

Any atomizing humidifier carries both water and dissolved minerals into the air. As the liquid water droplets disperse and evaporate, the minerals form a fine dust carried with the air. This material can fall out of suspension in slow airstreams and accumulate inside the duct system. It is best to filter the air after an atomizing humidifier to remove fine, entrained mineral dust. This can be removed with medium-efficiency air filters.

Appendix

Resources

This list of component manufacturers represents some of the specialized companies serving the HVAC industry. This list is, however, not complete. These companies have equipment able to serve most applications. Feel free to contact these manufacturers for product information or for the name, address, and phone number of your local sales representative. In most cases, they can help with specific applications and to determine the correct unit to fit a given need.

Air Conditioning and Cooling

Chillers

Air Technology Systems, Inc.
1572 Tilco Dr.
Frederick, MD 21701
(301) 620-2033

Baltimore Aircoil Co.
P.O. Box 7322
Baltimore, MD 21227
(410) 799-6200

SnyderGeneral Corp.
McQuay Commercial Products Group
13600 Industrial Park Blvd.
P.O. Box 1551
Minneapolis, MN 55440
(612) 553-5330

The Trane Company
3600 Pammel Creek Rd.
LaCrosse, WI 54601
(608) 787-2000

York International Corp.
P.O. Box 1592-36BA
York, PA 17405-1592
(717) 771-7890

Heat recovery

Governair Corp.
4811 N. Sewell Ave.
Oklahoma City, OK 73118
(405) 525-6546

Paul Mueller Co.
P.O. Box 828
Springfield, MO 65801
(800) 641-2380

SnyderGeneral Corp.
McQuay Commercial Products Group
13600 Industrial Park Blvd.
P.O. Box 1551
Minneapolis, MN 55440
(612) 553-5330

The Trane Company
3600 Pammel Creek Rd.
LaCrosse, WI 54601
(608) 787-2000

Ice storage systems

Calmac Mfg. Corp.
101 W. Sheffield Ave.
Englewood, NJ 07631
(201) 569-0420

Paul Mueller Co.
P.O. Box 828
Springfield, MO 65801
(800) 641-2380

SnyderGeneral Corp.
McQuay Commercial Products Group
13600 Industrial Park Blvd.
P.O. Box 1551
Minneapolis, MN 55440
(612) 553-5330

The Trane Company
3600 Pammel Creek Rd.
LaCrosse, WI 54601
(608) 787-2000

Package

Rooftop

Carrier Air Conditioning
Div. of Carrier Corp.
P.O. Box 4808
Syracuse, NY 13221
(315) 432-6000

Governair Corp.
4811 N. Sewell Ave.
Oklahoma City, OK 73118
(405) 525-6546

Lennox Industries Inc.
4908 Lindberg Dr.
Addison, TX 75244
(214) 233-9214

Mammoth Inc.
13120-B County Rd. 6
Minneapolis, MN 55441
(612) 559-2711

SnyderGeneral Corp.
McQuay Commercial Products Group
13600 Industrial Park Blvd.
P.O. Box 1551
Minneapolis, MN 55440
(612) 553-5330

The Trane Company
Commercial Systems Group
3600 Pammel Creek Rd.
LaCrosse, WI 54601
(608) 787-2000

PTAC

Carrier Air Conditioning
Div. of Carrier Corp.
P.O. Box 4808
Syracuse, NY 13221
(315) 432-6000

G.E. Co., G.E. Appliances
Appliance Park
Louisville, KY 40225
(800) 626-2004

*SnyderGeneral Corp.
McQuay Commercial Products Group
13600 Industrial Park Blvd.
P.O. Box 1551
Minneapolis, MN 55440
(612) 553-5330

*SnyderGeneral also has their AAF Commercial Products Group at the same address and telephone number for these products.

Penn Ventilator Co., Inc.
Red Lion & Gantry Roads
Philadelphia, PA 19115
(215) 677-1647

The Trane Company
Unitary Products Group
Guthrie Highway
P.O. Box 1008
Clarksville, TN 37040-1008
(615) 648-5951

Split systems

Central

Carrier Air Conditioning
Div. of Carrier Corp.
P.O. Box 4808
Syracuse, NY 13221
(315) 432-6000

Coolheat Inc.
P.O. Box 638
Linden, NJ 07036
(908) 925-4473

Governair Corp.
4841 N. Sewell Ave.
Oklahoma City, OK 73118
(405) 525-6546

Mammoth Inc.
13120-B County Rd. 6
Minneapolis, MN 55441
(612) 559-2711

SnyderGeneral Corp.
McQuay Commercial Products Group
13600 Industrial Park Blvd.
P.O. Box 1551
Minneapolis, MN 55440
(612) 553-5330

Modular

Carrier Air Conditioning
Div. of Carrier Corp.
P.O. Box 4808
Syracuse, NY 13221
(315) 432-6000

Coolheat Inc.
P.O. Box 638
Linden, NJ 07036
(908) 925-4473

Hitachi America Ltd.
Industrial & Air Conditioning Div.
220 White Plains Rd.
Tarrytown, NY 10591
(914) 631-0600

Intercity-Products Corp.
1368 Heil-Quaker Rd.
P.O. Box 3005
Lavergne, TN 37086-1985
(615) 793-0459

Mitsubishi Electronics America, Inc.
HVAC Division
6100 Atlanta Blvd.
Norcross, GA 30071-1305
(404) 368-4823

SnyderGeneral Corp.
McQuay Commercial Products Group
13600 Industrial Park Blvd.
P.O. Box 1551
Minneapolis, MN 55440
(612) 553-5330

Toshiba America Consumer Products Inc.
Air Conditioning Division
1010 Johnson Dr.
Buffalo Grove, IL 60089
(708) 541-9400

Components

Condensers

Air cooled

Coolheat Inc.
P.O. Box 638
Linden, NJ 07036
(908) 925-4473

Miller-Picking
Rt. 403 S.
P.O. Box 130
Johnstown, PA 15907
(814) 479-4023

SnyderGeneral Corp.
McQuay Commercial Products Group
13600 Industrial Park Blvd.
P.O. Box 1551
Minneapolis, MN 55440
(612) 553-5330

Super Radiator Co.
451 Southlake Blvd.
Richmond, VA 23236
(804) 794-2887

The Trane Company
Unitary Products Group
Guthrie Highway
P.O. Box 1008
Clarksville, TN 37040-1008
(615) 648-5951

Evaporative

Alfa Laval Thermal
5400 International Trade Dr.
Richmond, VA 23231
(804) 236-1303

Coolheat Inc.
P.O. Box 638
Linden, NJ 07036
(908) 925-4473

Mammoth Inc.
13120-B County Rd. 6
Minneapolis, MN 55441
(612) 559-2711

Filter dryers

Henry Valve Co.
3215 North Ave.
Melrose Park, IL 60160
(708) 344-1100

Parker Hannifin Corp.
Refrigeration & Air Conditioning Div.
100 Dunn Rd.
Lyons, NY 14489-9763
(315) 946-4891

Sight glasses

AC&R Components Inc.
Henry Valve Co.
701 S. Main St.
Chatham, IL 60128
(217) 983-2408

Liquid line strainers

AC&R Components Inc.
Henry Valve Co.
701 S. Main St.
Chatham, IL 60128
(217) 983-2408

Parker Hannifin Corp.
Refrigeration & Air Conditioning Div.
100 Dunn Rd.
Lyons, NY 14489-9763
(315) 946-4891

Refrigerant hot gas bypass valves

Eaton Corp.
Appliance & Specialty Controls Div.
191 E. North Ave.
Carol Stream, IL 60188
(708) 260-3145

Staefa Control System, Inc.
8515 Miraini Dr.
San Diego, CA 92126
(619) 530-1000

Airflow and duct specialties

Coils

Direct expansion

Coolheat Inc.
P.O. Box 638
Linden, NJ 07036
(908) 925-4473

SnyderGeneral Corp.
McQuay Commercial Products Group
13600 Industrial Park Blvd.
P.O. Box 1551
Minneapolis, MN 55440
(612) 553-5330

Super Radiator Co.
451 Southlake Blvd.
Richmond, VA 23236
(804) 794-2887

USA Coil & Air Inc.
P.O. Box 578
Devault, PA 19432
(800) 872-2645

Steam and water

Coil Company, Inc.
125 S. Front St.
Colwyn, PA 19023
(800) 523-7590

Heatcraft, Inc.
Heat Transfer Div.
1000 Heatcraft Dr.
P.O. Box 1457
Grenada, MS 38902-1457
(800) 225-4328

SnyderGeneral Corp.
McQuay Commercial Products Group
13600 Industrial Park Blvd.
P.O. Box 1551
Minneapolis, MN 55440
(612) 553-5330

Super Radiator Co.
451 Southlake Blvd.
Richmond, VA 23236
(804) 794-2887

USA Coil & Air Inc.
P.O. Box 578
Devault, PA 19432
(800) 872-2645

Duct sound absorbers

Certain Teed Corp.
HVAC Insulation Group
P.O. Box 860
Valley Forge, PA 19482
(610) 341-7949

Knauf Fiber Glass
240 Elizabeth St.
Shelbyville, IN 46176
(800) 825-4434

Schuller International, Inc.
P.O. Box 5108
Denver, CO 80217
(303) 978-4900

Duct hangers

Grinnell Corp.
3 Tyco Park
Exeter, NH 03833
(603) 778-9200

United McGill Corp.
Airflow Group
One Mission Park
P.O. Box 7
Groveport, OH 43125-0007
(614) 836-9981

Dampers

Fire

Greenheck
P.O. Box 410
Schofield, WI 54476-0410
(715) 359-6171

Penn Ventilator Co., Inc.
Red Lion & Gantry Roads
Philadelphia, PA 19115
(215) 677-1647

Ruskin Manufacturing
3900 Dr. Greaves Rd.
Kansas City, MO 64030
(816) 761-7476

Manual

Greenheck
P.O. Box 410
Schofield, WI 54476-0410
(715) 359-6171

Johnson Controls, Inc.
507 E. Michigan St.
P.O. Box 433
Milwaukee, WI 53201-0423
(414) 274-4000

Penn Ventilator Co., Inc.
Red Lion & Gantry Roads
Philadelphia, PA 19115
(215) 677-1647

Ruskin Manufacturing
3900 Dr. Greaves Rd.
Kansas City, MO 64030
(816) 761-7476

Motor operated

Gordon-Platt Energy Group, Inc.
Strother Airport
P.O. Box 650
Winfield, KS 67156-0850
(800) 638-6940

Greenheck
P.O. Box 410
Schofield, WI 54476-0410
(715) 359-6171

Penn Ventilator Co., Inc.
Red Lion & Gantry Roads
Philadelphia, PA 19115
(215) 677-1647

The Trane Company
Commercial Systems Group
3600 Pammel Creek Rd.
LaCrosse, WI 54601
(608) 787-2000

Outdoor air intake

Acme Engineering & Manufacturing Corp.
1820 N. York
P.O. Box 978
Muskogee, OK 74402
(918) 682-7791

Greenheck
P.O. Box 410
Schofield, WI 54476-0410
(715) 359-6171

Penn Ventilator Co., Inc.
Red Lion & Gantry Roads
Philadelphia, PA 19115
(215) 677-1647

The Trane Company
Commercial Systems Group
3600 Pammel Creek Rd.
LaCrosse, WI 54601
(608) 787-2000

Smoke

Greenheck
P.O. Box 410
Schofield, WI 54476-0410
(715) 359-6171

Johnson Controls, Inc.
507 E. Michigan St.
P.O. Box 433
Milwaukee, WI 53201-0423
(414) 274-4000

Penn Ventilator Co., Inc.
Red Lion & Gantry Roads
Philadelphia, PA 19115
(215) 677-1647

Ruskin Manufacturing
3900 Dr. Greaves Rd.
Kansas City, MO 64030
(816) 761-7476

Diffusers

* Acutherm
1766 Sabre St.
Hayward, CA 94545
(510) 785-0510

*This company specializes in automatic temperature control diffusers.

Hart & Cooley Inc.
500 E. 8th St.
Holland, MI 49423
(618) 392-7855

Krueger Division of AirSystem Components
P.O. Box 5486
Tucson, AZ 87503-9990
(602) 622-7601

The Trane Company
Commercial Systems Group
3600 Pammel Creek Rd.
LaCrosse, WI 54601
(608) 787-2000

Filters

High efficiency

Airguard Industries, Inc.
P.O. Box 32578
Louisville, KY 40232-2578
(505) 969-2304

Honeywell, Inc.
Commercial Buildings Group
Honeywell Plaza
Minneapolis, MN 55408
(612) 870-2747

SnyderGeneral Corp.
AAF Commercial Products Group
13600 Industrial Park Blvd.
P.O. Box 1551
Minneapolis, MN 55440
(612) 553-5330

HEPA

The Birdwell Co.
3708 Greenhouse Rd.
Houston, TX 77084-5512
(800) 237-2095

Mars Sales Co., Inc.
14716 S. Broadway St.
Gardena, CA 90248
(800) 421-1266

Purolator Products Air Filtration Co.
Commercial/Industrial Division
207 Johnston Pkwy.
P.O. Box 940
Kenly, NC 27542
(800) 843-0116

Throwaway

Airguard Industries, Inc.
P.O. Box 32578
Louisville, KY 40232-2578
(505) 969-2304

Fiberbond Corp.
110 Menke Rd.
Michigan City, IN 46360
(219) 879-4541

Purolator Products Air Filtration Co.
Commercial/Industrial Division
207 Johnston Pkwy.
P.O. Box 940
Kenly, NC 27542
(800) 843-0116

Louvers and registers

Acme Engineering & Manufacturing Corp.
1820 N. York
P.O. Box 978
Muskogee, OK 74402
(918) 682-7791

Greenheck
P.O. Box 410
Schofield, WI 54476-0410
(715) 359-6171

Penn Ventilator Co., Inc.
Red Lion & Gantry Roads
Philadelphia, PA 19115
(215) 677-1647

Ruskin Manufacturing
3900 Dr. Greaves Rd.
Kansas City, MO 64030
(816) 761-7476

Tuttle & Bailey
Hart & Cooley, Inc.
500 E. 8th St.
Holland, MI 49423
(616) 392-7855

Static pressure regulators

Honeywell, Inc.
Commercial Buildings Group
Honeywell Plaza
Minneapolis, MN 55408
(612) 870-2747

Johnson Controls, Inc.
507 E. Michigan St.
P.O. Box 433
Milwaukee, WI 53201-0423
(414) 274-4000

Turning vanes

PEPCO
50 Tannery Rd.
Building 3
Branchburg, NJ 08876
(908) 534-6111

Schuller International, Inc.
P.O. Box 5108
Denver, CO 80217
(800) 654-3103

Tuttle & Bailey
Hart & Cooley, Inc.
500 E. 8th St.
Holland, MI 49423
(616) 392-7855

Vents for pressure relief and intake air

Acme Engineering & Manufacturing Corp.
1820 N. York
P.O. Box 978
Muskogee, OK 74402
(918) 682-7791

Greenheck
P.O. Box 410
Schofield, WI 54476-0410
(715) 359-6171

Schuller International, Inc.
P.O. Box 5108
Denver, CO 80217
(800) 654-3103

Cooling Towers

Baltimore Aircoil Co.
P.O. Box 7322
Baltimore, MD 21227
(410) 799-6200

IMECO, Inc.
3820 Illinois Hwy. 26 S.
Polo, IL 61064
(815) 946-2351

The Marley Cooling Tower Co.
5800 Foxridge Dr.
Mission, KS 66202
(913) 362-1818

Water treatment

Ashland Chemical Co.
Division of Ashland Oil, Inc.
Drew Water Services
One Drew Plaza
Boonton, NJ 07005
(201) 263-7800

Calgon Corp.
Subsidiary of Merck & Co., Inc.
Route 60 at Campbells Run Rd.
P.O. Box 1346
Pittsburgh, PA 15230
(412) 777-8000

Nalco Chemical Co.
One Nalco Center
Naperville, IL 60563-1198
(708) 305-1000

Dehumidifiers

Adsorption

Rotors

U.S. Rotors Inc.
1430 Progress Way #106
Sykesville, MD 21784
(410) 795-0002

Systems

Air Technology Systems, Inc.
1572 Tilco Dr.
Frederick, MD 21701
(301) 620-2033

Kathabar Systems
P.O. Box 791
New Brunswick, NJ 08903
(908) 356-6000

Munters, DryCool
16825 I-35 N.
Selma, TX 78154
(210) 651-5018

Refrigerant

Air Technology Systems, Inc.
1572 Tilco Dr.
Frederick, MD 21701
(301) 620-2033

Desert Air Corp.
8300 W. Sleske Ct.
Milwaukee, WI 53223
(414) 357-7400

Nesbitt, Mestek, Inc.
4850 Rhawn St.
Philadelphia, PA 19136
(215) 331-5555

Ducts

Flexible

The Flexhaust Company
11 Chestnut St.
Amesbury, MA 01913
(508) 388-9700

Hart & Cooley, Inc.
500 E. 8th St.
Holland, MI 49423
(616) 392-7855

Nonmetallic

Knauf Fiber Glass
240 Elizabeth St.
Shelbyville, IN 46176
(800) 825-4434

Schuller International, Inc.
P.O. Box 5108
Denver, CO 80217
(800) 654-3103

United McGill Corp.
Airflow Group
One Mission Park
P.O. Box 7
Groveport, OH 43125-0007
(614) 836-9981

Metal round prefabricated

Quickdraft
Div. of C.A. Litzler Co.
1525 Perry Dr. S.W.
P.O. Box 80659
Canton, OH 44708
(216) 477-4574

United McGill Corp.
Airflow Group
One Mission Park
P.O. Box 7
Groveport, OH 43125-0007
(614) 836-9981

Fans and Blowers

Axial, centrifugal, and propeller fans

BarryBlower
Div. of SnyderGeneral Corp.
13600 Industrial Park Blvd.
P.O. Box 1551
Minneapolis, MN 55440
(612) 553-5330

Buffalo Forge Co.
490 Broadway
Buffalo, NY 14204
(716) 847-5121

Greenheck
P.O. Box 410
Schofield, WI 54476-0410
(715) 359-6171

Hartzell Fan, Inc.
910 S. Downing St.
P.O. Box 919
Piqua, OH 45356-0919
(513) 773-7411

New York Blower Co.
7660 Quincy St.
Willowbrook, IL 60521
(708) 655-4881

Penn Ventilator Co., Inc.
Red Lion & Gantry Roads
Philadelphia, PA 19115
(215) 677-1647

The Trane Company
Commercial Systems Group
3600 Pammel Creek Rd.
LaCrosse, WI 54601
(608) 787-2000

Replacement centrifugal fan wheels

Continental Fan Mfg. Co.
2296 Kenmore Ave.
Buffalo, NY 14207
(716) 842-0670

Loren Cook Co.
2015 E. Dale St.
P.O. Box 4047
Springfield, MO 65808
(417) 689-6474

Greenheck
P.O. Box 410
Schofield, WI 54476-0410
(715) 359-6171

New York Blower Co.
7660 Quincy St.
Willowbrook, IL 60521
(708) 655-4881

Heat Exchangers

Air

Coil runaround

Coolheat Inc.
P.O. Box 638
Linden, NJ 07036
(908) 925-4473

SnyderGeneral Corp.
McQuay Commercial Products Group
13600 Industrial Park Blvd.
P.O. Box 1551
Minneapolis, MN 55440
(612) 553-5330

The Trane Company
Commercial Systems Group
3600 Pammel Creek Rd.
LaCrosse, WI 54601
(608) 787-2000

Heat pipe and plate and frame

Air Enterprises
735 Glaser Pkwy.
Akron, OH 44306
(216) 794-9770

Des Champs Laboratories, Inc.
P.O. Box 220
Natural Bridge Station, VA 24579
(703) 291-1111

Governair Corp.
4841 N. Sewell Ave.
Oklahoma City, OK 73118
(405) 525-6546

Rotary

Conserv*A*Therm Corp.
460 Hillside Ave.
P.O. Box 698
Hillside, NJ 07205
(908) 688-0304

Governair Corp.
4841 N. Sewell Ave.
Oklahoma City, OK 73118
(405) 525-6546

Water

Plate and frame

Alfa Laval Thermal
5400 International Trade Dr.
Richmond, VA 23231
(804) 236-1303

Baltimore Aircoil Co.
P.O. Box 7322
Baltimore, MD 21227
(410) 799-6200

ITT Fluid Handling
8200 N. Austin Ave.
Morton Grove, IL 60053
(708) 966-3700

Paul Mueller Co.
P.O. Box 828
Springfield, MO 65801
(800) 641-2830

Shell and tube

Bryan Steam Corp.
P.O. Box 27
Peru, IN 46970
(317) 473-6651

Coolheat Inc.
P.O. Box 638
Linden, NJ 07036
(908) 925-4473

ITT Bell & Gossett
Unit of ITT Fluid Technology Corp.
8200 N. Austin Ave.
Morton Grove, IL 60053
(708) 966-3700

Tube and tube

Coolheat Inc.
P.O. Box 638
Linden, NJ 07036
(908) 925-4473

Paul Mueller Co.
P.O. Box 828
Springfield, MO 65801
(800) 641-2830

Heat Pumps

Air to air

Carrier Air Conditioning
Div. of Carrier Corp.
P.O. Box 4808
Syracuse, NY 13221
(315) 432-6000

Friedrich Air Conditioning Co.
4200 N. Pan Am Expy.
San Antonio, TX 78219
(210) 225-2000

G.E. Co., G.E. Appliances
Appliance Park
Louisville, KY 40225
(800) 626-2004

Governair Corp.
4841 N. Sewell Ave.
Oklahoma City, OK 73118
(405) 525-6546

The Trane Company
Commercial Systems Group
3600 Pammel Creek Rd.
LaCrosse, WI 54601
(608) 787-2000

Water source

Carrier Air Conditioning
Div. of Carrier Corp.
P.O. Box 4808
Syracuse, NY 13221
(315) 432-6000

SnyderGeneral Corp.
McQuay Commercial Products Group
13600 Industrial Park Blvd.
P.O. Box 1551
Minneapolis, MN 55440
(612) 553-5330

The Trane Company
Commercial Systems Group
3600 Pammel Creek Rd.
LaCrosse, WI 54601
(608) 787-2000

Heat Recovery Units

Refrigerant to water

Paul Mueller Co.
P.O. Box 828
Springfield, MO 65801
(800) 641-2830

The Trane Company
Commercial Systems Group
3600 Pammel Creek Rd.
LaCrosse, WI 54601
(608) 787-2000

USA Coil & Air Inc.
P.O. Box 578
Devault, PA 19432
(800) 872-2645

Humidifiers

Atomizing

Armstrong International, Inc.
816 Maple St.
Three Rivers, MI 49093
(616) 273-1415

Ellis & Watts
Division of Dynamics Corp. of America
4400 Willow Lake Ln.
Batavia, OH 45103
(513) 752-9000

Nortec Industries, Inc.
P.O. Box 698
Ogdensburg, NY 13669
(315) 425-1255

Steam

Armstrong International, Inc.
816 Maple St.
Three Rivers, MI 49093
(616) 273-1415

Nortec Industries, Inc.
P.O. Box 698
Ogdensburg, NY 13669
(315) 425-1255

Sussman Electric Boilers
Division of Sussman Automatic Corp.
43-20 34th St.
Long Island City, NY 11101
(718) 937-4500

Heating

Boilers

Electric

Bryan Steam Corp.
P.O. Box 27
Peru, IN 46970
(317) 473-6651

Electro-Steam Generator Corp.
1000 Bernard St.
Alexandria, VA 22314-1299
(800) 634-8177

Reimers Electra Steam, Inc.
4407 Martinsburg Pike
Clearbrook, VA 22624
(703) 662-3811

Williams & Davis Boilers, Inc.
P.O. Box 539
Hutchins, TX 75141
(214) 225-2365

Gas

Bryan Steam Corp.
P.O. Box 27
Peru, IN 46970
(317) 473-6651

Fulton Companies
Fulton Boiler Works, Inc.
Port & Jefferson Sts.
Pulaski, NY 13142
(315) 298-5121

Parker Boiler Co.
5930 Bandini Blvd.
Los Angeles, CA 90040
(213) 727-9800

Teledyne-Laars
20 Industrial Way
Rochester, NH 03867
(603) 335-6300

Burners

Cleaver-Brooks
P.O. Box 421
Milwaukee, WI 53201
(414) 359-0600

Gordon-Platt Energy Group
Division of Aqua Chem, Inc.
351 21st St.
Monroe, WI 53566
(608) 329-3166

Webster Engineering & Manufacturing Co., Inc.
619 Industrial Rd.
P.O. Box 748
Winfield, KS 67156
(316) 221-7464

Duct heaters

Electric

INDEECO
425 Hanley Industrial Ct.
St. Louis, MO 63144
(800) 243-8162

The Trane Company
Commercial Systems Group
3600 Pammel Creek Rd.
LaCrosse, WI 54601
(608) 787-2000

Gas

Modine Manufacturing Co.
1500 DeKoven Ave.
Racine, WI 53403
(414) 636-1200

Rupp Industries, Inc.
Temp Air Division
11550 Rupp Dr.
Burnsville, MN 55337
(612) 894-3000

The Trane Company
Commercial Systems Group
3600 Pammel Creek Rd.
LaCrosse, WI 54601
(608) 787-2000

Oil

Hart & Cooley, Inc.
500 E. 8th St.
Holland, MI 49423
(619) 392-7855

The Trane Company
Commercial Systems Group
3600 Pammel Creek Rd.
LaCrosse, WI 54601
(608) 787-2000

Electric radiant ceilings

Marley Electric Heating
470 Beauty Spot Rd. E.
Bennettsville, SC 29512
(803) 479-4121

Furnaces

Inter-City Products
1368 Heil-Quaker Rd.
P.O. Box 3005
Lavergne, TN 37086-1985
(615) 793-0459

Mammoth Inc.
13120-B County Rd. 6
Minneapolis, MN 55441
(612) 559-2711

Reznor/Thomas & Betts
1555 Lynnfield Rd., Suite 250
Memphis, TN 38119
(800) 695-1901

The Trane Company
Commercial Systems Group
3600 Pammel Creek Rd.
LaCrosse, WI 54601
(608) 787-2000

Hydronic Specialties

Air eliminators

Amtrol, Inc.
1400 Division Rd.
West Warwick, RI 02893-2300
(401) 884-6300

Armstrong Pumps, Inc.
93 East Ave.
North Tonawanda, NY 14120
(716) 693-8813

ITT Bell & Gossett
Unit of ITT Fluid Technology Corp.
8200 N. Austin Ave.
Morton Grove, IL 60053
(708) 966-3700

Spirotherm Inc.
Marketing & Sales
429 Kay Ave.
Addison, IL 60101
(708) 543-5850

Backflow preventers

Cla-Val Co.
P.O. Box 1325
Newport Beach, CA 92659-0325
(714) 548-2201

Grinnell Corp.
3 Tyco Park
Exeter, NH 03833
(603) 778-9200

Zurn Industries
Hydromechanics Division
1801 Pittsburgh Ave.
Erie, PA 16512

ITT Bell & Gossett
Unit of ITT Fluid Technology Corp.
8200 N. Austin Ave.
Morton Grove, IL 60053
(708) 966-3700

Parker Boiler Co.
5930 Bandini Blvd.
Los Angeles, CA 90040
(213) 727-9800

Sparco Inc.
65 Access Rd.
Warwick, RI 02886
(401) 738-4290

Radiation—baseboard, convection, and fin tube

Burnham Corp.
Hydronics Division
P.O. Box 3079
Lancaster, PA 17604
(717) 397-4701

Coolheat Inc.
P.O. Box 638
Linden, NJ 07036
(908) 925-4473

* Runtal North America Inc.
Runtal Radiators Div.
187 Neck Rd.
Ward Hill, MA 08135
(508) 373-1666

*This company specializes in architecturally integrated convectors and radiators.

The Trane Company
Commercial Systems Group
3600 Pammel Creek Rd.
LaCrosse, WI 54601
(608) 787-2000

Weil-McLain, a United Dominion Co.
500 Blain St.
Michigan City, IN 46360-2388
(219) 879-6561

Expansion tanks

Amtrol, Inc.
1400 Division Rd.
West Warwick, RI 02893-2200
(800) 726-6962

Armstrong Pumps, Inc.
93 East Ave.
North Tonawanda, NY 14120
(716) 693-8813

Pipe coverings

Adhesives

Armstrong World Industries, Inc.
P.O. Box 3001
Lancaster, PA 17604
(717) 396-4093

Loctite Corp.
Permatex Industrial Division
705 N. Mountain Rd.
Newington, CT 06111
(800) 641-7376

Pittsburgh Corning Corp.
800 Presque Isle Dr.
Pittsburgh, PA 15239
(412) 327-6100

Vapor barriers

IMCOA
4325 Murray Ave.
Halston City, TX 76117
(817) 485-5290

Vimasco Corp.
Plant Rd.
P.O. Box 516
Nitro, WV 25143-0516
(304) 755-3328

Expansion joints

Hyspan Precision Products, Inc.
1685 Brandywine Ave.
Chula Vista, CA 91911
(619) 421-1355

Thermo Tech Inc.
6701 Stapleton Dr. N.
Denver, CO 80216
(303) 322-0181

Plastic radiant pipe

Heatlink USA Inc.
89 54th St. S.W.
Grand Rapids, MI 49548
(616) 532-4266

PEPCO
50 Tannery Rd.
Building 3
Branchburg, NJ 08876
(908) 534-6111

Rehau Inc.
Heating & Sanitary
1501 Edwards Ferry Rd.
P.O. Box 1706
Leesburg, VA 22075
(703) 777-5255

Prefabricated radiant panels

Embassy Industries Inc.
300 Smith St.
Farmingdale, NY 1173
(516) 694-1800

Pumps

Water circulation

Armstrong Pumps, Inc.
93 East Ave.
North Tonawanda, NY 14120
(716) 693-8813

Dunham-Bush, Inc.
Dunham Division
811 E. Main St.
Marshalltown, IA 50158
(515) 752-4291

ITT Bell & Gossett
Unit of ITT Fluid Technology Corp.
8200 N. Austin Ave.
Morton Grove, IL 60053
(708) 966-3700

Weinman Pumps, a Burke Pumps, Inc. Co.
584 Commerce Rd.
P.O. Box 1364
Conway, AR 72032
(501) 329-9811

Condensation removal (from coils and pans)

Alyan Pump Co.
303 S. 69th St.
Upper Darby, PA 19082

Skidmore
Subs. of Vent-Rite Valve Corp.
1875 Dewey Ave.
P.O. Box 8583
Benton Harbor, MI 49022
(616) 925-8812

Valves

Balancing and flow control

Amtrol, Inc.
1400 Division Rd.
West Warwick, RI 02893-2300
(800) 726-6962

Armstrong Pumps, Inc.
93 East Ave.
North Tonawanda, NY 14120
(716) 693-8813

*Griswold Controls
2803 Barranca Rd.
Irvine, CA 92714
(714) 559-6000

*This company specializes in automatic hydronic flow control valves.

Landis & Gyr Powers, Inc.
1000 Deerfield Parkway
Buffalo Grove, IL 60089
(708) 215-1000

Mixing (thermostatic) for temperature regulation

Leonard Valve Co.
1360 Elmwood Ave.
Cranston, RI 02910
(401) 461-1200

Powers Process Controls
Unit of Mark Controls Corp.
3400 Oakton St.
Chicago, IL 60076
(708) 673-6700

T & S Brass & Bronze Works
2 Saddleback Cove
P.O. Box 1088
Travelers Rest, SC 29690
(803) 834-4102

Solenoid

Cla-Val Co.
P.O. Box 1325
Newport Beach, CA 92659-0325
(714) 548-2201

PEPCO
50 Tannery Rd.
Building 3
Branchburg, NJ 08876
(908) 534-6111

Two and three way

Automatic Switch Co. (ASCO)
50-60 Hanover Rd.
Florham Park, NJ 07932
(201) 966-2000

Henry Valve Co.
3215 North Ave.
Melrose Park, IL 60160
(708) 344-1100

Siebe Environmental Controls
Barber-Colman/Robertshaw Products
1354 Clifford Ave.
Loves Park, IL 61132
(815) 637-3000

Zone

*California Economizer
5731 McFadden Ave.
Huntington Beach, CA 92649
(714) 898-9963

*This company specializes in supplying complete multiple-zone systems.

Kreuter Manufacturing Co.
19476 Industrial Dr.
P.O. Box 497
New Paris, IN 46553
(219) 831-5250

Siebe Environmental Controls
Barber-Colman/Robertshaw Products
1354 Clifford Ave.
Loves Park, IL 61132
(815) 637-3000

Instruments and Equipment

Air

Flow and speed

Alnor Instrument Co.
7555 N. Linder Ave.
Skokie, IL 60077
(708) 677-3500

Dwyer Instruments Inc.
P.O. Box 373
Michigan City, IN 46360
(219) 879-8000

Shortridge Instruments, Inc.
7855 E. Redfield Rd.
Scottsdale, AZ 85260
(602) 991-6744

Pressure—filter losses

Airguard Industries, Inc.
P.O. Box 32578
Louisville, KY 40232-2578
(505) 969-2304

Dwyer Instruments Inc.
P.O. Box 373
Michigan City, IN 46360
(219) 879-8000

Psychrometers

Brooklyn Thermometer Co., Inc.
90 Verdi St.
Farmingdale, NY 11735
(516) 694-7610

Omega Engineering, Inc.
One Omega Dr.
P.O. Box 4047
Stanford, CT 06907
(203) 359-1660

Water flow

Armstrong Pumps, Inc.
93 East Ave.
North Tonawanda, NY 14120
(716) 693-8813

Carlon Meter Co., Inc.
1710 Eaton Dr.
Grand Haven, MI 49417
(800) 253-3098

Omega Engineering, Inc.
One Omega Dr.
P.O. Box 4047
Stanford, CT 06907
(203) 359-1660

Onicon, Inc.
2161 Logan St.
Clearwater, FL 34625
(813) 447-6140

Refrigeration

Charging

AES-Ntron, Inc.
Neutronics, Inc.
456 Creamery Way
Exton, PA 19341
(610) 524-8800

Penguin Refrigeration, Inc.
3407 Avenue East
Arlington, TX 76011
(800) 232-2124

Gauge sets

Dresser Industries
Instrument Division
250 E. Main St.
Stratford, CT 06497-5145
(800) 328-8258

Weskler Instruments Corp.
80 Mill Rd.
P.O. Box 808
Freeport, NY 11520-0808
(516) 623-0100

Refrigerant recovery

National Refrigerants, Inc.
11401 Roosevelt Blvd.
Philadelphia, PA 19154
(800) 262-0012

Penguin Refrigeration, Inc.
3407 Avenue East
Arlington, TX 76011
(800) 232-2124

The Trane Company
Commercial Systems Group
3600 Pammel Creek Rd.
LaCrosse, WI 54601
(608) 787-2000

Insulation

Equipment

Knauf Fiber Glass
240 Elizabeth St.
Shelbyville, IN 46176
(800) 825-4434

Molded Acoustical Products of Easton, Inc.
3 Danforth Dr.
Easton, PA 18042-8993
(610) 253-7135

Pittsburgh Corning Corp.
800 Presque Isle Dr.
Pittsburgh, PA 15239
(412) 327-6100

Thermal Ceramics
P.O. Box 923
Augusta, GA 30903
(706) 796-4200

Pittsburgh Corning Corp.
800 Presque Isle Dr.
Pittsburgh, PA 15239
(412) 327-6100

Schuller International, Inc.
P.O. Box 5108
Denver, CO 80217
(800) 654-3103

Pipe, tubing, and duct

Knauf Fiber Glass
240 Elizabeth St.
Shelbyville, IN 46176
(800) 825-4434

Appendix

B

Standards and Codes

There are many codes that influence the HVAC requirements for buildings. Almost every building in the United States is covered under a building code, usually adopted at the state level and possibly modified at the municipal level.

Institutional occupancies, particularly health care and school buildings, often have their own codes. These codes might also include sections on proper heating and cooling, or isolation of toxic substances from the occupants. For example, some codes limit the use of toxic antifreeze solutions used in hydronic systems. Other requirements commonly include special fire barriers needed where pipes, ducts, or cables penetrate fire walls. It can be very difficult to learn all of these codes prior to bidding on a job.

The best way to ascertain any special code requirements is to ask the building owner or administrator after the job has been won but before starting actual construction. They might not know of the actual mechanical requirements, but they should be able to direct you to the agency involved (for example, a department of health). Ask for the name of the inspector that usually visits the building and contact that person directly. They should be able to advise you of any special requirements before work starts.

Building Codes

Each state or locality generally has its own building codes, and which code is used can be determined by a phone call to your local building inspector's office. Although most codes have their specific clauses and caveats, they are usually based on one of three national codes. These codes are modified somewhat by state or local governments, but usually are adopted substantially intact. Feel free to contact these code agencies if you have any questions.

Contractors should obtain a copy of the building code that is used in their locality. Contact your state or municipal offices to order a copy. The underlying code writing agencies can also provide copies of their source documents. If the local inspector says that their codes are the same as one of the national codes, it might be less expensive to order it directly from that agency. The code agencies are

Building Officials and Code Administrators International, Inc. (BOCA)
4051 W. Flossmore Rd.
Country Club Hills, IL 60478-5795

BOCA publishes the *National Building Code* and the *National Mechanical Code*. These are primarily used in the northeast and north central areas of the United States.

International Conference of Building Officials (ICBO)
5360 S. Woekman Mill Rd.
Whittier, CA 90601

ICBO publishes the *Uniform Building Code* and the *Uniform Mechanical Code*. These are used in the western United States.

Southern Building Code Congress International (SBCCI)
900 Monclair Rd.
Birmingham, AL 35213-1206

SBCCI publishes the *Standard Building Code* and the *Standard Mechanical Code* used in the southern United States.

National Fire Protection Agency

The NFPA is a research agency that investigates fire safety and how it can be enhanced. The NFPA is not a code agency, but the code agencies use NFPA standards as a basis for building codes. Several NFPA standards directly impact the work of HVAC technicians and it is important to be familiar with these.

National Fire Protection Agency standards are the source of almost every HVAC code used in the United States, and they are often quoted or referenced directly. Even if particular standards are not part of the building code, installations compatible with NFPA standards demonstrate that the technician did the best job possible.

Standards are available covering all aspects of fire protection. These include industry-specific requirements that do not include HVAC work, as well as standards that deal directly with heating and cooling. The standards that most HVAC technicians should be familiar with include

31	Oil Burning Equipment
54	National Fuel Gas Code
58	LP-Gas Storage
70	National Electrical Code
70A	Dwelling Electrical Code
90A	Air Conditioning Systems
90B	Warm Air Heating and Air Conditioning
92A	Smoke Control Systems
96	Vapor Removal for Cooking Equipment
101	Life Safety Code
204M	Smoke and Heat Venting
211	Chimneys, Fireplaces and Vents
214	Water Cooling Towers
8501	Single Burner Boiler-Furnaces

Copies of each of these, and a list of other standards, are available directly from the NFPA. They can be contacted at

National Fire Protection Agency (NFPA)
1 Battery March Park
P.O. Box 9101
Quincy, MA 02269
(617) 770-3000

American Society of Heating, Refrigeration, and Air Conditioning Engineers

The ASHRAE is not a code agency. They research the science behind the HVAC field and endeavor to perfect heating and cooling system designs and installations. To this end, they publish standards describing the best methods to use. While the NFPA is generally concerned with fire safety, the ASHRAE works to improve all aspects of HVAC system design.

The ASHRAE standards are sometimes quoted in building and agency codes, especially in the areas of ventilation requirements. Heating and cooling contractors should be familiar with these standards, especially if their local codes require them. Standards most often cited by codes include

Standard 15	Describes equipment room ventilation, layout, and refrigeration system installations.
Standard 34	Categorizes refrigerants by toxicity and flammability. Describes the potential for refrigerant leaks into occupied spaces and sets limits on the amount of refrigerant allowed.
Standard 62	Specifies the amount of outside air ventilation to be supplied to occupied spaces. Many newer codes cite the 1989 edition of this standard calling for 20 cfm per occupant. Higher ventilation amounts might be required in smoking lounges and other areas that contain irritants.

Copies can be obtained directly from the ASHRAE. They can be contacted at

The American Society of Heating, Refrigeration, and Air Conditioning Engineers, Inc.
1791 Tullie Circle NE
Atlanta, GA 30329
(404) 636-8400

Appendix

C

SI (Metric) Units

All of the calculations and measures used in this book are convertible to SI units, which can be desirable when calculating heat flows directly in SI units and to convert IP (English) specifications to SI scales. Converting to a common measurement system when estimating is essential to ensure accuracy.

This appendix provides two SI conversions: Units of measure and equations. Unit conversions allow individual terms and specifications to be changed to metric, while the equations allow heat transfer and other calculations to be made directly in SI units.

Unit Conversions

The following table shows multiplication factors required to change IP units to their corresponding SI measures. Multiplying the IP unit by its factor will give a result in the SI unit shown. If it is necessary to convert an SI unit to its corresponding IP measurement, divide the SI number by the same factor. The result is the IP unit.

Unit	Factor	SI Unit
BTU	1.055	kJ
BTU/(hr × ft × deg. F)	1.731	W/(m × deg. C)
BTU × in./(hr × ft^2 × deg. F)	0.144	W/(m × deg. C)
BTU/hr	0.293	W
BTU/(hr × ft^2 × deg. F)	5.68	W/(m^2 × deg. C)
BTU/(lb × deg. F) {specific heat}	4.19	kJ/(kg × deg. C)
EER	0.293	COP

Unit	Factor	SI Unit
ft	0.3048	m
ft/min.	0.00508	m/sec
ft/sec	0.3048	m/sec
ft of water {head pressure}	2.99	kPa
ft^2	0.0929	m^2
$ft^2 \times hr \times$ deg. F/Btu {R-value}	0.176	($m^2 \times$ deg. C)/W
ft^3	28.3	L
ft^3	0.0283	m^3
ft^3/min.	0.472	L/sec
gal	3.79	L
gal	0.00379	m^3
gal/min.	0.0631	L/sec.
horsepower {motor}	0.746	kW
in. of mercury {head pressure}	3.38	kPa
in. of water {head pressure}	249	Pa
in./100 ft {thermal expansion}	0.833	mm/m
$in.^2$	645	mm^2
mile/hr	1.61	km/hr
mile/hr	0.447	m/sec
pound {weight or mass}	0.454	kg
pound/ft^2	4.88	kg/m^2
pound/ft^2	47.9	Pa
pound/ft^3 {density}	16	kg/m^3
pound/gal	120	kg/m^3
pound/$in.^2$	6.89	kPa
therm	105.5	MJ
ton {12,000 Btu/hr}	3.52	kW

Equations

Many of the IP equations used to find values for heating and cooling system estimations have corresponding SI forms. Generally, only a multiplying constant must be adjusted to allow direct estimations in SI units.

The following are most of the equations used in this book. Both the original IP and the "converted" SI versions are given. The equations are listed by the chapter in which they were introduced.

Chapter 1

The amount of heat stored in a substance is

$$\text{Heat} = \text{Mass} \times \text{Specific heat} \times \text{Temperature}$$

This basic relationship remains unchanged when all of the units are metric. The equation with all SI units is

$$\text{kJ} = \text{kg} \times \left(\frac{\text{kJ}}{(\text{kg} \times \text{deg. C})} \right) \times \text{deg. C}$$

Where the specific heat of a substance is given in IP units, Btu / (lb × deg. F), the equation becomes

$$\text{kJ} = 4.19 \times \text{kg} \times \left(\frac{\text{Btu}}{(\text{lb} \times \text{deg. F})}\right) \times \text{deg. C}$$

Power and energy are related by the equation

$$\text{Power} = \frac{\text{Energy}}{\text{Time}}$$

The SI units for this equation are

$$\text{Watts} = \frac{\text{Joules}}{\text{sec}}$$

Chapter 2

The rate of conductive heat flow is related to temperature difference and object size with the equation

$$q = k \times \left(\frac{A}{L}\right) \times T_d$$

The SI terms are

q = heat flow, watts
k = heat conductivity, watts/sec/m/deg. C
A = area through which the heat is flowing, m^2
L = length of material the heat is flowing through, m
T_d = (High temp – Low temp), deg. C

If the IP value for the conductivity (k) term is known, all of the other units can be SI if the equation is given a constant:

$$\text{W} = 1.731 \times k \left(\frac{\text{Btu}}{(\text{hr/ft/deg. F})}\right) \times \frac{\text{m}^2}{\text{m}} \times \text{deg. C}$$

Resistance to heat flow (R-value) is defined as

$$R = \frac{L}{k}$$

The "pure" SI terms of this equation are

L = length of the heat flow path or thickness of the material, m

k = Thermal conductivity, $\dfrac{\text{W}}{(\text{m} \times \text{deg. C})}$

If given an R-value in its IP units, multiply as shown to get the SI units:

$$R_{SI} = 0.176 \times R_{IP}$$

The flow of heat is related to temperature and R-value with the equation

$$\text{Heat flow} = \text{Area} \times \frac{T_d}{R_{total}}$$

Again, this equation is fine if all of the units are consistently IP or SI. If IP values for the total R are available, use the equation

$$W = 5.67 \times m^2 \times \frac{T_d}{R_{IP\ total}}$$

The IP equation for sensible heat flow associated with the movement of standard air is

$$\text{Btu/hr} = 1.08 \times \text{cfm} \times T_d$$

This can be converted to SI units as

$$W = 4.12 \times \text{L/sec} \times T_d$$

Note that all of the units are metric. No IP unit conversions are included.

The IP equation for the total heat flow caused by the movement of standard air is

$$\text{Total Btu/hr} = 4.5 \times \text{cfm} \times \text{Enthalpy difference}$$

That equation's SI equivalent is

$$W = 14 \times \text{L/sec} \times \text{Enthalpy difference}_{SI}$$

The enthalpy should be expressed as kJ/kg. If IP enthalpy values (with units of Btu/lb) are known, use the equation

$$W = 32.5 \times \text{L/sec} \times \text{Enthalpy difference}_{IP}$$

Estimating the number of air changes per hour is done with the IP equation

$$\text{ACH} = 60 \times \frac{\text{cfm}}{\text{Space volume, ft}^3}$$

The SI unit equation is

$$\text{ACH} = 3.6 \times \frac{\text{L/sec}}{\text{Space volume, m}^3}$$

Heat flow associated with water flow is given in the IP equation

$$\text{Btu/hr} = 498 \times \text{gpm} \times T_d,\ \text{deg. F}$$

The SI equation for water borne heat transfer is

$$W = 4160 \times \text{L/sec} \times T_d,\ \text{deg. F}$$

Chapter 3

Some heat flows are discussed as a function of the length of a perimeter foundation or slab and the temperature difference. The heat flow and the rest of the terms of the equation can be converted to SI units by multiplying the given factor by 1.75. For example, the heat flow through an uninsulated slab edge is

$$0.8 \text{ Btu/ft per degree F of perimeter edge}$$

The SI equation is

$$0.8 \times 1.75 = 1.4 \text{ W/m per degree C of perimeter edge}$$

Solar heat gains through windows are determined by the equation

$$\text{Btu/hr} = 125 \times \text{Area, ft}^2 \times \text{SC}$$

The SI equation is

$$\text{W} = 394 \times \text{Area, m}^2 \times \text{SC}$$

Heat gains from a motor in a fan's airstream are found by the equation

$$\text{Btu/hr} = \text{Motor hp} \times 2040$$

The SI calculation is very simple. About 80% of the motor's electrical power will enter the air as heat, so the heat load is

$$\text{W} = 800 \times \text{Motor kW}$$

Chapter 4

Velocity pressure (as inches of water) is related to air speed (fpm) by

$$\text{pressure} = \left(\frac{\text{Speed}}{4005}\right)^2$$

The SI relationship is

$$\text{kPa} = 103 \times \text{m/sec}^2$$

Air speed is related to volume and duct size by

$$\text{ft/min} = 144 \times \frac{\text{cfm}}{\text{Area, in.}^2}$$

SI units follow the equation

$$\text{m/sec} = 1000 \times \frac{\text{L/sec}}{\text{Area, mm}^2}$$

The relationship of a circular duct's area to its diameter is the same when using consistent SI units:

$$\text{Area,}_{\text{mm}^2} = 0.785 \times \text{Diameter, mm}^2$$

The size of a duct to reach a target velocity is found with the equation

$$\text{Diameter, inches} = \sqrt{\frac{183 \times \text{cfm}}{\text{ft/min}}}$$

The same relationship for an SI design is expressed with

$$\text{Diameter, mm} = 100 \times \sqrt{\frac{\text{L/sec}}{\text{m/sec}}}$$

Chapter 5

Pressure drop along a pipe flowing water is found with the two IP equations:

$$\text{psi} = 0.0807 \times f \times \text{Length, ft} \times \frac{\text{Velocity, ft/sec}^2}{\text{Diameter, in.}}$$

$$\text{psi} = 0.0135 \times f \times \text{Length, ft} \times \frac{\text{Flow, gpm}}{\text{Diameter, in.}^5}$$

The equivalent SI equations are

$$\text{kPa} = 499 \times f \times \text{Length, m} \times \frac{\text{Velocity, m/sec}^2}{\text{Diameter, mm}}$$

$$\text{kPa} = 51{,}100{,}000 \times f \times \text{Length, m} \times \frac{\text{Flow, L/sec}}{\text{Diameter, mm}}$$

Water flowing through n number of parallel, same-sized tubes has a velocity found with

$$\text{ft/sec} = 0.408 \times \frac{\text{gpm}}{(\text{Diameter, inches} \times n)}$$

This relationship in SI units is

$$\text{m/sec} = 50 \times \frac{\text{L/sec}}{(\text{Diameter, mm} \times n)}$$

Control valves should be sized for an application based on their required pressure drop in a full-flow condition. The factor used to rate the capacity of the valve is its C_v. The IP formula is:

$$C_v = \frac{\text{gpm}}{\sqrt{\text{Pressure, lb/in.}^2}}$$

An SI C_v exists, found by dividing the full flow in liters per second by the square root of the pressure drop (in kPa). However, the value found with this formula is not the same as the IP C_v. To convert between the two (and determine an IP C_v value from SI conditions) use the formula

$$C_{vIP} = 41.6 \times \frac{\text{L/sec}}{\sqrt{\text{Pressure, kPa}}}$$

This will allow the use of IP-rated valves where the existing flow and pressure drops are known in SI units.

Water temperature increase caused by pumping is found with the IP equation

$$\text{Temp rise, deg. F} = \text{Head, ft. water} \times \frac{(1 - \text{Efficiency})}{(778 \times \text{Efficiency})}$$

Expressed in SI units, the equation becomes

$$\text{Temp rise, deg. C} = 0.239 \times \text{Head, Pa} \times \frac{(1 - \text{Efficiency})}{\text{Efficiency}}$$

Horsepower required to pump a given amount of water through a system is found with the IP equation

$$\text{hp} = \frac{(\text{gpm} \times \text{Head, ft. water} \times \text{Specific gravity})}{(3960 \times \text{Efficiency})}$$

The amount of power in metric SI units is found with the equivalent equation

$$\text{kW} = 0.1 \times \text{L/sec} \times \text{Head, kPa} \times \frac{\text{Specific gravity}}{\text{Efficiency}}$$

The suction side pressure on a pump is found by estimating its NPSH. In IP units, the equation is

NPSH, ft. water = 33.8 + Static head – Friction head – Vapor pressure

The same equation using kPa is

NPSH, kPa = 11.3 + Static head – Friction head – Vapor pressure

Note that all of the terms must be expressed in kPa. Multiply pressures expressed as feet of water by 2.99 to get their equivalent kPa values.

Chapter 8

Coefficient of performance is found by dividing an air conditioner's cooling power by the electrical power it uses. In the text, both were expressed as Btu/hr but the equation is the same if both are expressed as watts. Because SI units rate both heat flow and electrical power in watts, it is inherently an

SI equation. If a system is rated by its energy efficiency ratio it can be converted to COP by the equation

$$\text{COP} = 0.293 \times \text{EER}$$

Proper airflow for purging a mechanical room is related to the amount of refrigerant in the room's largest system with the equation

$$\text{cfm} = 100 \times \sqrt{\text{Mass, lb}}$$

The proper SI airflow is

$$\text{L/sec} = 31.8 \times \sqrt{\text{Mass, kg}}$$

Chapter 9

Water source heat pump systems require a proportional water flow to heat flow capacity to keep water temperatures within bounds. The text gave an example of 2.5 gpm per thousand Btu/hr of heat pump capacity. The IP equation is

$$\text{Pump gpm} = \frac{\text{Total Btu}}{(1000 \times 2.5)}$$

To find the proper SI pumping capacity, use the equation

$$\text{Pump L/sec} = \frac{\text{Total W}}{(4640 \times 2.5)}$$

Storage tank capacity is related to flow rate and desired time to store water. The IP equation to find the tank volume is

$$\text{Capacity, gal} = 60 \times \text{Pump gpm} \times \text{hr}$$

The same SI equation to find tank capacity is

$$\text{Capacity, L} = 3600 \times \text{Pump L/sec} \times \text{hr}$$

Chapter 11

The example estimating the required capacity of a humidifier used IP units throughout the text. The number of pounds of water per cubic foot of air, the airflow in cfm, and the final answer in gallons per day capacity can all be converted to SI units. To change the water's density in air to kilograms per cubic meter, use the equation

$$\text{kg/m}^3 = 16 \times \text{lb/ft}^3$$

Multiply the CFM air flow by 0.472 to get liters per second, or

$$\text{L/sec.} = 0.0472 \times \text{CFM}$$

Finally, convert the gallons per day to liters per day with the formula

$$\text{liters} = 0.264 \times \text{gallons}$$

Index

A

Illustration page numbers are in **boldface**.

D

E

I

K

L

S

T

U

V

W